Errata

ANALYSIS AND SOFTWARE OF CYLINDRICAL MEMBERS

Edited by

Wai-Fah Chen and Shouji Toma

ISBN: 0-8493-8282-3

Page 13, Equation 2.12 should read:

$$B = \frac{(\sigma_i/\sigma_y - 1)\sigma_y}{e^{\frac{\sigma_i/\sigma_y}{\sigma_i/\sigma_y - 1}}} = \frac{\mu - 1}{e^{\mu/(\mu-1)}}\sigma_y$$

Page 56, the fifth definition from the bottom should read:

θ_{y1}, θ_{y2} angle to define yielded range

Page 59, the equation in the middle should read:

$$W_m = \frac{M_{pc}}{p} - W_i \quad \text{or} \quad W_m = \frac{M_{mc}}{p} - W_i$$

Page 125, Equation 4.15 should read:

$$\epsilon_o = \left[\frac{P}{2Etr} - r\Phi\left(\cos\theta_y - \cos\theta_{y2}\right) - \epsilon_y\left(\theta_{y1} - \theta_{y2}\right)\right] \times \frac{1}{\theta_{y1} + \theta_{y2}}$$

$$\text{for} \begin{cases} \epsilon_o - r\Phi \leq -\epsilon_y \\ \epsilon_o + r\Phi \geq -\epsilon_y \end{cases}$$

Page 150, Equation 4.34 should read:

$$M_E = \frac{P}{k\sin\frac{kL}{2}}\left[\frac{Q}{2P}\left(\sin ka \sin\frac{kL}{2} + \cos ka \cos\frac{kL}{2} - \cos\frac{kL}{2}\right) + \theta_c - \theta_E \cos\frac{kL}{2}\right]$$

ANALYSIS and SOFTWARE of CYLINDRICAL MEMBERS

New Directions in Civil Engineering

SERIES EDITOR: W. F. CHEN *Purdue University*

Published and Forthcoming Titles

Response Spectrum Method in Seismic Analysis and Design of Structures
Ajaya Kumar Gupta *North Carolina State University*

Stability Design of Steel Frames
W. F. Chen *Purdue University* and E. M. Lui *Syracuse University*

Concrete Buildings: Analysis for Safe Construction
W. F. Chen *Purdue University* and K. H. Mossallam *Ministry of Interior, Saudi Arabia*

Stability and Ductility of Steel Structures under Cyclic Loading
Yuhshi Fukumoto *Osaka University* and George C. Lee *State University of New York at Buffalo*

Unified Theory of Reinforced Concrete
Thomas T. C. Hsu *University of Houston*

Flexural-Torsional Buckling of Structures
N. S. Trahair *University of Sydney*

Advanced Analysis of Steel Frames: Theory, Software, and Applications
W. F. Chen *Purdue University* and S. Toma *Hokkaigakuen University*

Water Treatment Processes: Simple Options
S. Vigneswaran *University of Technology, Sydney* and C. Visvanathan *Asian Institute of Technology*

Simulation-Based Reliability Assessment for Structural Engineers
Pavel Marek *San José State University,* Milan Guštar *Prague* and Thalia Anagnos *San José State University*

Analysis and Software of Cylindrical Members
W. F. Chen *Purdue University* and S. Toma *Hokkaigakuen University*

Fracture Mechanics of Concrete
Zdeněk P. Bažant *NorthwesternUniversity* and Jaime Planas *Technical University, Madrid*

Artificial Intelligence and Expert Systems for Engineers
C.S. Krishnamoorthy *Indian Institute of Technology* and S. Rajeev *Indian Institute of Technology*

Introduction to Environmental Geotechnology
Hsai-Yang Fang *Lehigh University*

Buckling of Thin-Walled Structures
J. Rhodes *University of Strathclyde*

Fracture Processes of Concrete
Jan G.M. van Mier *Delft University of Technology*

Limit Analysis and Concrete Plasticity (2nd Edition)
M.P. Nielsen *Technical University of Denmark*

Aseismic Testing of Building Structures
Bo-Long Zhu *Tongji University, Shanghai*

Winter Concreting
B.A. Krylov *Design and Technology Institute (NIIZhB), Moscow*

Contaminated Soils and Sediments
Raymond N. Yong *McGill University*

ANALYSIS and SOFTWARE of CYLINDRICAL MEMBERS

Wai-Fah Chen
Shouji Toma

CRC Press is an imprint of the
Taylor & Francis Group, an **informa** business

CRC Press
Taylor & Francis Group
6000 Broken Sound Parkway NW, Suite 300
Boca Raton, FL 33487-2742

First issued in paperback 2019

ISBN-13: 978-0-367-44873-8 (pbk)
ISBN-13: 978-0-8493-8282-6 (hbk)

Visit the Taylor & Francis Web site at
http://www.taylorandfrancis.com

and the CRC Press Web site at
http://www.crcpress.com

Preface

Accurate information regarding the behavior of structural members throughout the entire range of loading up to ultimate load and including the post-buckling and cyclic inelastic behaviors is essential for engineers to make a realistic assessment of strength and risk of their structures. This book attempts to provide such basic information for structural engineers in the case of cylindrical members as used in offshore structures.

The book is intended primarily for practicing structural engineers familiar with the processes of design and construction of offshore structures, but requiring more information on the performance and specific design criteria of these members, and for students of structural engineering requiring a broader knowledge of the behavior of structural members other than those commonly available under the general heading of "Steel Design."

The scope of the book is indicated by the contents. It sets out initially to describe concisely the concept, the assumptions, and the basic formulation steps, goes on to show how these theoretical formulations are implemented into computers and how the programs are documented into user-friendly manuals, and finally outlines the solution procedures of some typical sample problems using the software developed. These solutions may be used in the ultimate processes of design and performance analysis of offshore structural engineering.

January, 1995

W. F. Chen
S. Toma

Contents

1 INTRODUCTION 1
S. Toma and W. F. Chen
1.1 General 1
1.2 Circular Cylindrical Members 1
1.3 Analytical Approach 3
1.4 Organization of the Book 3
1.5 Computer Software 4

2 CENTRALLY AND ECCENTRICALLY COMPRESSED COLUMNS 5
S. Toma and W. F. Chen
2.1 Introduction 5
2.2 Behavior and Strength of Short Columns 7
2.2.1 Moment-Curvature Relation 7
2.2.2 Tangent Stiffness Method 7
2.2.3 Initial Imperfections 14
2.2.4 Effect of Hydrostatic Pressure 16
2.2.5 Computer Implementation 23
2.2.6 Numerical Results 25
2.3 Behavior and Strength of Long Columns 34
2.3.1 Newmark's Method 34
2.3.2 Out-of-Straightness 37
2.3.3 Computer Implementation 38
2.3.4 Numerical Results 39
2.3.5 Effect of Hydrostatic Pressure 44
2.4 User's Manual for the Program NEWMARK 44
2.5 Sample Calculations 49
2.5.1 Short-Column Analysis 49
2.5.2 Long-Column Analysis 51
References 53

3 APPROXIMATE ANALYSIS OF BEAM-COLUMNS 55
S. Toma and W. F. Chen
Notations 55
3.1 Introduction 56
3.2 Load-Deflection Relation (P-w Curve) 60

3.2.1 Assumed Deflection Method 60
3.2.2 Deflection Functions 60
3.2.3 Moment-Thrust-Curvature Relationships (M-P-Φ Curve).. 61
3.3 Load-Shortening Relation (P-Δ Curve) 61
3.4 Elastic Analysis 63
3.4.1 Basic Concept 63
3.4.2 Load-Deflection Relation (P-w Curve) 63
3.4.3 Load-Shortening Relation (P-Δ Curve) 67
3.5 Plastic Hinge Method 68
3.5.1 Basic Concept 68
3.5.2 Load-Deflection Relation (P-w Curve) 69
3.5.3 Load-Shortening Relation (P-Δ Curve) 70
3.6 Modified Plastic Hinge Method 70
3.6.1 Basic Concept 70
3.6.2 Load-Deflection Relation (P-w Curve) 70
3.6.3 Load-Shortening Relation (P-Δ Curve) 71
3.7 Exact-Moment Curvature Method 72
3.7.1 Basic Concept 72
3.7.2 Closed-Form Expression of M-P-Φ 73
3.7.3 Load-Deflection Relation (P-w Curve) 76
3.7.4 Load-Shortening Relation (P-Δ Curve) 78
3.8 Average Flow Moment Method 81
3.8.1 Basic Concept 81
3.8.2 Average Flow Moment 83
3.8.3 Load-Deflection Relation (P-w Curve) 85
3.8.4 Load-Shortening Relation (P-Δ Curve) 85
3.9 Computer Implementation 86
3.10 User's Manual for the Program ADMCOL 86
3.11 Sample Calculations 87
3.12 Numerical Results 89
3.12.1 Plastic Hinge and Modified Plastic Hinge Methods 89
3.12.2 Exact Moment-Curvature Method 89
3.12.3 Average Flow Moment Method 99
3.12.4 Conclusions 99
References 104

4 CYCLIC BEHAVIOR AND MODELING 105
S. Toma and W. F. Chen
Notations 105
4.1 Introduction 106
4.2 Cyclic Behavior of Short Tube 108
4.2.1 Basic Concept 108
4.2.2 Stress-Strain Relation 108
4.2.3 Exact M-P-Φ and P-M-ϵ_o Relations 111
4.2.4 Computer Implementation 111

4.2.5 User's Manual for the Program MPCYCL 111
4.2.6 Numerical Results 115
4.2.7 Closed Form Expressions of M-P-Φ Curves 115
4.2.8 Closed Form Expressions of P-M-ϵ_o Curves 123
4.3 Cyclic Analysis of Pin-Ended Columns 126
4.3.1 Basic Concept—Newmark's Method 126
4.3.2 Computer Implementation 126
4.3.3 User's Manual for the Program BMCYCL 129
4.3.4 User's Manual for the Program APCYCL 133
4.3.5 Numerical Results 137
4.4 Cyclic Analysis of Fixed-Ended Columns 142
4.4.1 Basic Concept — Hinge-by-Hinge Method 142
4.4.2 Compressive Axial Force (Stages 1 to 4) 144
4.4.3 Tensile Axial Force (Stages 5 to 9) 150
4.4.4 Load-Shortening Relation 156
4.4.5 Computer Implementation 158
4.4.6 User's Manual for the Program FIXCYCL 158
4.4.7 Numerical Results 163
References 165

5 ANALYSIS CONSIDERING LOCAL BUCKLING EFFECTS 167
I. S. Sohal and W. F. Chen
5.1 Introduction 167
5.2 Kinematic Model for Cross Sectional Distortion 168
5.3 M-P-Φ Analysis of Section 169
5.3.1 Pre-Local-Buckling Analysis 169
5.3.2 Post-Local-Buckling Analysis 169
5.3.3 Reversed Loading Analysis 170
5.3.4 Closed Form Expressions for a Complete Cycle of Loading 172
5.4 Load-Deflection Analysis of Members 178
5.4.1 Modified Assumed Deflection Method 179
5.4.2 Relation for Elastic Regime 181
5.4.3 Relation for Primary Yield Regime 183
5.4.4 Relation for Secondary Yield Regime 183
5.4.5 Relation for Post-Local Buckling Regime 183
5.4.6 Relation for Reversed Loading Regime 184
5.5 Load-Shortening Analysis of Member 187
5.6 Computer Implementation 188
5.7 User's Manual for the Program BRACE 191
5.8 Solution of Sample Examples 193
5.8.1 Effects of Local Buckling on a Pin-Ended Column 193
5.8.2 Effects of Diameter-to-Thickness Ratio on Fixed-Ended Column 201
5.8.3 Effects of Slenderness Ratio on Fixed-Ended Column 201

5.8.4 Effects of End Moments on Pin-Ended Beam-Column.... 203
5.8.5 Effects of Lateral Loads on Pin-Ended Beam-Column.... 204
References 207

6 ANALYSIS CONSIDERING DENT DAMAGE EFFECTS.... 209
L. Duan and W. F. Chen
Notations.... 209
6.1 Introduction 210
6.2 M-P-Φ Relationships for Dented Cylindrical Sections 212
6.2.1 Undented Cylindrical Sections 212
6.2.2 Dented Cylindrical Sections 215
6.2.3 M-P-ϵ_o Expressions.... 221
6.3 Member Analysis Considering Dent Damage Effect.... 225
6.3.1 General Description.... 226
6.3.2 Numerical Procedure.... 228
6.3.3 Load-Shortening Relations.... 233
6.4 Computer Implementation 233
6.4.1 Program BCDENT 233
6.4.2 Structure of Program BCDENT 234
6.5 User's Manual for BCDENT 237
6.5.1 Input Data Organization.... 237
6.5.2 Input Data Formats 238
6.5.3 Operation of BCDENT on an IBM-PC.... 243
6.5.4 Examples 243
6.6 Solutions of Undented Cylindrical Member Behavior 258
6.6.1 Undented Columns without Local Buckling.... 258
6.6.2 Undented Columns with Local Buckling.... 258
6.7 Solutions of Dented Cylindrical Member Behavior 259
6.7.1 Pin-Ended Dented Columns.... 260
6.7.2 Dented Beam-Columns 261
References 266

7 ANALYSIS OF INTERNALLY GROUT-REPAIRED DAMAGED MEMBERS 269
J. M. Ricles
7.1 Introduction 269
7.2 M-P-ϕ Analysis.... 273
7.3 Member Analysis 278
7.4 Computer Implementation 283
7.5 User's Manual 292
7.6 Solutions for Member Behavior 295
References 303

Index 305

1: Introduction

S. Toma
Department of Civil Engineering, Hokkai-Gakuen University, Sapporo, Japan

W. F. Chen
School of Civil Engineering, Purdue University, West Lafayette, Indiana

1.1 GENERAL

The circular cylindrical members are used extensively in the construction of offshore oil-production platforms (Fig. 1.1). This is because cylindrical members have larger torsional rigidity than open section members and show a superior behavior for structural stability. In addition to the stability strength, the circular shape reduces the flow resistance more than any other shape. Hence, it is the best shape for structural members used in offshore structures.

There are several design specifications written specifically for offshore platforms, including the one issued by the American Petroleum Institute (API) and the others by the bureaus of shipping industries such as the American Bureau of Shipping (ABS) and Det Norske Veritus (DNV) in Norway. These specifications dictate the strength of circular tubes and provide the design guidelines for engineers. Furthermore, the circular tubes with small diameter are often used for roof truss of inland framed structures. In many countries, the building codes also provide the rules for the design of cylindrical members.

This book provides the analytical means for the study of circular tube members as used in offshore structures. Many different analytical procedures will be described in the book, ranging from very sophisticated analytic methods to a rather crude approximate procedure for the monotonic loading, including the post-buckling behavior and the cyclic loading. These methods are useful for the study of the structural behaviors of cylindrical members.

Recently, square-shaped cylindrical members with thick walls have become common in onshore building structures due to their superiority in structural properties and constructional convenience. However, only the circular tubes are described in this book.

The strength of steel frames with cylindrical members is often controlled by the connections. The connections are usually the weak spots of the structures, but the proper design of these connections is beyond the scope of this book.

1.2 CIRCULAR CYLINDRICAL MEMBERS

Circular tubes can be categorized by their manufacturing process. They are: (1) hot-rolled cylindrical members, also called "seamless pipe," made in a similar manner

0-8493-8282-3/96/$0.00+$.50

FIGURE 1.1 Fixed offshore platform.

to the structural shapes; (2) fabricated cylindrical members welded longitudinally to form a can after a plate is bent and then welded transversely to make a long member by connecting cans; and (3) electrically seamed members made by first cold-forming with straightening and bending the coiled plate and then welding longitudinally.

Cylindrical members with different sizes and manufacturing processes will have different structural properties. Fabricated cylindrical members with large diameter and relatively thick plate are used in offshore structures. Manufactured hot-rolled members with small diameter (less than 350 mm) and with thick plate (up to 30 mm) are usually used for tubing of the oil production facility rather than structural members. Electrically seamed members with smaller diameter and thinner plate are used for secondary members rather than primary members.

Since the analytical procedures described in this book consider all possible imperfections, such as initial deflection and residual stresses, they can be applied directly to any types of tubular members by simply modifying the sectional shape, such as square tube, or the pattern of residual stress distributions.

1.3 ANALYTICAL APPROACH

One of the purposes of structural analysis is to trace the load-deflection (or force-displacement) relations of the structure, consisting of several basic steps ranging from the analysis of a small part (material property) to the entire system of the structure. The stress-strain relation of the material at any point is the most fundamental load-deformation relation required in any structural analysis. In this book, the elastic–perfectly plastic type of stress-strain relation is used.

The next step is to develop the relationship of the sectional forces to the curvature or axial strain. This is obtained by integrating the stress-strain relation throughout the entire cross section, where results are expressed by the moment-curvature relation (M-P-Φ curve) and the thrust-axial strain relation (P-M-ϵ_0 curve). In this book, a wide range of analytical procedures is described, ranging from a very rigorous integration process to several levels of approximation in order to study the M-P-Φ and P-M-ϵ_0 relations. The tangent stiffness method is used for the rigorous analysis.

The third step is the member behavior, which is obtained by the integration of the sectional load-deformation relations, M-P-Φ and P-M-ϵ_0 curves, along its member length. The member behavior is expressed by the relation of axial and/or lateral forces to axial strain and/or lateral deflection. In this book, Newmark's method is used in the numerical integration along the member.

The last step is to analyze the load-deformation relation of the whole structural system. The integration of all member behaviors gives the system behavior. However, the study of the structural system behavior is outside the scope of this book. Only the load-deformation behaviors of members are treated.

Although many analytical methods such as finite element approach, shell approach, etc., can be applied for cylindrical members, the beam-column approach for a bar element is generally adequate for practical design work and is therefore used in this book. Cylindrical members may be treated as a perfect circular shape, but they may be dented by local buckling or by the collision of ships, which reduces the member strength significantly. These effects will be discussed in this book.

1.4 ORGANIZATION OF THE BOOK

This book consists of an introduction and six chapters, each having a different type of member modeling and analytical means. In Chapter 2, the strength and load-deformation behavior of the cylindrical members subjected to axial force is analyzed by using the rigorous numerical approach, i.e., Newmark's method. Also, the effect of hydrostatic pressure is investigated for members in deep water closed inside for air/water tightness.

Chapter 3 describes an approximate analysis procedure for which several levels of simplifications are introduced. For the first simplification, the deflection shape is assumed to be a simple function, and the analysis is reduced to an one-degree-of-freedom system. When the closed form expressions for M-P-Φ curves are used instead of the numerical integration as described in Chapter 2, the analysis provides

TABLE 1.1 Computer Programs

Program Names (Directory)	Type of Analysis	Chapters
1. NEWMARK (NEW)	Monotonic member behaviors (Newmark integration method)	Chapter 2
2. ADMCOL (ADM)	Monotonic member behaviors (assumed deflection method)	Chapter 3
3. MPCYCL (MPC)	Cyclic M-P-Φ curves	Chapter 4
4. BMCYCL (BMC)	Cyclic member behaviors (exact, pin-ended)	Chapter 4
5. APCYCL (APC)	Cyclic member behaviors (approximate, pin-ended)	Chapter 4
6. FIXCYCL (FIX)	Cyclic member behaviors (approximate, fixed-ended)	Chapter 4
7. BRACE (BRA)	Local buckling effects	Chapter 5
8. BCDENT (BCD)	Dent damage effects	Chapter 6
9. DGROUT (DGR)	Grout-repaired damaged members	Chapter 7

Notes: The floppy disk is formatted by MS-DOS 1.44MB. In each directory, the FORTRAN source program and the input data and corresponding output results for sample problems are stored. The programs are compatible with personal computers or workstations.

a level of accuracy similar to that of Chapter 2. The most drastic simplification is made in the plastic hinge solution in which the elastic–perfectly plastic type of M-P-Φ curve is used. This is also compared with other analyses in this chapter.

Since offshore structures are subjected to repeated wave action, the cyclic behavior of steel cylindrical members is important and is discussed in Chapter 4. Here, the material is assumed to be the elastic–perfectly plastic type, as in other chapters. The cyclic load-deformation behaviors for the section and for the member are obtained by introducing some further simplifications.

Local buckling will occur when the thickness of the tube wall is relatively thin and the bending moment is large. Chapter 5 describes the effect of local buckling on the sectional and the member behaviors. In Chapter 6, the reduction in strength is studied for members dented by ship collision or by other causes. This is important for assessing the reduction in strength and the necessity for repairing offshore structures.

Finally, Chapter 7 discusses the analysis of the damaged members that are repaired by grouting internally. Grout repair is used extensively in practical offshore applications, and it is important to know the performance of such members.

The analyses described in each chapter are first presented theoretically, followed by the development of computer software. This will be explained next.

1.5 COMPUTER SOFTWARE

The computer programs for all these analyses were developed using FORTRAN language, and their source lists are supplied in the attached floppy disk with sample problems. The computer implementation and the input manual are given in each chapter. A list of the programs developed is summarized in Table 1.1.

The reader is encouraged to first run the sample problems in the book to learn the input manual of the programs and then to run their own problems to understand the theory and design of cylindrical members more deeply.

2: Centrally and Eccentrically Compressed Columns

S. Toma
Department of Civil Engineering, Hokkai-Gakuen University, Sapporo, Japan

W. F. Chen
School of Civil Engineering, Purdue University, West Lafayette, Indiana

2.1 INTRODUCTION

There are many types of cylindrical members, such as hot-rolled tube, spiral welded tube, cold-formed thin tube, fabricated tube, etc. Among them, a fabricated cylindrical column, as commonly used in offshore structures (Fig. 2.1) contains imperfections that are far more complicated than the hot-rolled members, such as channel, angle, and wide flange, because they involve a more complicated manufacturing process: (1) rolling from a steel plate to form a cylinder and (2) welding of the longitudinal seam of the cylinder to form a "can." This results in significant circumferential and longitudinal residual stresses as well as out-of-roundness of the cross section. Furthermore, the transverse welding of the "cans" to form a long column results in a significant out-of-straightness of the column. Therefore, analytical procedures developed for fabricated tube involve these complicated factors and can be applied to any other types of cylindrical members.

A realistic design for axially loaded fabricated cylindrical columns must consider the fact that an actual cylindrical column is geometrically and materially imperfect and is frequently subjected to bending moments resulting from unavoidable end eccentricities and lateral forces. Thus, all fabricated cylindrical columns must be designed as beam-columns.

The beam-column analysis can be divided into two steps: (1) the moment-curvature behavior of a short column and (2) the load-deflection behavior of a long column. The moment-curvature relationship is of prime importance in the analysis of any long beam-column. The slope of this relation curve gives the required stiffness of a beam-column. For an elastic cross section, the sectional stiffness, EI, is a constant. This presents no difficulties for solutions. However, for an elastic–plastic cross section, the moment-curvature relationship becomes nonlinear. The value of EI in this case can be considered as the current slope between moment M and curvature, Φ, which depends on the magnitude of moment M.

0-8493-8282-3/96/$0.00+$.50

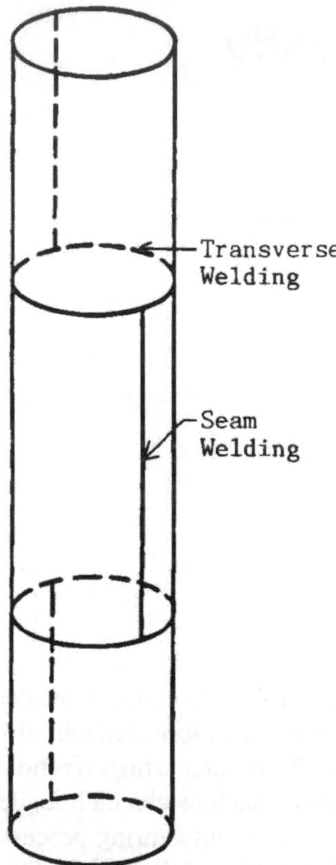

FIGURE 2.1 Fabricated cylindrical steel column as used in offshore structures.

The existence of residual stresses and out-of-roundness will affect the slope of a moment-curvature relationship. As expected, the longitudinal residual stress will cause the cross section to reach the elastic-plastic regime much earlier than without the effect of longitudinal residual stress.

In this chapter, the moment-curvature behavior of a short, fabricated cylindrical column is first studied considering the effects of residual stresses and out-of-roundness caused by forming and welding. This is followed by the study of elastic–plastic behavior of long columns, considering the effect of out-of-straightness.

In particular, the effect of hydrostatic pressure will be described. For deepwater platforms, an additional important consideration for tubular members used in offshore structures is the interaction of axial and bending stresses, with compressive hoop stresses caused by the external hydrostatic pressure. Due to the reversible nature of storm forces, this axial stress can be tensile as well as compressive. As with other factors, the external pressure may significantly reduce the axial load-carrying capacity of actual tubes to less than that predicted by the conventional design column curves.

As reported in the reference (Chen and Ross, 1977), two types of failure modes were observed in the 10 full-size column tests: local buckling, recognized by the checkerboard pattern of the cross section distortion, and overall column-type buckling, characterized by extensive yield of a "can" and relatively little cross-section distortion. In this chapter, only overall buckling is analyzed, and the effect of local buckling will be studied in Chapter 5. Shell theory related to local buckling is not considered in this chapter.

2.2 BEHAVIOR AND STRENGTH OF SHORT COLUMNS

2.2.1 Moment-Curvature Relation

The basic relation required in any long beam-column analysis is the relation between moment and curvature for a given thrust. This relationship reflects the structural response of a "can" to such loadings. The individual and combined effects of longitudinal and circumferential residual stresses, and out-of-roundness of the tube on the behavior and strength of a fabricated "can" can be assessed by studying the moment-curvature-thrust relationships. This relationship is used as the basic building block for the analysis of long columns to be described in subsequent sections.

The tangent stiffness method (Santathadaporn and Chen, 1972) is applied here to obtain moment-curvature relationship for a fabricated cylindrical steel column segment subjected to axial compression and external hydrostatic pressure combined with biaxial bending moment. In this part of analysis, both the initial out-of-roundness of the tube and the longitudinal and circumferential residual stresses are considered, using Tresca yield criterion. In particular, allowance has been made for relieving circumferential residual stresses due to applied longitudinal loads, which cause "Poisson's ratio effect" on circumferential residual stress. Considering the interaction of residual stresses and hoop stresses caused by hydrostatic pressure, and magnified by initial out-of-roundness, a computer program was developed to provide numerical results.

2.2.2 Tangent Stiffness Method

Basic Concept

The tangent stiffness method is effective in numerically obtaining the non-linear moment-curvature-thrust relationship when the cross section is partially yielded. The matrix formulation developed in this method is convenient for computer solution. A theoretical method for analysis of a short biaxially loaded H-column was developed using the tangent stiffness approach (Santathadaporn and Chen, 1972). This method was extended here to include the interaction between longitudinal and circumferential residual stresses and circumferential hoop stresses.

For a biaxially loaded column, the appropriate set of generalized stresses are bending moments M_x and M_y and axial force P (Fig. 2.2). The corresponding set of generalized strains are bending curvatures Φ_x and Φ_y and axial strain ϵ_o.

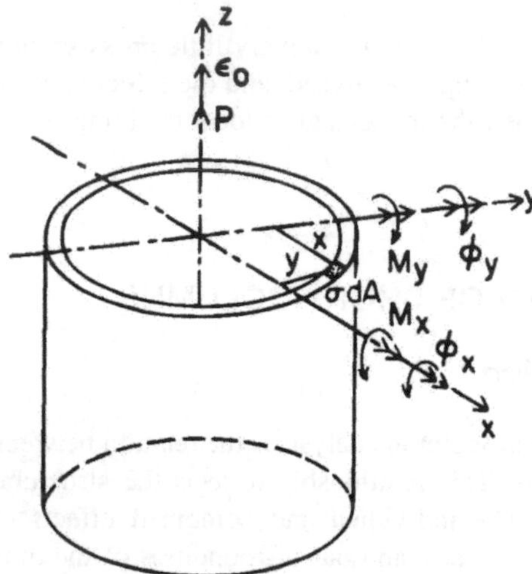

FIGURE 2.2 Column segment under axial load, external hydrostatic pressure, and biaxial bending moment.

$$\{f\} = \begin{Bmatrix} M_x \\ M_y \\ P \end{Bmatrix} \tag{2.1}$$

$$\{x\} = \begin{Bmatrix} \Phi_x \\ \Phi_y \\ \epsilon_0 \end{Bmatrix} \tag{2.2}$$

Assuming plane sections remain plane after bending and introducing the effective modulus concept, we have

$$M_x = \int \sigma y dA = \int E_{eff}(y\Phi_x - x\Phi_y + \epsilon_0 + \epsilon_r) y dA \tag{2.3a}$$

$$M_y = -\int \sigma x dA = -\int E_{eff}(y\Phi_x - x\Phi_y + \epsilon_0 + \epsilon_r) x dA \tag{2.3b}$$

$$P = \int \sigma dA = \int E_{eff}(y\Phi_x - x\Phi_y + \epsilon_0 + \epsilon_r) dA \tag{2.3c}$$

in which E_{eff} = effective Young's modulus
A = area of cross section
ϵ_r = residual strains in the longitudinal direction, and
x,y = coordinates

The generalized stresses and strains are shown in positive directions in Fig. 2.2. The orientation of axes x and y are defined by its relative position to the longitudinal weld as marked arbitrarily in Fig. 2.1. The objective of this analysis is to calculate the deformation history {x} of the cross section corresponding to a given path of loading history {f}.

Since plastic behavior depends highly on the previous load history of the structure, it is possible to establish analytically only the relationship between the infinitesimal generalized stress increments {df} or $\{\dot{f}\}$ and the corresponding infinitesimal generalized strain increments {dx} or $\{\dot{x}\}$. The following relationship between $\{\dot{f}\}$ and $\{\dot{x}\}$ can be derived for a particular column cross section:

$$\begin{Bmatrix} \dot{M}_x \\ -\dot{M}_y \\ \dot{P} \end{Bmatrix} = \begin{bmatrix} Q_{11} & Q_{12} & Q_{13} \\ Q_{21} & Q_{22} & Q_{23} \\ Q_{31} & Q_{32} & Q_{33} \end{bmatrix} \begin{Bmatrix} \dot{\Phi}_x \\ \dot{\Phi}_y \\ \dot{\epsilon}_0 \end{Bmatrix} \tag{2.4}$$

or

$$\{\dot{f}\} = [Q]\{\dot{x}\} \tag{2.5}$$

In Eq. (2.4), Q_{ij} is defined as

$$\begin{aligned} Q_{11} &= \int E_{eff} y^2 dA & Q_{12} = Q_{21} &= \int E_{eff} xy dA \\ Q_{22} &= \int E_{eff} x^2 dA & Q_{13} = Q_{31} &= -\int E_{eff} y dA \\ Q_{33} &= \int E_{eff} dA & Q_{23} = Q_{32} &= -\int E_{eff} x dA \end{aligned} \tag{2.6}$$

and Eq. (2.5) can be rewritten in the form

$$\{\dot{x}\} = [Q]^{-1}\{\dot{f}\} \tag{2.7}$$

For the special case when the entire cross section is elastic, it can be seen from Eq. (2.6) that

$$Q_{ij} = 0 \qquad (i \neq j) \tag{2.8}$$

For a partially yielded section, Eq. (2.8) no longer holds. None of the elements of the tangent stiffness matrix will be zero, except when the section is completely yielded. Unlike the elastic problems, the matrix [Q] of the partially yielded section is a function of current state of stress and strain as well as the properties of the material and the cross section. For an elastic–perfectly plastic material under uniaxial state of stress, the value of E is zero in the yielded zone. Therefore, only the area of the elastic core will contribute to the integration in Eq. (2.6). This implies that further increment of external forces is resisted by the remaining elastic area of the section only. This is the case for most commonly used column cross sections. However, for the case of fabricated cylindrical cross sections used in offshore

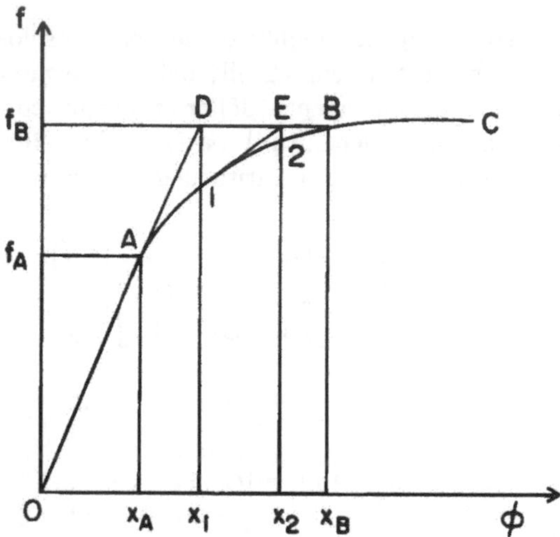

FIGURE 2.3 Tangent stiffness procedure.

structures, the yielding of a material element is caused by both circumferential and longitudinal stresses. The value of E in Eq. (2.6) must therefore be replaced by an "effective" modulus E_{eff} in the regions of biaxial yielding. Further discussions on the value of E_{eff} will be given next. Once the tangent stiffness matrix [Q], corresponding to a given state of stress can be evaluated, it is a simple matter to find the path of generalized strains {x} for a given path of generalized stresses {f} through step-by-step incremental calculations.

The tangent stiffness method consists of finding successive solutions as the load on the column is increased in steps. This is shown in Fig. 2.3. Let $\{f_A\}$ and $\{x_A\}$ be the known state. Now compute the deformation in state "B" when the prescribed force is $\{f_B\}$. The intermediate curvature state $\{x_1\}$ is first calculated by using the current tangent stiffness at state "A." This results in a modification of the previous stiffness associated with state "A" to the current tangent stiffness associated with state "1." If the increment of force is small, the first estimate of the increment of deformation from Eq. (2.7) is quite accurate and subsequent correction will not be necessary. Even with a large incremental force, the solution will generally converge within just a few cycles of iteration. The unbalanced force resulting from each iterative cycle is always smaller than the previous one and diminishes rapidly.

Effective Young's Modulus

An important facet of the present analysis is that elemental stresses in two directions (i.e., longitudinally and circumferentially) are linked by a Poisson's ratio effect. As the longitudinal stresses are applied (whether by application of axial load or bending moment) the circumferential residual stresses and hoop stresses will be affected, i.e., an increase in compression longitudinal stress produces a circumferential stress that increases circumferential compression stress and decreases circumferential ten-

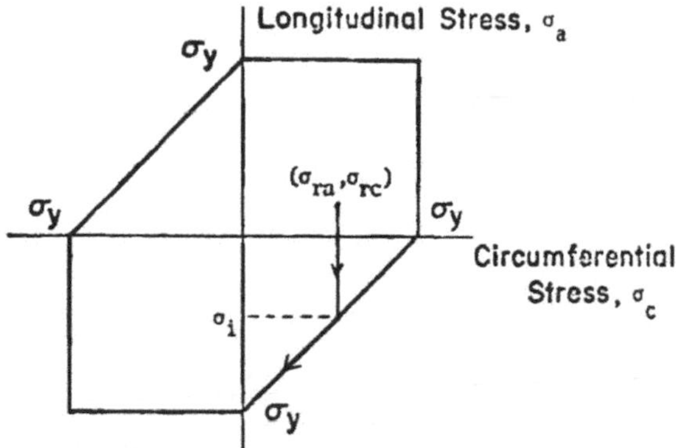

FIGURE 2.4 Tresca yield criterion.

sion stress. This interaction causes problems in the analysis when either the longitudinal stress is tensile and the circumferential stress is compressive or vice versa, because in these regimes the Tresca yield curves are at 45° to the biaxial principal stress axes (Fig. 2.4). Not only will this interaction affect the stress state at which the element yields, but it will obviously affect the subsequent loading history of each particular element.

When both circumferential and longitudinal stresses are either in tension or compression, the Tresca yield condition indicates that the limiting stress of an element is the uniaxial yield stress σ_y, beyond which the element can assume no more load, but merely deforms plastically. However, in the tension-compression regimes of the Tresca yield diagram, the element stress condition is such that the elemental stress state can change while the element still remains on the yielded curve, but merely shifts its position on the sloping lines of the Tresca yield diagram. For this case, the limiting longitudinal stress reaches, eventually, the uniaxial yield stress σ_y of the material. In other words, an element which has yielded in either of the tension-compression zones of the Tresca yield diagram does have some resistance for an increased section curvature.

A rigorous plasticity solution requires the generalized strain vector to be normal to the yield surface (flow rule)(Chen and Atsuta, 1976). However, an exact solution of this type is complicated. In order to find a relatively simple, yet reasonably accurate solution to this impasse, the concept of effective modulus E_{eff} is adopted herein. In the elastic range, the element is assumed to have a constant elastic modulus E, until the element yields at a longitudinal stress σ_i. We also know that at the uniaxial yield stress σ_y the effective E value for a perfect plastic material is zero. These two conditions were linked with a straight line variation, as shown in Fig. 2.5, to provide a reasonable estimate of decreasing elemental resistance to an increased applied longitudinal loading in the tension-compression regimes of the Tresca yield diagram. Thus, the effective Young's modulus varies linearly from E at the first yield stress σ_i to zero at the uniaxial yield value σ_y. The linear variation of Young's modulus in the plastic range results in the expression

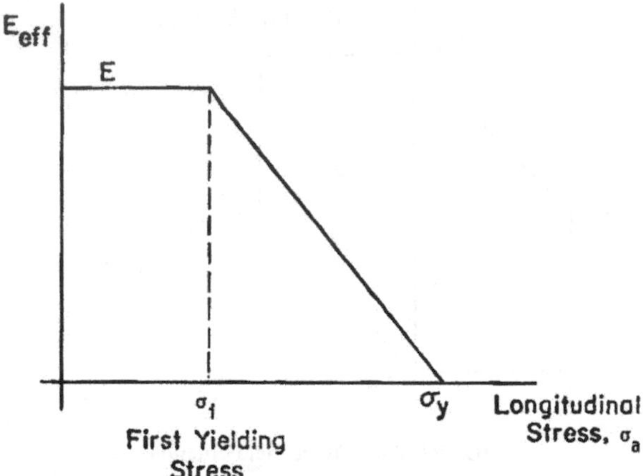

FIGURE 2.5 Effective Young's modulus idealization for biaxial stress condition.

$$E_{eff} = \frac{d\sigma}{d\epsilon} = \frac{E}{\sigma_i - \sigma_y}(\sigma - \sigma_y) \tag{2.9}$$

Stress-Strain Relationship

The tube is assumed to be in the plane stress state. When the entire cross section is in the elastic range, the two-dimensional elastic stress-strain relationship can be applied. However, in the present analysis, since only the axial strain increment $\dot{\epsilon}_a$ is prescribed, it is necessary to assume the relationship between the circumferential strain change $\dot{\epsilon}_c$ caused by the axial strain increment $\dot{\epsilon}_a$. It is known that the actual interaction between $\dot{\epsilon}_c$ and $\dot{\epsilon}_a$ lies between the two extreme conditions: (1) uniaxial behavior where the circumference of the tube is allowed to expand or contract freely in the radial direction and (2) plane strain behavior where the circumference of the tube is restrained completely from radial expansion or contraction. For the tubular columns as used in offshore structures, it is assumed that the tube can expand or contract freely so that in the elastic range the circumferential stresses σ_c will not be affected by the axial stress changes σ_a. In the biaxial stress space (σ_a, σ_c), this means that the stress point moves in the direction parallel to the σ_a axis (Fig. 2.4). As the axial stresses σ_a are increased as the result of the axial load combined with biaxial bending, the stress point located initially at the residual stresses σ_{rc} and σ_{ra} moves vertically upward until it reaches the Tresca yield curve, then moves in the direction toward its corresponding final point, σ_y.

As described previously, after the initial yielding of an element, the linear variation of the Young's modulus is assumed, as shown in Fig. 2.5 (see Eq. (2.9)).

The integration of Eq. (2.9) results in

$$\sigma = Be^{\frac{E\epsilon}{\sigma_i - \sigma_y}} + \sigma_y \tag{2.10}$$

in which B is the integration constant.

Using the initial yield values, that is

$$\sigma = \sigma_i = E\epsilon_i \quad \text{at} \quad \epsilon = \epsilon_i \tag{2.11}$$

the constant B can be expressed in terms of these initial yield values

$$B = \frac{(\sigma_i/\sigma_y - 1)\sigma_y}{e^{\frac{\sigma_i/\sigma_y}{\sigma_i/\sigma_y - 1}}} = \frac{\mu - 1}{e^{\mu/(\mu-1)}}\sigma_y \tag{2.12}$$

in which

$$\mu = \frac{\sigma_i}{\sigma_y} \tag{2.13}$$

Substituting the value of B, Eq. (2.12), into Eq. (2.10), the stress-strain relationship in the plastic range is obtained.

$$\sigma = \sigma_y\left\{1 - (1 - \mu)e^{\frac{\mu - \zeta_\epsilon}{1-\mu}}\right\} \tag{2.14}$$

in which

$$\zeta_\epsilon = \frac{\epsilon}{\epsilon_y} \tag{2.15}$$

As an example, the elastic and plastic stress-strain relationships are plotted in Fig. 2.6 for the following set of typical values

$$\sigma_y = 34 \text{ ksi}, \qquad \epsilon_y = 1.1332 \times 10^{-3}$$

$$\sigma_i = 15 \text{ ksi}, \qquad \epsilon_i = 0.5 \times 10^{-3}$$

It is of interest to note that for an elastic–perfectly plastic material, the biaxial stress interaction between the longitudinal and circumferential stresses results in a work-hardening type of material behavior with the initial yield value of σ_i = 15 ksi and ultimate strength of σ_y = 34 ksi. The slope of the stress-strain curve gives the effective tangent modulus E_{eff} of the tube.

In the present analysis, each element is treated independently and each contributes to the total sectional stiffness as individual; the interaction among the elements is not considered. To account for this interaction of elements, a more rigorous approach such as shell analysis or finite element method is needed in addition to the incremental theory of plasticity. This will make the problem exceedingly complicated.

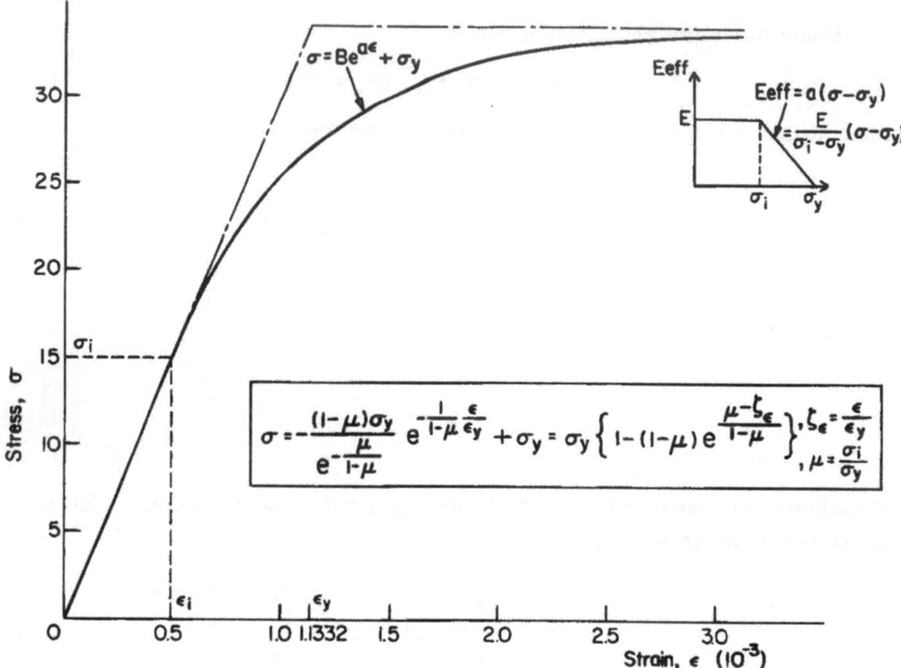

$$\sigma = -\frac{(1-\mu)\sigma_y}{e^{-\frac{\mu}{1-\mu}}} e^{-\frac{1}{1-\mu}\frac{\epsilon}{\epsilon_y}} + \sigma_y = \sigma_y\left\{1-(1-\mu)e^{\frac{\mu-\zeta_\epsilon}{1-\mu}}\right\}, \zeta_\epsilon = \frac{\epsilon}{\epsilon_y}, \mu = \frac{\sigma_i}{\sigma_y}$$

FIGURE 2.6 Stress-strain relationship in the elastic and plastic ranges.

2.2.3 Initial Imperfections

Residual Stress Distribution

There are two major types of residual stresses existing in fabricated tubular columns—longitudinal and circumferential residual stresses. During the forming process by which a flat plate is converted to a cylinder, significant circumferential residual stresses are induced that vary through the wall thickness of the cylinder. The longitudinal welding process induces an additional longitudinal residual stress into the "can." The measured residual stresses were reported in the references (Chen and Ross, 1977) and are summarized next.

The longitudinal residual stress distribution is shown in Fig. 2.7. This distribution represents an average value through the wall thickness. The curved solid line through the experimental results has the form predicted by Marshall (1970), and the dotted lines are a straight-line approximation suggested as a simplified alternative.

In any such distribution, the resultant axial force must of course be zero. It is particularly noteworthy that longitudinal residual stresses appear to go through a zero point at distances from the weld equal to multiples of the column radius.

In Fig. 2.8, the circumferential residual stresses through the wall thickness are given. If a section through the column wall is taken as shown in Fig. 2.8a, then a typical result is shown in Fig. 2.8b. Since the results between the various test sites around the circumference do not differ significantly, the suggested circumferential residual stress distribution through the wall thickness is shown in Fig. 2.8c.

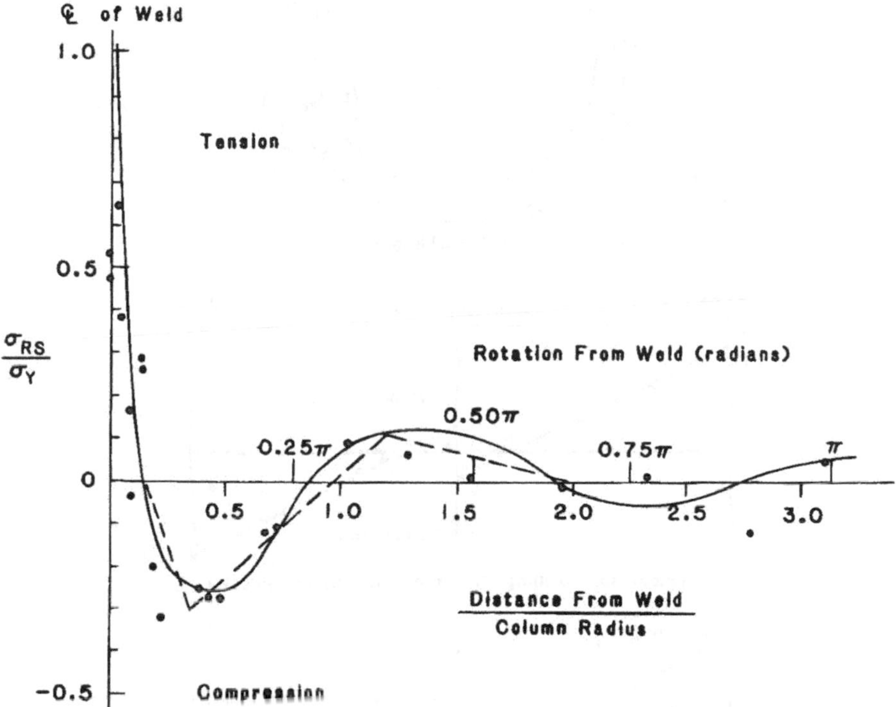

FIGURE 2.7 Longitudinal residual stress distribution obtained from slicing method.

The idealized residual stress distributions in axial and circumferential directions of the tube are shown in Fig. 2.9. These linearized distributions are adopted for the present analysis for steel tubes with the yield value of $\sigma_y = 36$ ksi.

Out-of-Roundness

The effect of out-of-roundness in the tubes on the behavior and strength of cylindrical columns is included. Such imperfections may be caused by either initial fabrication of each "can" and also by subsequent welding of cans to form the cylinders (see Fig. 2.1). Herein, a geometrical imperfection is introduced in the form of

$$w = w_i \cos 2\theta \tag{2.16}$$

where w_i is the initial maximum out-of-roundness as shown in Fig. 2.10, and θ is the angle around the circumference. If the maximum and minimum diameters of the tube are denoted by D_{max} and D_{min}, respectively, the initial deflection w_i can be expressed in the form

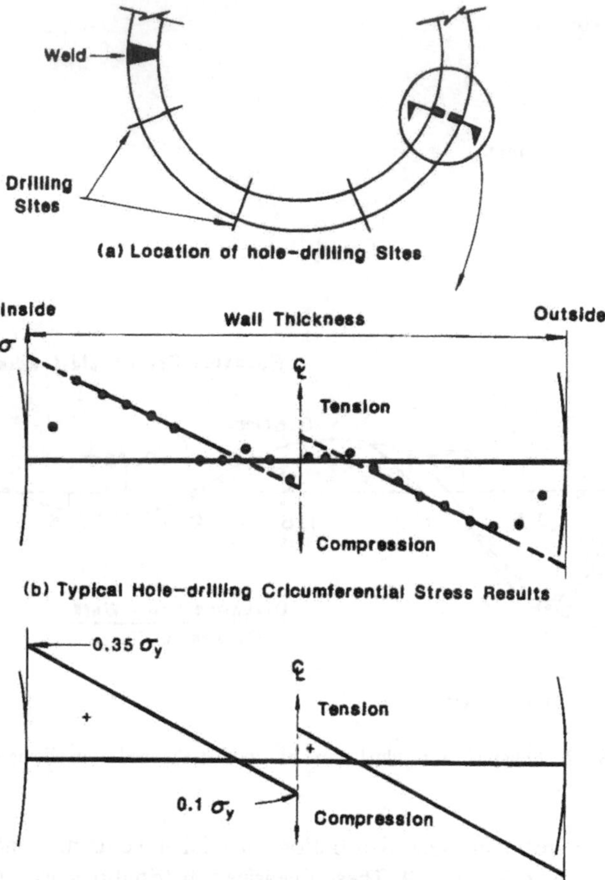

FIGURE 2.8 Circumferential residual stress pattern.

$$\frac{D_{max} - D_{min}}{D} = \frac{4w_i}{D} = 0.01 \tag{2.17}$$

where we have assumed the maximum out-of-roundness to be 1%. It is to be noted that measurements of the out-of-roundness of the ten fabricated tubular specimens reported in the earlier work (Chen and Ross, 1977) indicate that, in general, there is less than 1% difference between two perpendicular diameters at all positions along the column length. This 1% out-of-roundness adopted in the present analysis is the maximum tolerance specified by API (API, 1972) for allowable fabrication imperfections for out-of-roundness.

2.2.4 Effect of Hydrostatic Pressure

Amplification of Out-of-Roundness

The introduction of external hydrostatic pressure, Q_{cr}, to the imperfect cylinder results in an increased out-of-roundness, which in turn produces a significant bending

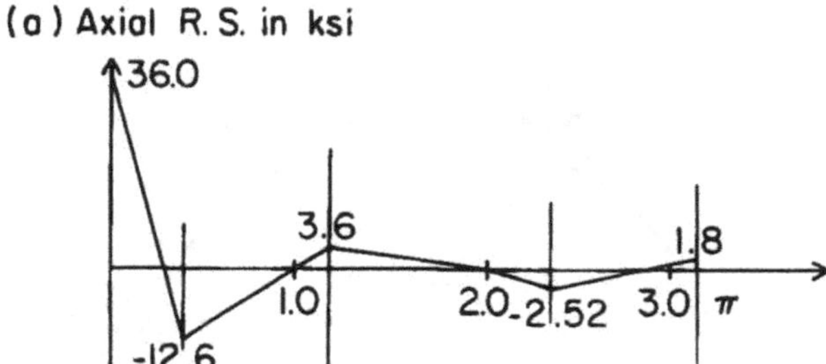

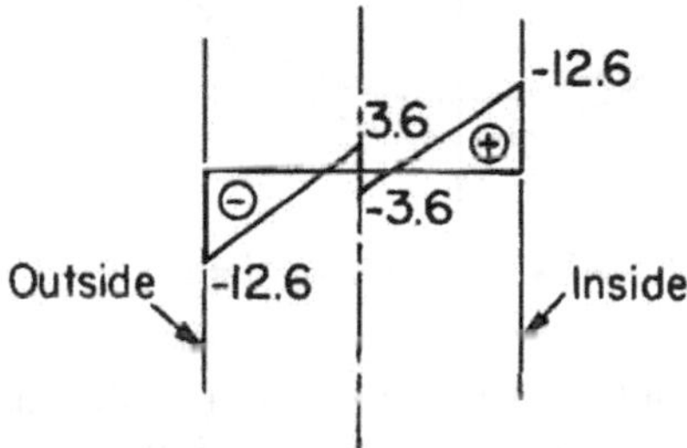

FIGURE 2.9 Residual stress distributions adopted in the present analysis.

stress in the circumference of the wall in addition to the uniform hoop compression. Herein, the total magnified geometrical imperfection is assumed to have the form

$$w = \frac{w_i \cos 2\theta}{1 - Q/Q_{cr}} \tag{2.18}$$

where Q_{cr} is the elastic buckling pressure for a perfect long cylinder. This critical pressure has the value (see, for example, Timoshenko and Gere, 1961).

$$Q_{cr} = \frac{2E(t/D)^3}{1 - \nu^2} \tag{2.19}$$

where t = wall thickness
D = mean diameter of the tube
ν = Poisson's ratio

As an example, Fig. 2.11 shows the out-of-roundness contours as amplified by various levels of external hydrostatic pressures Q/Q_{cr} = 0, 0.2, 0.4, 0.6 and 0.8.

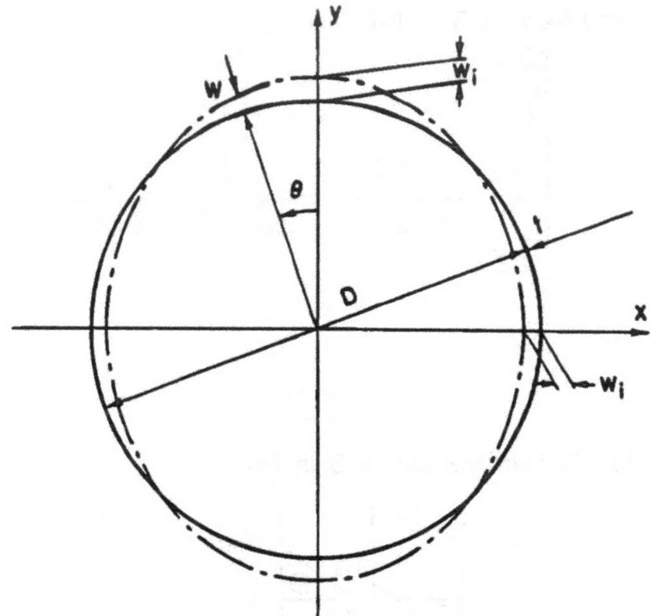

FIGURE 2.10 Out-of-roundness.

For a 15-in. tube with an initial imperfection of 1%, the maximum amplified deflection can be as large as 0.2 inch when $Q/Q_{cr} = 0.8$ (see Fig. 2.11a, not to scale). Figure 2.11b shows the comparison of the contour of a perfect cylinder with those out-of-roundness contours to the same scale.

Stress-Paths

For a cylindrical segment subjected to axial force and bending moment as well as hydrostatic pressure, the moment-curvature curves are obtained in two stages: the tangent stiffness method is first applied to calculate the stress distribution in the tube subjected to a given hydrostatic pressure. This part of the solution determines the initial state of stress in the tube before the axial compression and bending moment are applied to the tube. This is then followed by the application of tangent stiffness method again to obtain the moment-curvature solutions corresponding to a specified hydrostatic pressure Q and axial compression P.

Figure 2.12 shows the typical stress paths for four elements through the wall thickness of the tube. The initial residual stresses existing in the wall are denoted by the four points A1, A2, A3, and A4. After the application of external hydrostatic pressure (i.e., $Q/Q_{cr} = 0.4$), these points move horizontally to the left and the corresponding points are now denoted by B1, B2, B3, and B4. As the axial stresses σ_a are increased as a result of applying axial load combined with biaxial bending, these stress points move vertically downward until they reach the Tresca yield curve. After initial yielding, the stress points trace the Tresca yield curve moving in the

(a)

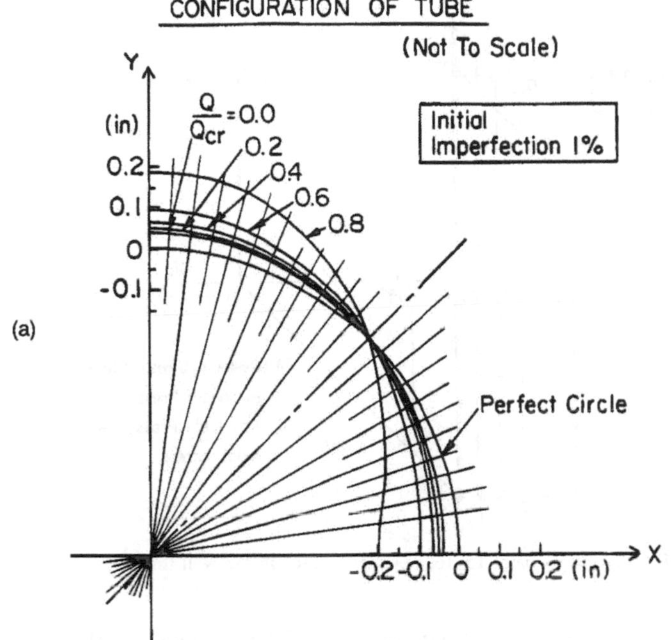

(b)

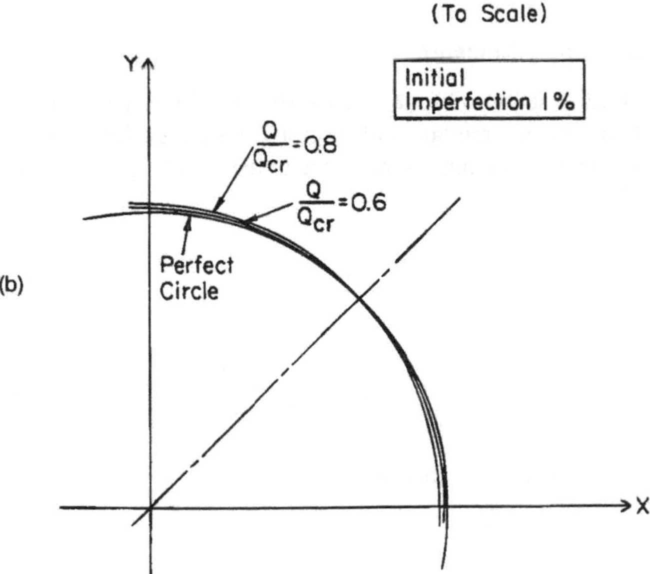

FIGURE 2.11 (a) Out-of-roundness amplified by hydrostatic pressure (D = 15 in.). (b) Comparison of perfect cylinder with out-of-roundness contours (D = 15 in.).

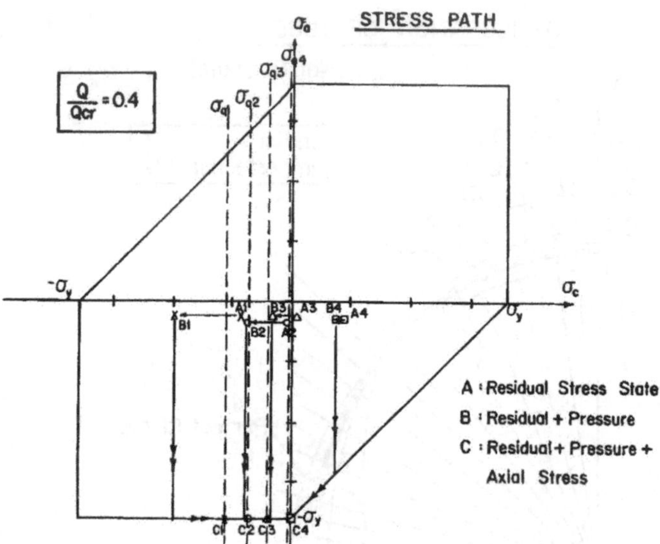

FIGURE 2.12 Stress paths for four elements through the wall thickness.

direction toward their corresponding final points C1, C2, C3, and C4 with the values σ_{q1}, σ_{q2}, σ_{q3}, and σ_{q4}, respectively, which are the stresses in equilibrium with external hydrostatic pressure. The calculation of these stresses follows.

Stresses Due to Hydrostatic Pressure

From equilibrium considerations only, it can be shown (Timoshenko and Gere, 1961) that the resultant circumferential axial force P_q and circumferential bending moment M_q per unit length of column segment due to hydrostatic pressure, Q, have the values (Fig. 2.13)

$$P_q = -Q\frac{D}{2} \tag{2.20}$$

$$M_q = P_q w = -Q\frac{D}{2}\frac{w_i \cos 2\theta}{1 - Q/Q_{cr}} \tag{2.21}$$

In the elastic range, the maximum elastic stresses through the wall thickness can be expressed as the sum of average axial stress σ_{ave} and bending stress σ_b

$$\sigma_{q_{\max}} = \sigma_{ave} \pm \sigma_b = \frac{QD}{t}\left[-\frac{1}{2} \pm \frac{3}{t}\frac{w_i \cos 2\theta}{1 - Q/Q_{cr}}\right] \tag{2.22}$$

In the elastic–plastic range, Eq. (2.22) no longer holds. The stress distributions

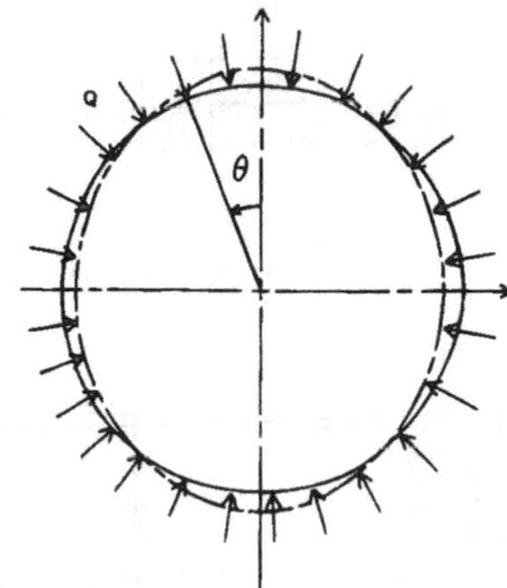

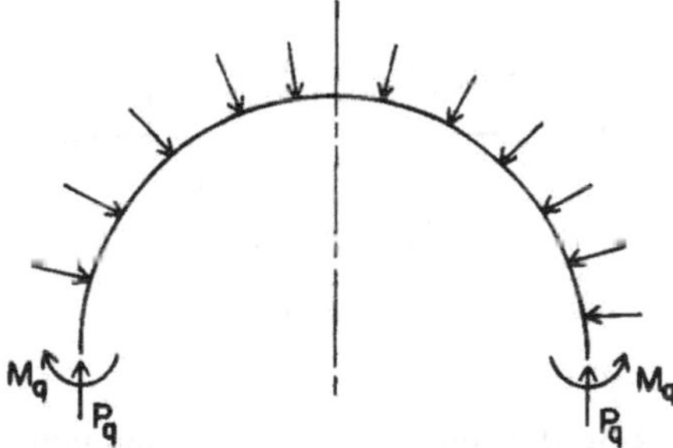

FIGURE 2.13 Axial force and bending moment due to hydrostatic pressure.

in the plastic tube due to a given external hydrostatic pressure, Q, can now be determined by the application of the tangent stiffness method described previously.

Here, as in the axial bending case, we assume that plane section through wall thickness remains plane after deformation. Thus, the strain ϵ_q at a point through the wall thickness of the tube can be expressed in a linear form as (Fig. 2.14a)

$$\epsilon_q = y\Phi_q + \epsilon_{0q} + \epsilon_r \tag{2.23}$$

and the strain rate equation is

$$\dot{\epsilon}_q = y\dot{\Phi}_q + \dot{\epsilon}_{0q} \tag{2.24}$$

since the residual strain, ϵ_r, is independent of time.

The equations of equilibrium in the circumferential direction are

$$\dot{M}_q = \int \dot{\sigma}_q y dA \tag{2.25}$$

$$\dot{P}_q = \int \dot{\sigma}_q dA$$

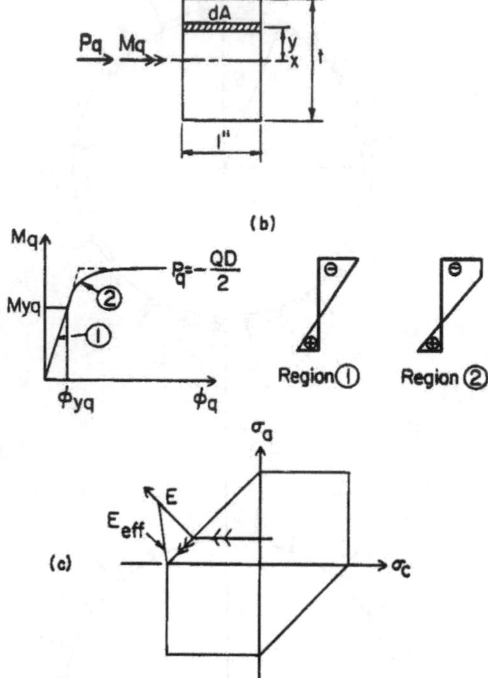

FIGURE 2.14 Calculations of stresses in the plastic range due to hydrostatic pressure.

Using the concept of effective modulus

$$\dot{\sigma}_q = E_{eff}\dot{\epsilon}_q \tag{2.26}$$

and the compatibility relation Eq. (2.24), Eq. (2.25) can be rewritten in the matrix form as

$$\begin{Bmatrix} \dot{M}_q \\ \dot{P}_q \end{Bmatrix} = \begin{bmatrix} \int E_{eff}y^2 dA & \int E_{eff}y dA \\ \int E_{eff}y dA & \int E_{eff} dA \end{bmatrix} \begin{Bmatrix} \dot{\Phi}_q \\ \dot{\epsilon}_{0q} \end{Bmatrix} \tag{2.27}$$

Figure 2.14b shows the elastic–plastic circumferential bending moment-curvature relation (M_q, Φ_q) corresponding to a given hydrostatic pressure Q. The corresponding stress path in the biaxial stress space (σ_c, σ_a) is shown in Fig. 2.14c. The effective Young's modulus and the stress-strain relation are described in Section 2.2.2.

A limit analysis solution for the pipeline hydrostatic collapse pressure was reported (Pan, 1978). This solution was obtained by simply equating the maximum elastic moments M_q as given by Eq. (2.21) to the fully plastic moment capacity M_{pc}

of the tube, including the effect of axial force P_q. The result can be written in the following form

$$\frac{Q/Q_y}{1 - Q/Q_{cr}}\left[\frac{D_{\max} - D_{\min}}{D}\right]\left[\frac{D}{t}\right] + \left[\frac{Q}{Q_y}\right]^2 = 1.0 \qquad (2.28)$$

where t is the thickness of the wall and Q_y is the yield pressure of a perfectly round pipe that has the value $\sigma_y 2t/D$.

Further, the elastic limit pressure at which the extreme fiber stress of Eq. (2.22) just reaches the yield stress σ_y has the form (Timoshenko and Gere, 1961)

$$\frac{Q}{Q_y} = \frac{1}{1 + \left[\frac{D_{\max} - D_{\min}}{D}\right]\left[\frac{D}{t}\right]\frac{3/2}{1 - Q/Q_{cr}}} \qquad (2.29)$$

These two extreme solutions along with the result of present computer analysis are compared in Fig. 2.15. In this figure, the solid line is the present computer solution using the tangent stiffness method as described here with four layers through the wall thickness. The tangent stiffness solution is seen slightly below that of the limit analysis solution. This is because the moment-curvature relationship ($M_q - \sigma_q$) used in the limit analysis is idealized as linearly elastic and perfectly plastic type as shown by the dashed lines in Fig. 2.14b.

2.2.5 Computer Implementation

Based upon the equations formulated, a computer program has been developed to provide numerical results. A brief flow chart is given in Fig. 2.16. The program is written in the language FORTRAN. The program, which is named "NEWMARK," includes short-column and long-column analysis. The program list and sample calculations are stored in the accompanying floppy disk.

First, the input data are read into the computer, such as material properties, member and element sizes, load increments, residual stresses, and imperfections. Second, the sectional characteristics and coordinates of each element are generated from the input data by the subroutine "ELEMENT." The subroutine "INITIAL" then assigns the residual stress distribution. This is followed by the subroutine "FORCE," which assigns the order of load increments, ΔM_x, ΔM_y and ΔP, and the subroutine "CURVE," which assigns the magnitude of the load increment for the solution.

The tangent stiffness iteration is carried out by the subroutine "REPEAT" with its convergence controlled by the subroutine "CONVRG." The subroutine "REPEAT" consists of four parts: the calculation of sectional stiffness matrix, its inverse matrix, strain increments (curvatures), and resultant stresses (force) of the section. The subroutine "YONGEF" calculates the effective Young's modulus, E_{eff},

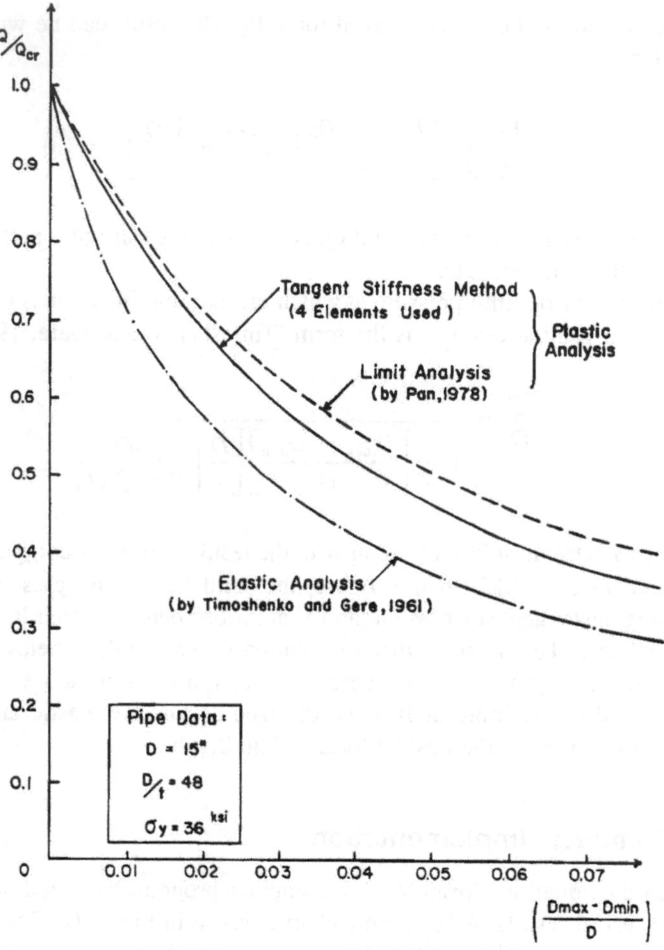

FIGURE 2.15 Comparison of pipe collapse pressure.

for each element in the cross section. This value is used for subsequent stress and stiffness matrix calculations.

The elements of the tangent stiffness matrix are evaluated numerically by dividing the cross section into finite elements as shown in Fig. 2.17 with four layers through the thickness of the wall. The effect of mesh sizes is studied by dividing the circumference of the tube into 48, 72, and 96 equal radial segments, which result in a total of 192, 288, and 384 elements, respectively, as shown in Fig. 2.17. The numerical results of this study are summarized and compared in Table 2.1. The dimensions, loadings, and material properties of the tube are also given in Table 2.1. It can be seen that the mesh with n = 192 gives a sufficiently accurate result and hence is adopted herein for the parameter studies of the moment-curvature-thrust relationships of the tubular segments in the subsequent sections.

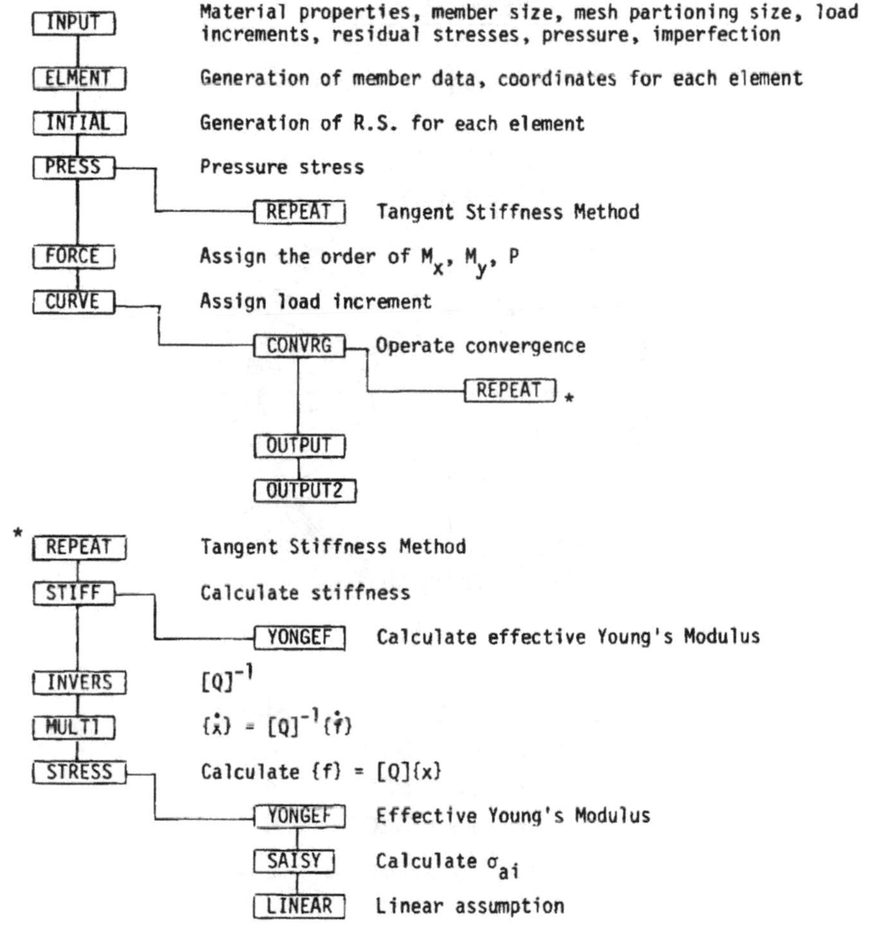

FIGURE 2.16 Computer program for moment-curvature relationship.

2.2.6 Numerical Results

Numerical studies have been performed for sectional behavior of tubular members using the program NEWMARK. Typical results are summarized in what follows.

Effect of Residual Stresses

Moment-curvature relations for a tubular cross section with both longitudinal and circumferential residual stresses are shown in Fig. 2.18. The residual stress distributions adopted for the curves are shown in Fig. 2.9. It can be seen that the residual stresses significantly reduce the stiffness of the cross section, especially near the region immediately beyond the initial yielding of a cross section. Also, it can be seen that as the axial load increases, the moment capacity of a cross section with

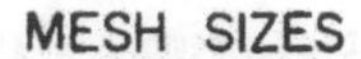

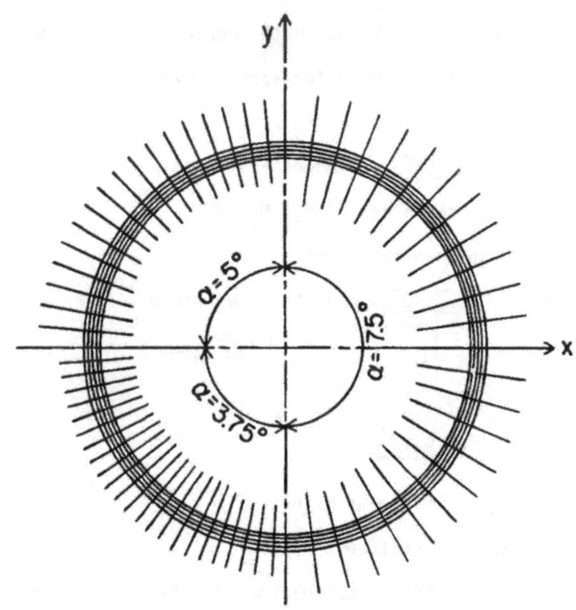

α		NTH		NLA
7.5°	x	48	= 360°	4
5°	x	72	= 360°	4
3.75°	x	96	= 360°	4

FIGURE 2.17 Tubular section divided into finite elements.

TABLE 2.1 Comparison of Curvature for Three Mesh Sizes
(unit: radian per unit length)

MESH \ M_y	1400 in-kip	1700 in-kip	2000 in-kip	2150 in-kip
48 × 4 = 192	1.216×10^{-4}	1.548×10^{-4}	2.890×10^{-4}	6.912×10^{-4}
72 × 4 = 288	1.215×10^{-4}	1.542×10^{-4}	2.881×10^{-4}	6.850×10^{-4}
96 × 4 = 384	1.214×10^{-4}	1.542×10^{-4}	2.870×10^{-4}	6.824×10^{-4}

Note: $M_x = 0$
$P = -103.7$ kip ($P/P_y = 0.2$)
Imperfection = 1%

$D = 15$ in $E = 30000$ ksi
$t = .312$ in $\sigma_y = 36$ ksi
$P_y = 518.3$ kip $\nu = 0.3$
$M_p = 2423.5$ in-kip

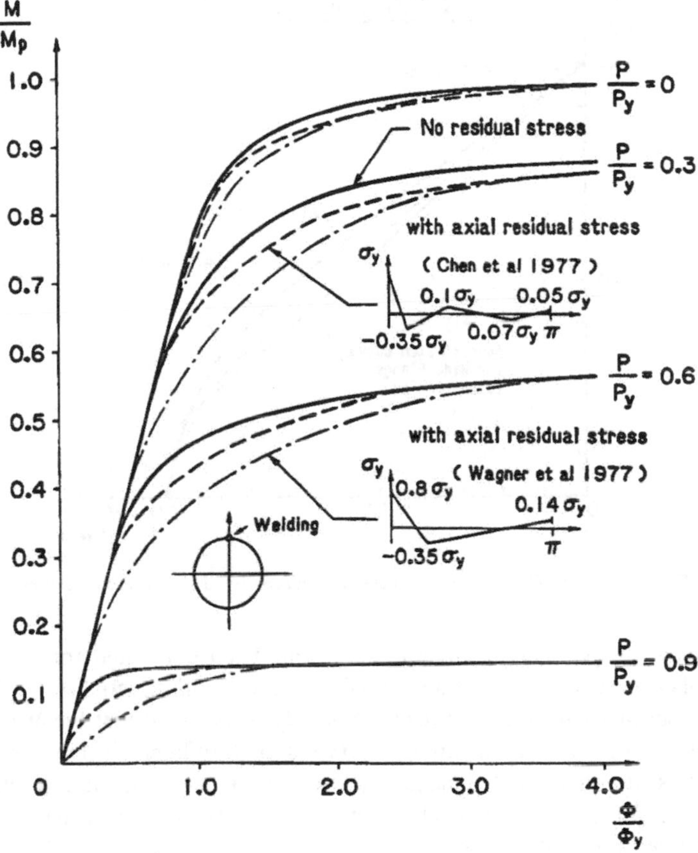

FIGURE 2.18 Comparison of M-P-Φ curves for residual stress distribution.

residual stresses approaches earlier to the maximum strength than without residual stresses.

The effect of different patterns of longitudinal residual stress distribution on the M-Φ-P curve is also shown in Fig. 2.18. With longitudinal residual stress distribution of the type adopted by Wagner et al. (1976), as shown in the inset of Fig. 2.18, their results are found to be considerably lower than the present results. This is because the longitudinal residual stress distribution used in the Wagner et al. analysis has a larger and more uniform compressive stress near the region away from the diametrical plane containing the longitudinal weld. Since in these regions the elements are far from the neutral axis, they result in an earlier yielding of the cross section.

Interaction Strength for Sectional Capacity

The maximum strength interaction curve between axial force, P/P_y, and bending moment, M/M_p, is shown as a solid line in Fig. 2.19 and is compared with that of

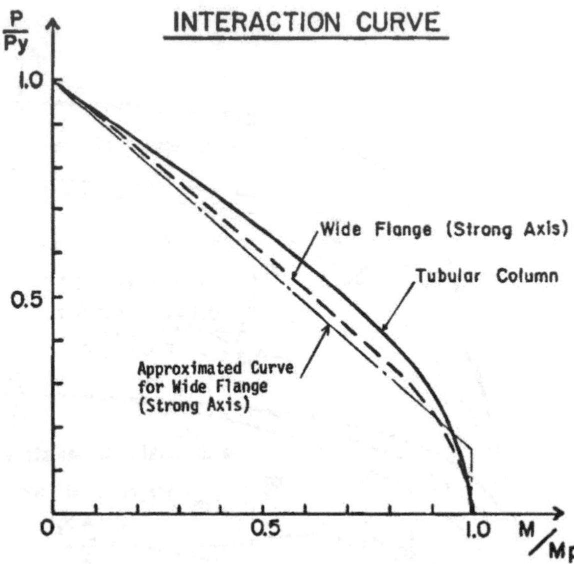

FIGURE 2.19 Comparison of interaction curves for different cross sections.

a wide flange shape. This interaction curve has been computed for an imperfect cylindrical cross section with both longitudinal and circumferential residual stresses as well as out-of-roundness described previously. It is found that the effect of out-of-roundness on the maximum strength curve is negligibly small. Further, since the residual stress do not affect the maximum strength of a cross section, it follows that the curve shown in Fig. 2.19 is also applicable for the case of a perfect circular section without residual stresses.

The interaction curve for a perfect circular cross section has been developed previously (ASCE, 1971) and its numerical results are shown in Fig. 2.20. The solid line in Fig. 2.19 corresponds to the special case of either $m_x = M_x/M_p = 0$ or $m_y = M_y/M_p = 0$ in Fig. 2.20. A good agreement is observed between the corresponding values of Fig. 2.19 and Fig. 2.20.

The theoretical interaction curve for a wide flange section bending about its strong axis as shown by the dashed line in Fig. 2.19 for the case $A_f/A_w = 1.0$ (A_f = one flange area and A_w = web area) can be approximated by the linear equations (ASCE, 1971)

$$\frac{M}{M_p} = 1.0 \quad \text{for} \quad 0 \le \frac{P}{P_y} \le 0.15 \tag{2.30}$$

$$\frac{M}{M_p} = 1.18\left(1 - \frac{P}{P_y}\right) \quad \text{for} \quad 0.15 < \frac{P}{P_y} \le 1.0 \tag{2.31}$$

A better approximation for the interaction curve of the wide flange shape shown in Fig. 2.19 has the form

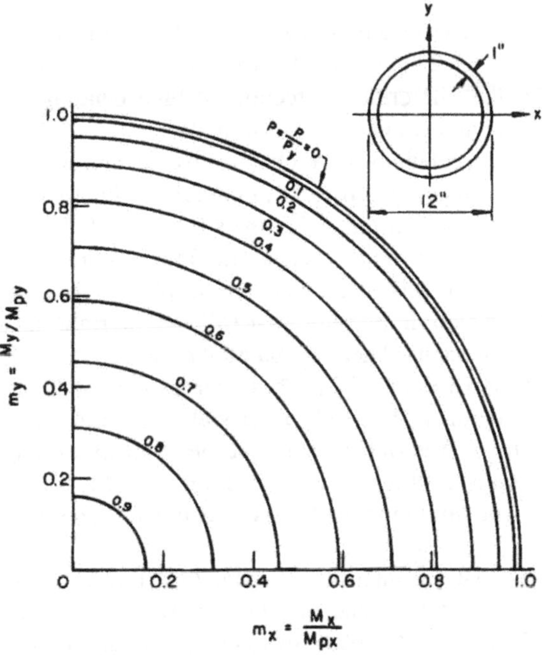

FIGURE 2.20 Interaction curves of hollow circular section.

$$\frac{M}{M_p} = 1.0 - 1.52\left(\frac{P}{P_y}\right)^2 \quad \text{for} \quad 0 \le \frac{P}{P_y} \le 0.5 \tag{2.32}$$

$$\frac{M}{M_p} = 1.24\left(1 - \frac{P}{P_y}\right) \quad \text{for} \quad 0.5 < \frac{P}{P_y} \le 1.0 \tag{2.33}$$

Using the same type of equations, the corresponding equations for a fabricated cylindrical cross section are obtained.

$$\frac{M}{M_p} = 1.0 - 1.18\left(\frac{P}{P_y}\right)^2 \quad \text{for} \quad 0 \le \frac{P}{P_y} \le 0.65 \tag{2.34}$$

$$\frac{M}{M_p} = 1.43\left(1 - \frac{P}{P_y}\right) \quad \text{for} \quad 0.65 < \frac{P}{P_y} \le 1.0 \tag{2.35}$$

Equations (2.34) and (2.35) give the maximum moment capacity of a tubular cross section in the presence of axial force.

Effect of Welding Location

As described previously, cylindrical members used in offshore structures are cold-formed from plate, welded along a single longitudinal seam, and used without any kind of stress relief. The resulting asymmetrical pattern of longitudinal residual stress, as shown in Fig. 2.7, and its effect on the moment-curvature behavior are examined here so that the critical direction of tube bending with respect to the location of longitudinal weld can be determined.

Moment-curvature relations for a tubular cross section with different directions of applied bending moment with respect to the longitudinal weld are shown in Fig. 2.21 for three cases of the constant axial loads $P/P_y = 0$, 0.3, and 0.6. The bending moment vector M, as shown in the insets of the figures follows the right-hand screw rule; that is, the moment acts in the diametrical plane, which is perpendicular to the moment vector shown in the sense that follows the right-hand screw rule. The moment-curvature curves has been computed by considering only the longitudinal residual stress, as shown in Fig. 2.9a. These curves are nondimensionalized by the fully plastic hinge moment, M_p, and initial yield curvature, Φ_y, of a perfect tube with no residual stress. It is well known that the intensity of axial force P reduces the plastic hinge moment of a tube cross section from M_p for $P = 0$ to M_{pc} for $P \neq 0$. This fully plastic limit moment M_{pc} is a sectional property, independent of the residual stress.

It can be seen that the intensity of axial load has a major influence on the determination of the critical plane of bending. However, in general the critical bending plane is found to be the plane that is perpendicular to the diametrical plane containing the longitudinal weld. In other words, the critical direction of bending moment is the one whose vector passes through the longitudinal weld. This conclusion, however, is not always correct when the intensity of moment is changed, but the differences among various orientations studied are found to be small.

Fortunately, the stiffness (i.e., slopes of the moment-curvature relations) and maximum moment capacity do not appear to differ much from the critical case described previously. For simplicity, we shall assume in the following parameter study that the critical bending moment is the one whose moment vector passes through the longitudinal weld.

Effect of Hydrostatic Pressure

Moment-curvature relations for a cylindrical cross section with residual stresses (both longitudinal and circumferential residual stresses) and geometric imperfection (with initial out-of-roundness of 1%) are shown in Fig. 2.22 for two values of external hydrostatic pressure, $Q/Q_{cr} = 0.4$ and 0.6. Theoretical assumptions for hydrostatic pressure are explained in Section 2.2.4. The intensity of axial load is varied from $P/P_y = 0$ to 0.8. For the case of a small axial load, $P/P_y = 0 \sim 0.2$, the intensity of hydrostatic pressure Q has a major influence on the values of initial yield moment, tangent stiffness, and maximum moment capacity of the tube. As the axial load increases to $P/P_y = 0.6$ and 0.8, apart from the case of a perfect tube, the moment-curvature relations are not affected by the presence of hydrostatic pressure.

Figure 2.23 shows the maximum strength interaction curves between the axial load P/P_y and bending moment M/M_p with hydrostatic pressure varying from $Q/Q_{cr} = 0$ to 0.8 and with 1% out-of-roundness. These curves have been computed

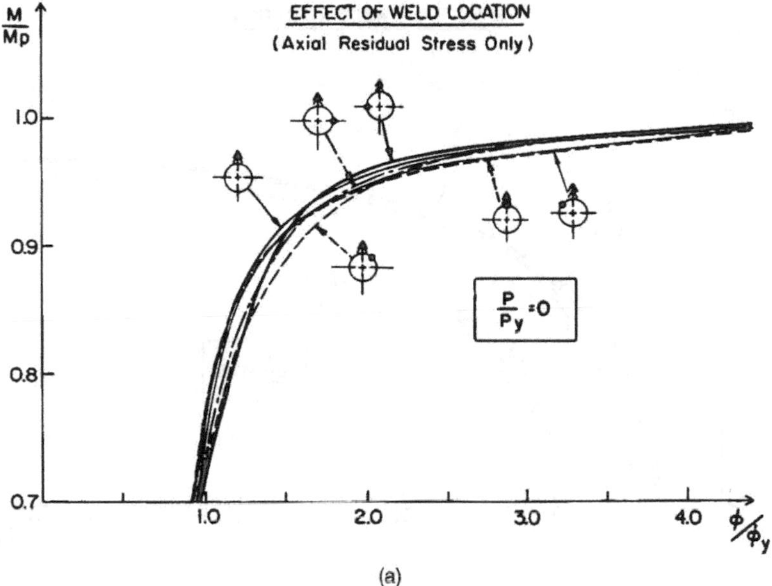

(a)

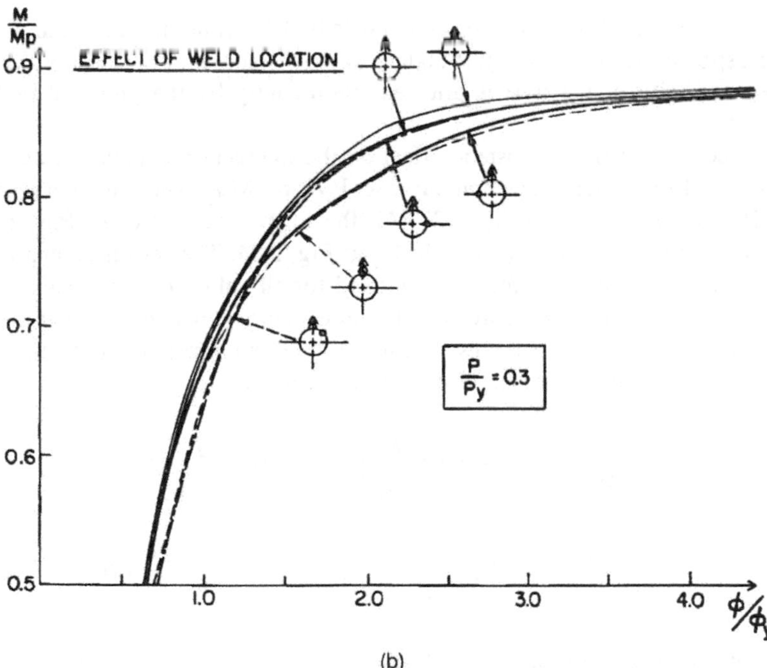

(b)

FIGURE 2.21 (a) Moment-curvature relations for tube bending with different weld location, $P/P_y = 0$. (b) moment-curvature relations for tube bending with different weld location, $P/P_y = 0.3$. (c) Moment-curvature relations for tube bending with different weld location, $P/P_y = 0.6$.

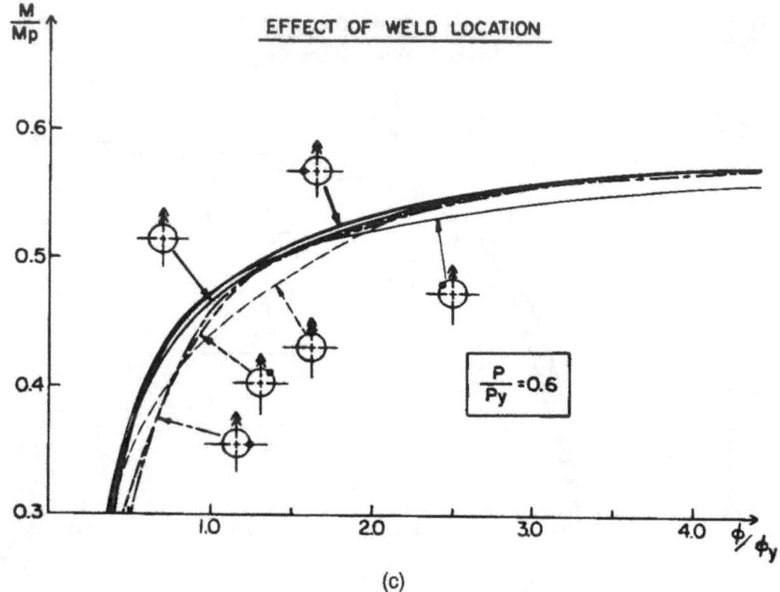

(c)

FIGURE 2.21 *Continued*

from an imperfect cylindrical cross section with both longitudinal and circumferential residual stresses. Here, in the region under low axial loads ($P/P_y \leq 0.5$), the maximum moment capacity of the tube is affected significantly by the presence of hydrostatic pressure.

To account for the hydrostatic effect on the interaction curves, we denote the end points of the exact interaction curve as P_{yQ} and M_{pQ}, where the curve crosses the P/P_y and M/M_p axes in Fig. 2.23. All the interaction curves of Fig. 2.23 are now replotted on axes of P/P_{yQ} vs. M/M_{pQ} in Fig. 2.24. The results shown in these replots are seen to form a rather narrow band for all values of Q/Q_{cr} varying from 0 to 0.8. Also shown in the figure are the average curves, which represent the band. The dotted representative curve as shown in Fig. 2.24 corresponding to 1% out-of-roundness can be approximated closely by the equations

$$\frac{M}{M_{pQ}} = 1.0 - 1.12\left(\frac{P}{P_{yQ}}\right)^2 \quad \text{for} \quad 0 \leq \frac{P}{P_{yQ}} < 0.65 \tag{2.36}$$

$$\frac{M}{M_{pQ}} = 1.5\left(1 - \frac{P}{P_{yQ}}\right) \quad \text{for} \quad 0.65 < \frac{P}{P_{yQ}} \leq 1.0 \tag{2.37}$$

These equations are also applicable to the case of 2% out-of-roundness, although the spreading of band is slightly wider than 1% out-of-roundness.

The fully plastic bending moment reduced for hydrostatic pressure, M_{pQ}, may be approximated empirically by the equation

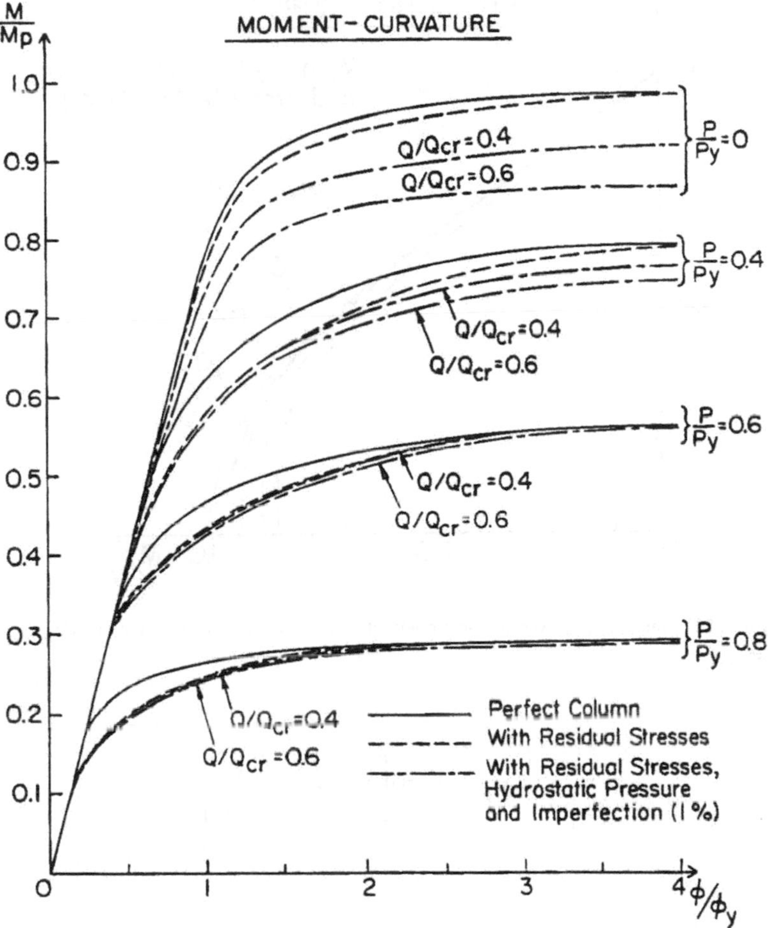

FIGURE 2.22 M-Φ-P-Q relations for an imperfect cylindrical cross section.

$$\frac{M_{pQ}}{M_p} = 1.0 - \frac{i_R + 3}{20}\left(\frac{Q}{Q_{cr}}\right) \tag{2.38}$$

in which i_R is the value of out-of-roundness of a cross section expressed in percent. Also, the reduced axial capacity of a cross section due to hydrostatic pressure may be approximated by the equation

$$\frac{P_{yQ}}{P_y} = 1.0 - 0.02\left(i_R \frac{Q}{Q_{cr}}\right)^2 \tag{2.39}$$

In Eqs. (2.38) and (2.39), the fully plastic moment M_p and the yield axial capacity P_y merely depend on cross section and can be calculated from the equations

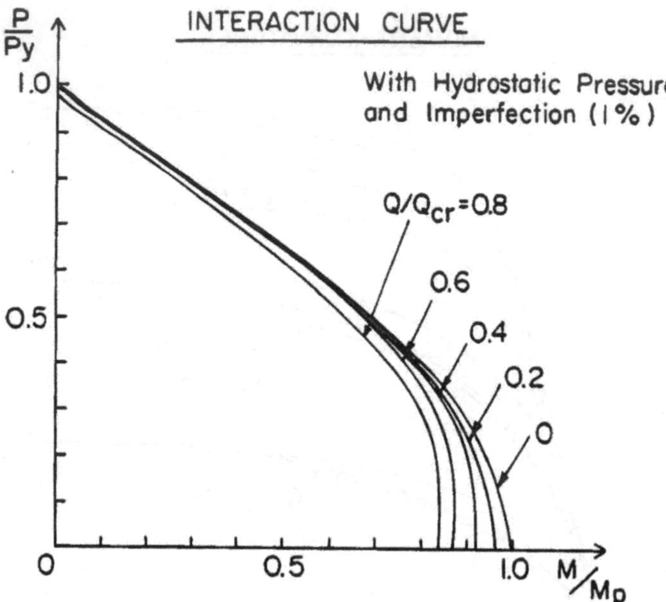

FIGURE 2.23 Interaction curves for an imperfect cross section with residual stresses (1% out-of-roundness).

$$M_p = \frac{\sigma_y}{6}D^3\left[1 - \left(1 - \frac{2t}{D}\right)^3\right] \tag{2.40}$$

and

$$P_y = \sigma_y A \tag{2.41}$$

It should be noted that the external hydrostatic pressure Q/Q_{cr} cannot exceed the pipe collapse pressure given in Fig. 2.15.

2.3 BEHAVIOR AND STRENGTH OF LONG COLUMNS

2.3.1 Newmark's Method

For the analysis of long beam-columns for which elastic theory may be applied, the governing differential equations may be solved rigorously by the use of formal mathematics. In the plastic or nonlinear range, however, the differential equations are often intractable and recourse must be made to numerical methods to obtain solutions. Newmark's method appears to be one of the most convenient numerical methods to obtain the maximum strength of a beam-column by first tracing the load

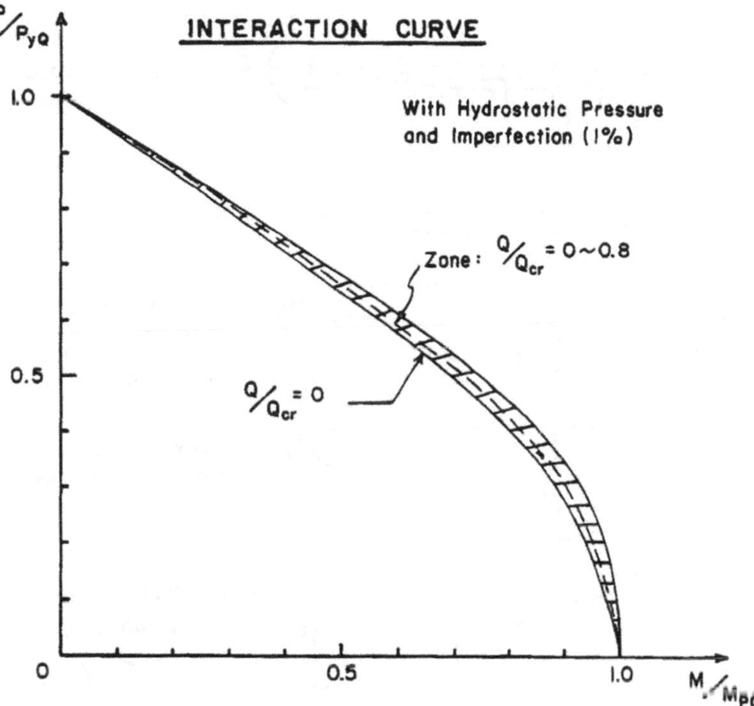

FIGURE 2.24 Interaction curves including the effect of hydrostatic pressure (1% out-of-roundness).

deflection (or load rotation) curve of a beam-column and then determining the peak point from this curve.

Newmark's integration method is a useful means of computing the deflected shape from a given curvature distribution. The moment-curvature-thrust relationship for the cross section must be known before applying this method.

Newmark's method requires iteration. The basic steps involved are shown in Fig. 2.25 and summarized below (Chen and Atsuta, 1976):

1. Assume a deflected shape.
2. Compute bending moments at all stations selected along the length of a member.
3. Compute curvatures at all stations from a known M-P-Φ relationship of the section.
4. Assume the distribution of curvature between two stations to be quadratic, and compute the contribution of the curvature to the slope at adjacent stations.
5. Compute slopes at all stations.
6. Integrate the slopes along the length of a member to obtain the new deflected shape.

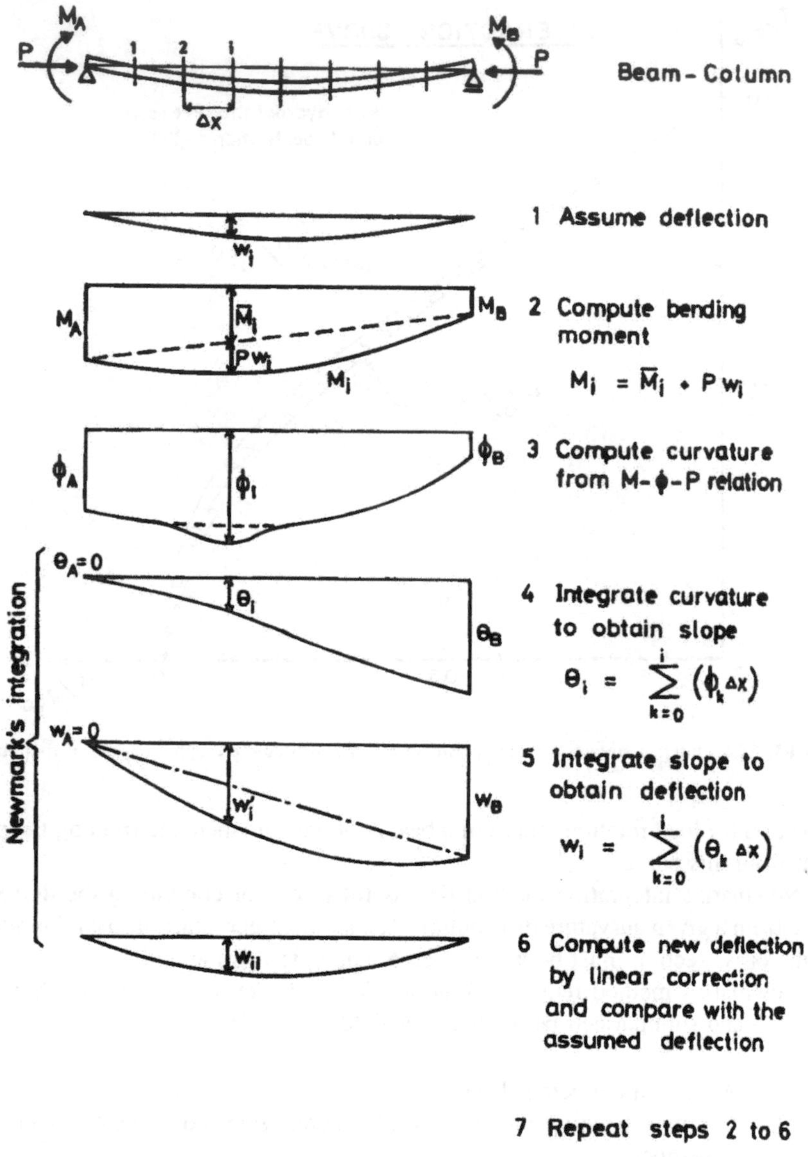

FIGURE 2.25 Beam-column analysis by Newmark's method.

7. Repeat steps (2) through (6) until the deflected shape converges to a prescribed tolerance.

The above procedures must be repeated for every increment of the axial load until the resultant deflection diverges, at which point the axial load exceeds the maximum strength of a beam-column (buckling load). Thus, the maximum strength

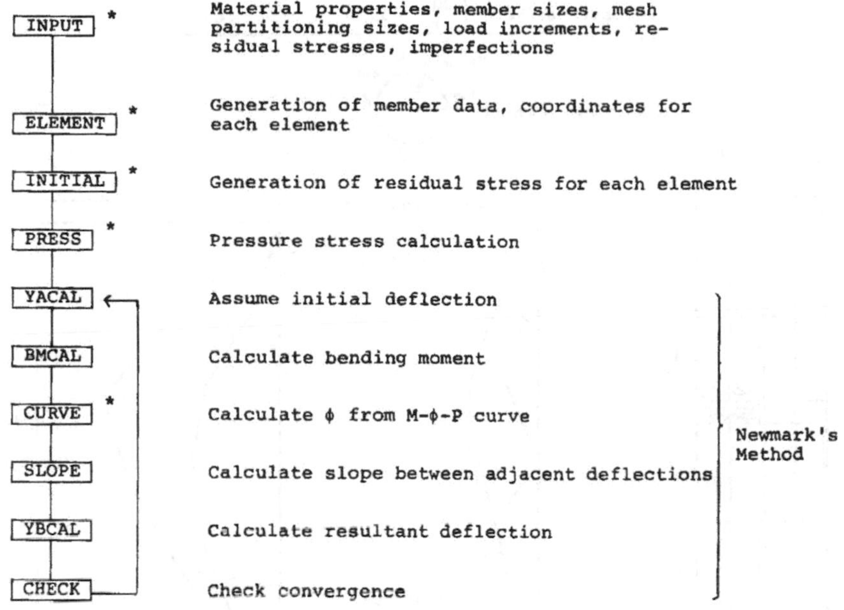

FIGURE 2.26 Computer program for long column analysis.

of a beam-column is obtained as the peak point of the axial load vs. the end rotation curve (or deflection) of a beam-column.

2.3.2 Out-of-Straightness

The effect of out-of-straightness on the buckling strength of a beam-column is considered here. Two types of out-of-straightness are considered: measured values for actual specimens and assumed values for general study.

Analytical studies of the results of tests on 10 full-scale specimens conducted at Lehigh University (Chen and Ross, 1977) were first made using the computer model developed. These studies provided the necessary confirmation of the validity of the analytical procedure. In the analysis, the measured initial deflection is approximated by piecewise straight lines connecting the "offsets" at adjacent stations. The out-of-straightness measured in the test and assumed in the calculation are shown by "dotted" and solid lines, respectively, in Fig. 2.27 to be analyzed later.

For general study, the column strength curves are developed, assuming the out-of-straightness to have the sinusoidal shape with maximum amplitudes of 0.1% and 0.2% of the column length. The maximum amplitude specified by API specifications is 0.1% (API, 1972).

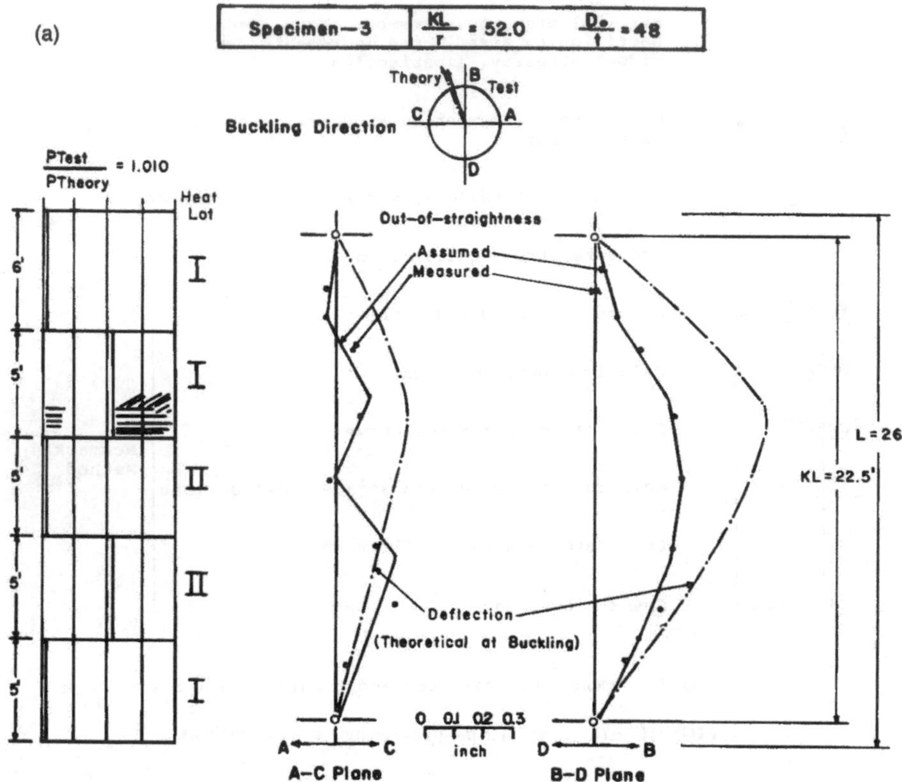

FIGURE 2.27 (a) Calculated deflection pattern at buckling (Specimen 3). (b) Calculated deflection pattern at buckling (Specimen 4).

2.3.3 Computer Implementation

A brief flow chart of the program for long-column analysis is shown in Fig. 2.26. The program NEWMARK for both short-column and long-column analysis is stored in the attached floppy disk. A sample calculation is also stored on the disk.

The convergence of the deflection in the program is specified when the difference between the assumed deflection and resulting deflection at each station is less than 1%. When the deflection diverges after a given increment of axial load, the mean value of the two last axial loads is taken as the buckling load of the column. This criterion of the convergence is found to be adequate for the analysis of ideal perfect circular columns.

The location of seam welding affects the direction of buckling and strength of the column in conjunction with the out-of-straightness. Therefore, Newmark's method for a long column is programmed to analyze in two perpendicular planes so that the buckling direction of a column can be determined.

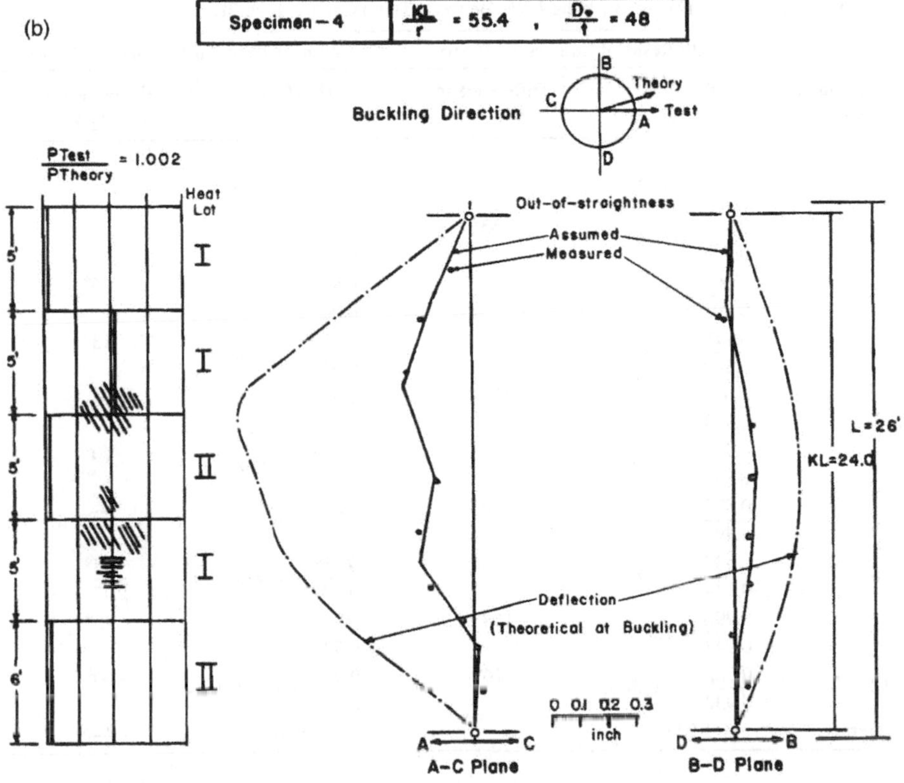

FIGURE 2.27 *Continued*

In the following numerical studies, five to seven elements are used for the study of Lehigh test specimens and eight elements are used for the development of a column strength curve.

2.3.4 Numerical Results

Numerical studies have been performed for member behaviors using the program NEWMARK. Typical results are summarized in the next section.

Analysis of Lehigh Tests

The Lehigh test results are analyzed here using the computer model developed (Toma and Chen, 1979). Table 2.2 shows the comparison between the theoretical values computed and the test results for the 10 Lehigh specimens. In the analysis, two cases are considered: with and without the effects of residual stresses. In general, it can be seen that the theoretical values with the effect of residual stresses are slightly more conservative than the experimental results. Material strain hardening

TABLE 2.2 Comparison of Buckling Load Between Theory and Lehigh Test

	With Residual Stresses in Theory		Without Residual Stresses in Theory	
Specimen Number	$(P_{test}/P_y)/(P_{theory}/P_y)$	Difference in Buckling Angle	$(P_{test}/P_y)/(P_{theory}/P_y)$	Difference in Buckling Angle
1	$\frac{0.981}{0.926} = 1.059$	54°	$\frac{0.981}{0.944} = 1.039$	28°
2	$\frac{0.988}{0.926} = 1.067$	53°	$\frac{0.988}{0.944} = 1.047$	30°
3	$\frac{0.900}{0.891} = 1.010$	4°	$\frac{0.900}{0.909} = 0.990$	6°
4	$\frac{0.910}{0.908} = 1.002$	17°	$\frac{0.910}{0.944} = 0.964$	13°
5	$\frac{1.010}{0.974} = 1.037$	171°	$\frac{1.010}{0.985} = 1.025$	—
6	$\frac{0.973}{0.964} = 1.009$	2°	$\frac{0.973}{0.985} = 0.988$	20°
7	$\frac{0.877}{0.891} = 0.984$	3°	$\frac{0.877}{0.944} = 0.929$	—
8	$\frac{0.938}{0.944} = 0.994$	46°	$\frac{0.938}{0.979} = 0.958$	5°
9	$\frac{0.990}{0.943} = 1.050$	164°	$\frac{0.990}{0.974} = 1.016$	45°
10	$\frac{0.968}{0.932} = 1.039$	16°	$\frac{0.968}{0.964} = 1.004$	60°
Average	1.025	—	0.996	—

probably contributes to this increased strength for tested specimens, which is not considered in the present computer model.

The buckling mode and the buckling direction are shown in Fig. 2.27. A specimen and its yielded pattern near buckling are shown to the left of the figures. KL, marked to the right of the figures, is the effective pinned-end length, which is used in the present theoretical analysis. Theoretical deflection curves in two orthogonal directions, A-C and B-D planes, are shown in the middle of the figures, from which the buckling direction and critical location of the column can be obtained.

The location of the critical "can" is affected by both initial out-of-straightness and strength of the material. For example, the critical section for specimens in Fig. 2.27a is located in the second "can" from the top, which has less strength than the "can" located at the center of the column; and this agrees fairly well with the observed buckling mode.

Column Strength Curves

Two theoretical column strength curves for a fabricated cylindrical column are computed considering the effects of longitudinal and circumferential residual stresses, 1% and 2% of out-of-roundness and 0.1% and 0.2% of out-of-straightness.

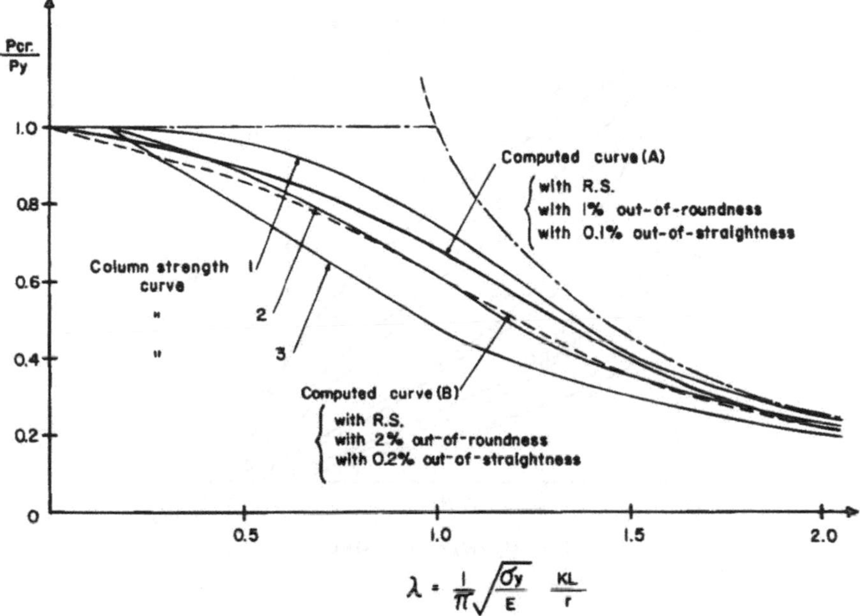

FIGURE 2.28 Comparison of cylindrical column strength curves with SSRC multiple column curves.

The two theoretical curves are compared with the SSRC multiple column curves (SSRC, 1988) in Fig. 2.28.

The computed column strength curve (A) in Fig. 2.28 lies between SSRC curves 1 and 2, which is close to the AISC LRFD design curve (1993) and the AISC plastic design curve (1989).

Comparing the computed curve (A) with the computed curve (B) in Fig. 2.28, it is seen that the increased imperfections have a significant weakening effect on the load-carrying capacity of a fabricated tube.

Long Columns Loaded Eccentrically at the Ends

The analysis of beam-columns subjected to eccentric loads applied at the ends is essentially the same as that of the axially loaded case. Herein, the imperfections and residual stresses adopted are 1% out-of-roundness, 0.1% out-of-straightness, and residual stress distributions of the type shown in Fig. 2.9. The number of stations used in the present Newmark's method is seven.

Interaction curves for the combinations of axial force and end moment that can be safely supported by the fabricated tubular column are shown in Fig. 2.29 for both perfect and imperfect columns.

The effect of imperfections and residual stresses on the maximum strength of a fabricated cylindrical column may be seen in these figures by comparing the solid curves with the dashed curves. It is seen that the difference between the perfect and

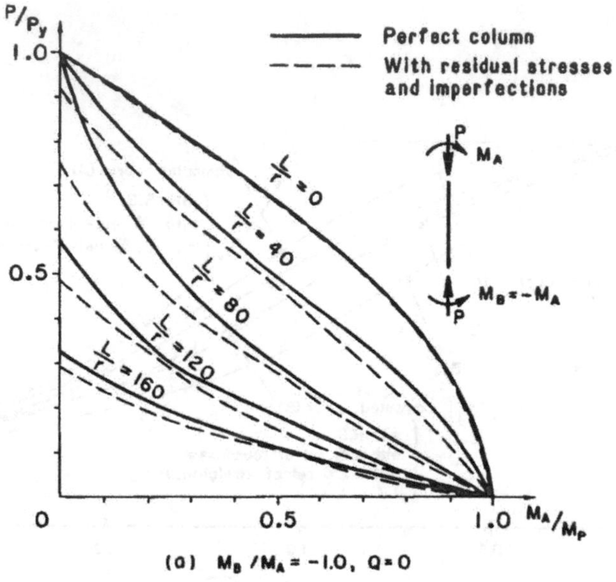

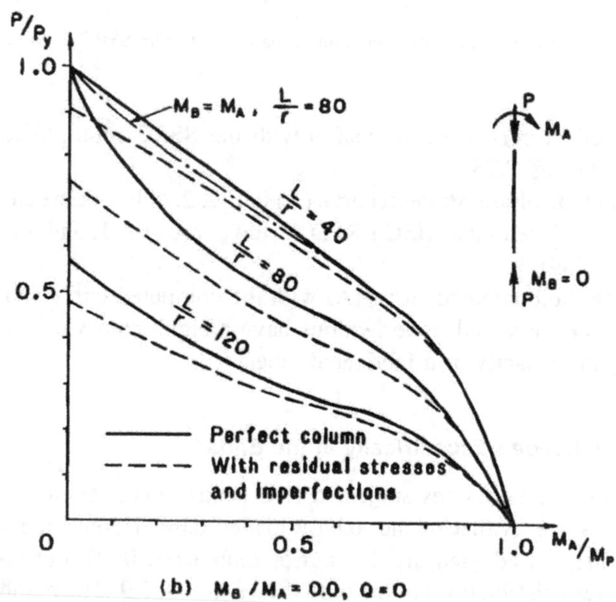

FIGURE 2.29 (a) Interaction curves eccentrically loaded columns, $M_B/M_A = -1.0$, $Q = 0$. (b) Interaction curves for eccentrically loaded columns, $M_B/M_A = 0.0$, $Q = 0$.

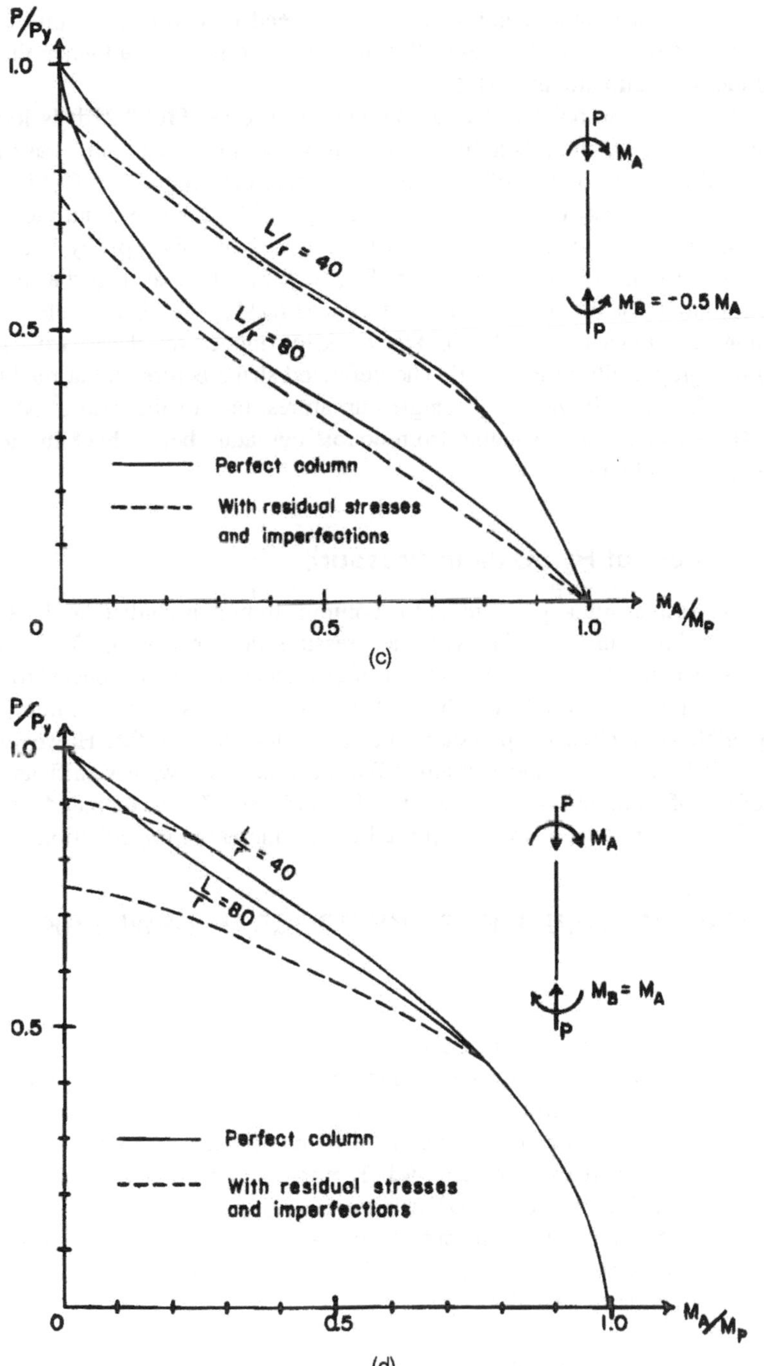

FIGURE 2.29 (c) Interaction curves for eccentrically loaded columns, $M_B/M_A = -0.5$, $Q = 0$. (d) Interaction curves for eccentrically loaded columns, $M_B/M_A = 1.0$, $Q = 0$.

imperfect curves is quite significant when the end eccentricity is small and the eccentricity ratio M_B/M_A decreases. The difference becomes small when the values of slenderness ratio are decreased.

If the inflection point in the double curvature case (Fig. 2.29d) is located at the center of the column, then the half length of the double curvature case may be considered the same as that of the single curvature case $M_B/M_A = 0$ (Fig. 2.29c). However, the comparison between the case $M_B = M_A$, $L/r = 80$ and the case $M_B = 0$, $L/r = 40$ in Fig. 2.29c shows that there is some discrepancy between the curves in the range of high axial force ($P.P_y \geq 0.5$). The reason is that the actual buckling mode for the double curvature case ($M_B/M_A = 1$) is a single curvature buckling, as shown in Fig. 2.30b. The buckling modes for these two cases are illustrated graphically in Fig. 2.30. The deflected shape before and at buckling for the case $M_B/M_A = 0$ are always single curvatures, but for the case of $M_B = M_A$ the deflected shapes are changing from double curvature before buckling to single curvature at buckling.

2.3.5 Effect of Hydrostatic Pressure

The effect of hydrostatic pressure on the column strength is studied, and the typical results are shown in Fig. 2.31, with the pressure varying from $Q/Q_{cr} = 0$ to 0.8 (Toma and Chen, 1983). In the case of out-of-straightness 0.1% and out-of-roundness 1% (or computed curve A), the effect of hydrostatic pressure on column strength is negligible for hydrostatic pressure up to the value $Q/Q_{cr} = 0.6$. However, when $Q/Q_{cr} = 0.8$, this hydrostatic pressure effect becomes somewhat significant. In the case of out-of-straightness 0.2% and out-of-roundness 2% (or computed curves B), this effect is seen to be slightly amplified by the increased imperfections.

2.4 USER'S MANUAL FOR THE PROGRAM NEWMARK

```
Line 1.
IND; Flag for type of analysis
       =1; Analysis of M-Φ curve (short column analysis,
           skip Lines 2 through 7)
       =2; Analysis of Newmark's method in the case of a
           symmetric column with respect to middle of the
           column (long column analysis)
       =3; Analysis of Newmark's method in the case of an
           asymmetric column along its length (long
           column analysis)

Line 2. (skip when IND = 1)
AL; Length of column (ft)
NEL; Number of elements for Newmark's method
```

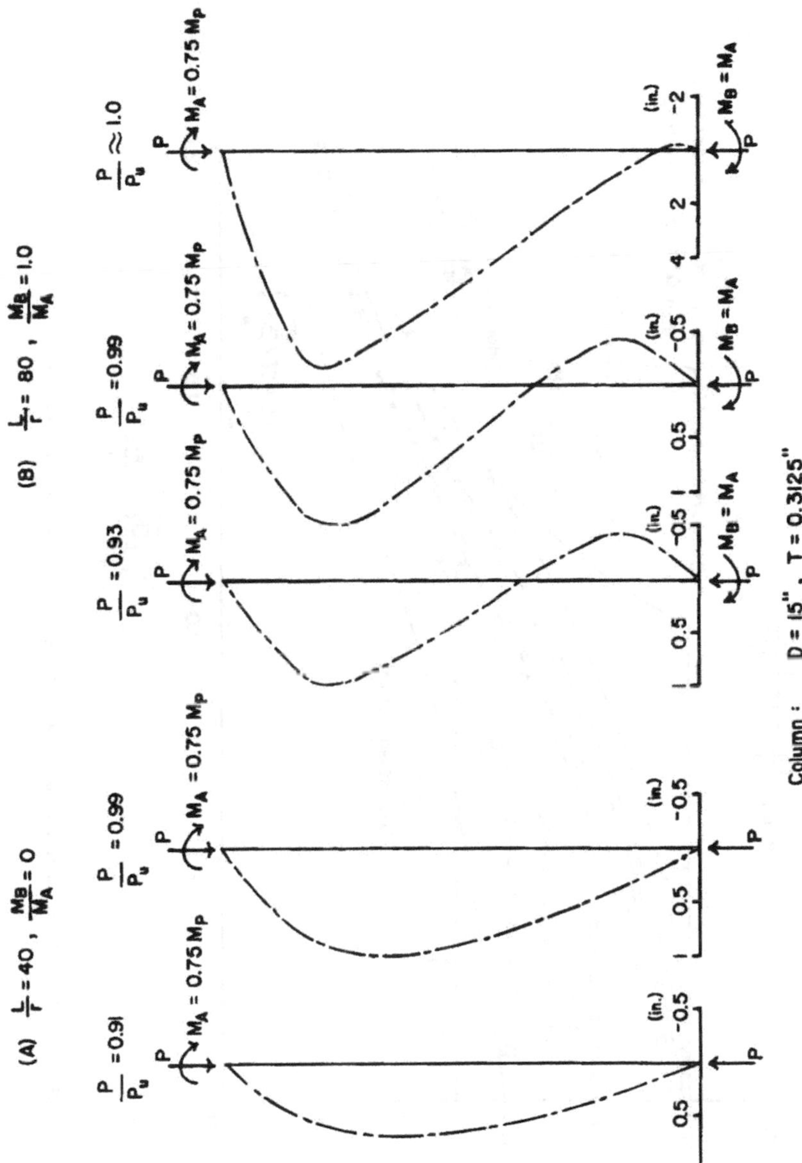

FIGURE 2.30 Buckling modes for $M_B/M_A = 0$ and 1.0.

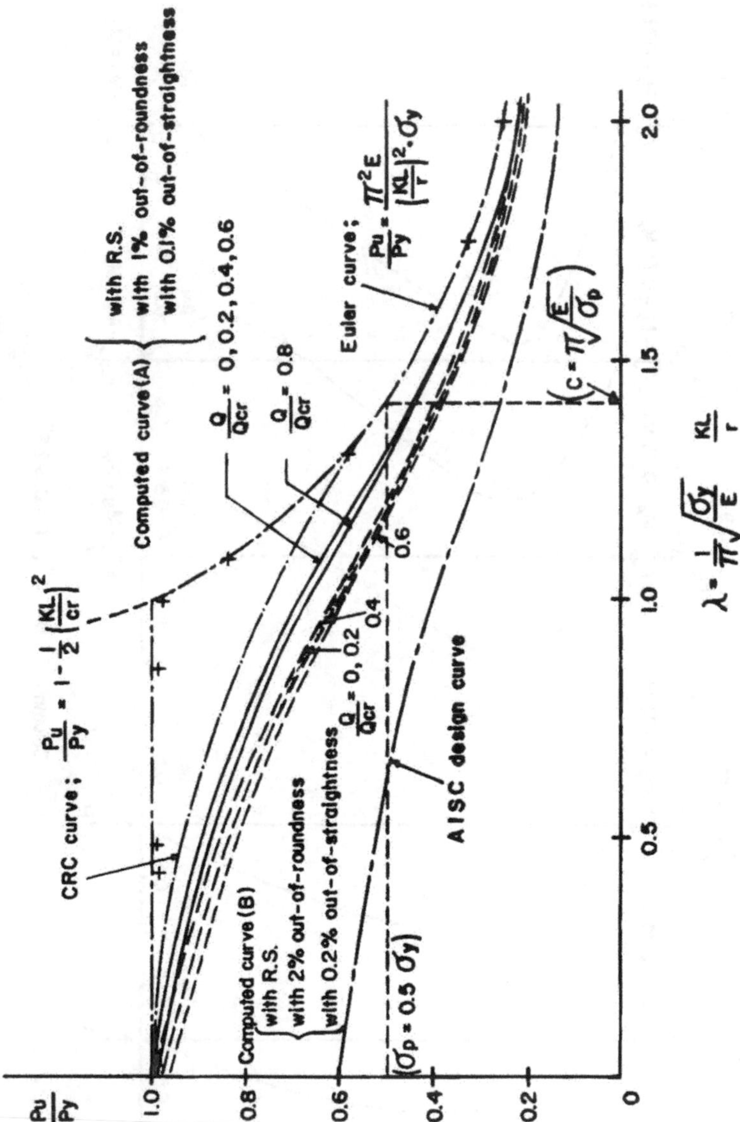

FIGURE 2.31 Effect of hydrostatic pressure on axial strength of fabricated tubular columns.

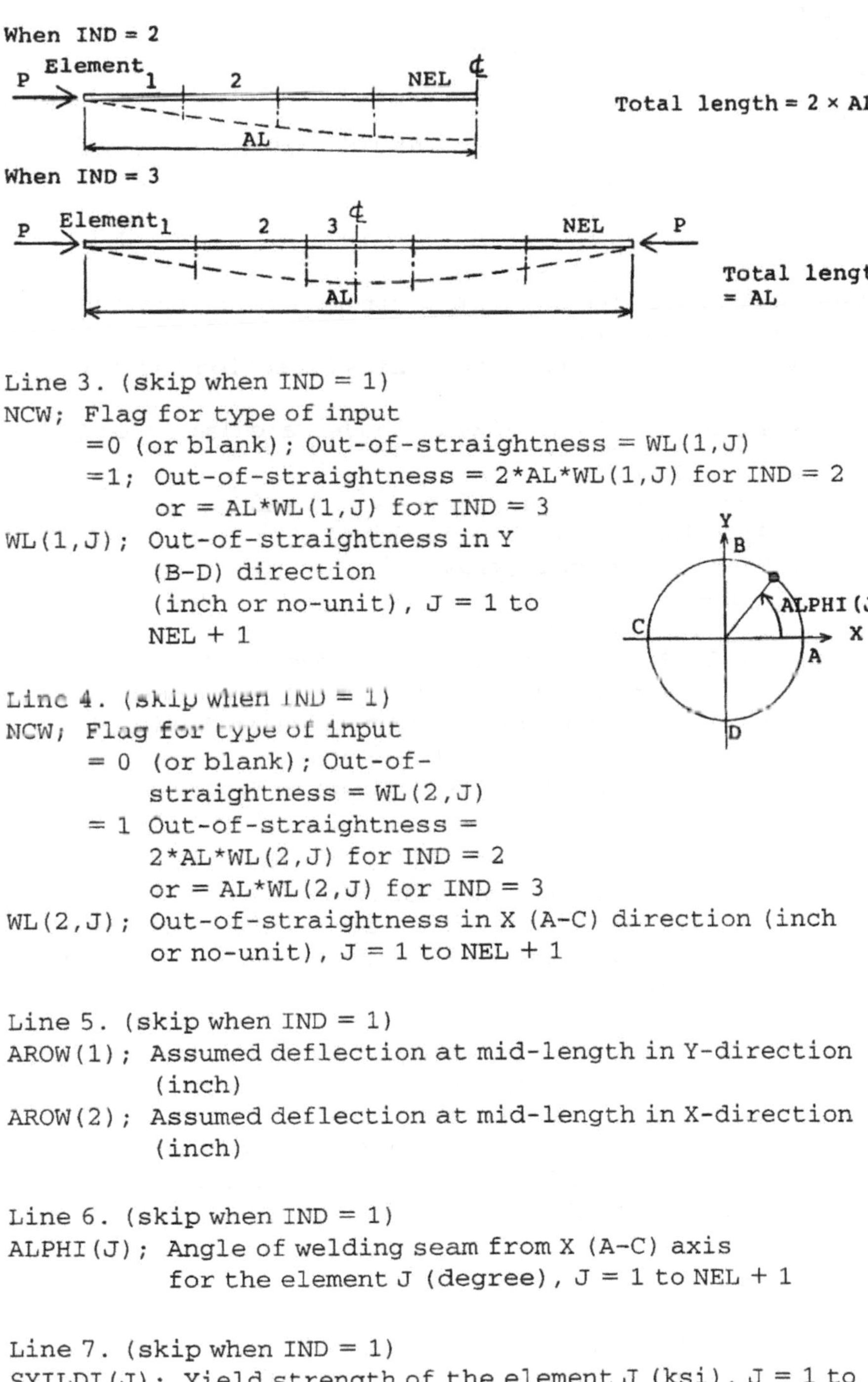


```
Line 3. (skip when IND = 1)
NCW; Flag for type of input
      =0 (or blank); Out-of-straightness = WL(1,J)
      =1; Out-of-straightness = 2*AL*WL(1,J) for IND = 2
          or = AL*WL(1,J) for IND = 3
WL(1,J); Out-of-straightness in Y
         (B-D) direction
         (inch or no-unit), J = 1 to
         NEL + 1

Line 4. (skip when IND = 1)
NCW; Flag for type of input
      = 0 (or blank); Out-of-
         straightness = WL(2,J)
      = 1 Out-of-straightness =
         2*AL*WL(2,J) for IND = 2
         or = AL*WL(2,J) for IND = 3
WL(2,J); Out-of-straightness in X (A-C) direction (inch
         or no-unit), J = 1 to NEL + 1

Line 5. (skip when IND = 1)
AROW(1); Assumed deflection at mid-length in Y-direction
         (inch)
AROW(2); Assumed deflection at mid-length in X-direction
         (inch)

Line 6. (skip when IND = 1)
ALPHI(J); Angle of welding seam from X (A-C) axis
          for the element J (degree), J = 1 to NEL + 1

Line 7. (skip when IND = 1)
SYILDI(J); Yield strength of the element J (ksi), J = 1 to
           NEL + 1
```

Line 8.
TITLE; Title of the calculation

Line 9.
SYILD; Yield strength (ksi), ignored when IND = 2 and 3 (see Line 7)
YONG; Young's modulus (ksi)
DIA; Diameter of pipe (in.)
TH; Thickness of pipe (in.)
NTH; Number of elements along circumference for M-Φ curve analysis
NLA; Number of elements in thickness direction for M-Φ curve analysis
ENU; Poisson's ratio (used only in the calculation of Q_{cr})

Line 10. (input in sequence for i = 1, 2 and 3)
FS(i); Starting applied forces (kips)
DF(i); Force increments (kips)
NF(i); Number of increments
i = 1; Axial force, P
i = 2; Bending moment with respect to Y axis, M_y (ignored when IND = 2 and 3)
i = 3; Bending moment with respect to X axis, M_x (ignored when IND = 2 and 3)
Note: Sign convention follows as shown in Figure 2.2.

Line 11.
NCTRL; Flag for equilibrium condition
= 1; the point in Tresca yield diagram stops at σ_y
= 3; the point in Tresca yield diagram stops at σ_i (lower bound theory)
NCTRL2; Flag for data output
NCTRL2; = 1; Output the reference data, coordinates of the elements, residual stresses, etc.
NCTRL2; = 0; Output no reference data
Note: When IND = 2 or 3, NCTRL2 is set as 0.
ALPH; Angle of the welding seam location from X (A–C) axis (ignored when IND = 2 and 3, see Line 6)

Line 12.
NRSTC; Number of input points for residual stress distribution in thickness direction
RLC(i); Location of the point (in.), RLC(i) = 0 to thickness

```
FLC(i); Residual stress in thickness direction (ksi),
         i = 1 to NRSTC
Note: Begin with external surface inward.

Line 13.
NRSTA; Number of input points for residual stress
        distribution along circumference
RLA(i); Location of the point (radian), RLA (i) = 0 to π
         (symmetry with respect to the weld seam)
FLA(i); Ratio of residual stress along circumference to
         yield stress, i = 1 to NRSTA

Line 14.
Q;      External pressure (ksi)
WI;     Initial out-of-roundness
        (in.)
BETA;   Angle of the direction of out-
        of-roundness
        from X-axis (degree)
```

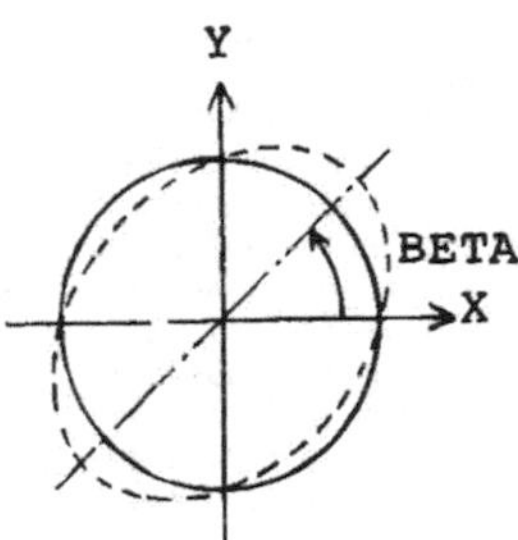

2.5 SAMPLE CALCULATIONS

2.5.1 Short-Column Analysis

INPUT DATA OF THE PROGRAM "NEWMARK" FOR A SHORT COLUMN PROBLEM

```
1
Sample Calculation for Short Column
3.500E+01 3.000E+04 1.000E+01     0.340     48     4          0.30
0.000E+00 1.083E+02  3  3.000E+02 5.000E+01 22 0.000E+00 0.000E+00 1
      1    0                 90
  4   0 -12.6 0.170  3.6 0.170  -3.6 0.340  12.6
  6   0       35 0.380 -12.6 1.180  3.6 1.980  0.0 2.360-2.520 3.142 1.80
    0.000E+00            0.000E+00        0
```

OUTPUT RESULT OF THE PROGRAM "NEWMARK" FOR A SHORT COLUMN PROBLEM

```
Sample Calculation for Short Column
  YIELD ST =    35.00    E = 30000.00    D =10.000    T = .340    NTHTA = 48    NT = 4
  POISSON RATIO =  .30

FORCE INCREMENT  (MX, MY, P)
      FORCE      INCREMENT      TIMES
         .00           .00          1
      300.00         50.00         22
         .00        108.30          3
 NCTRL = 1   NCTRL2 = 0
```

```
RESIDUAL STRESS
 ALPH = 90.000
  CIRCUM   .000-12.600    .170  3.600    .170 -3.600    .340 12.600
  AXIAL    .000 35.000    .380-12.600   1.180  3.600   1.980   .000   2.360 -2.520   3.142 1.800
 EXTERNAL PRESSURE   =        .00000E+00
 INITIAL OUT-OF-ROUNDNESS=  .00000E+00   BETA =    .00
DATA OF MEMBER
 AREA = .10318E+02 I = .12051E+03 S = .24101E+02 Z = .31740E+02 SHAPE FUNCTION = .13170E+01
 R =  .34174E+01
 M(YILD)= .84354E+03 M(PLASTIC)= .11109E+04 P(YILD)= .36114E+03 PHI(YIELD)= .23333E-03
 DA =    .55160E-01   .54214E-01   .53268E-01   .52322E-01
 RI =    .50000E+01   .49150E+01   .48300E+01   .47450E+01   .46600E+01

 QCR=   .27771E+01   STRESS DUE TO PRESSURE =   .00000E+00   .00000E+00
      MX          MY           P          XA(1)        XA(2)         XA(3)
-.67418E-05 .30000E+03  .11723E-04  -.30516E-05 .81915E-04   .49835E-05  ITERATION = 2
 .65616E-05 .35000E+03  .82412E-05  -.30516E-05 .95766E-04   .49835E-05  ITERATION = 1
 .12921E-04 .40000E+03  .11792E-04  -.30516E-05 .10962E-03   .49835E-05  ITERATION = 1
-.10149E-05 .45000E+03  .49126E-05  -.30516E-05 .12347E-03   .49835E-05  ITERATION = 1
 .20752E-04 .50000E+03  .54238E-05  -.30516E-05 .13732E-03   .49835E-05  ITERATION = 1
 .30421E-04 .55000E+03 -.42975E-05  -.30516E-05 .15117E-03   .49835E-05  ITERATION = 1
 .70970E-05 .60000E+03  .94460E-05  -.30516E-05 .16502E-03   .49835E-05  ITERATION = 1
 .97014E-05 .65000E+03 -.39902E-05  -.30516E-05 .17887E-03   .49835E-05  ITERATION = 1
 .16335E-02 .69982E+03 -.30614E-01  -.30516E-05 .19272E-03   .49835E-05  ITERATION = 1
 .51701E-04 .85000E+03  .99350E-05  -.30394E-05 .23621E-03   .66741E-05  ITERATION = 2
 .64344E-02 .89994E+03  .37737E-02  -.29949E-05 .25322E-03   .68463E-05  ITERATION = 2
 .65646E-01 .94951E+03  .12326E-01  -.27728E-05 .27851E-03   .90849E-05  ITERATION = 2
 .39070E+00 .99842E+03  .25087E+00  -.46937E-05 .32690E-03   .44009E-06  ITERATION = 2
-.29792E+00 .10482E+04  .12631E+00  -.12558E-04 .44622E-03  -.38015E-04  ITERATION = 2
-.96645E-01 .10999E+04  .43964E-03  -.33152E-04 .12294E-02  -.14103E-03  ITERATION = 4
**** STIFFNESS MATRIX IS SINGULAR
**** STIFFNESS MATRIX IS SINGULAR
-.84860E+00 .11092E+04  -.12825E+00 -.47750E-04 .38136E-02  -.21463E-03   ITERATION = 2
  SA(I,J) AND SC(I,J)
  1 -.3500E+02 -.3500E+02 -.3500E+02 -.3500E+02 -.8550E+01 -.4500E+00 .0000E+00 .0000E+00
  2 -.3500E+02 -.3500E+02 -.3500E+02 -.3500E+02 -.8550E+01 -.4500E+00 .0000E+00 .0000E+00
  3 -.3500E+02 -.3500E+02 -.3500E+02 -.3500E+02 -.8550E+01 -.4500E+00 .0000E+00 .0000E+00
  4 -.3500E+02 -.3500E+02 -.3500E+02 -.3500E+02 -.8550E+01 -.4500E+00 .0000E+00 .0000E+00
  5 -.3500E+02 -.3500E+02 -.3500E+02 -.3500E+02 -.8550E+01 -.4500E+00 .0000E+00 .0000E+00
  6 -.3500E+02 -.3500E+02 -.3500E+02 -.3500E+02 -.8550E+01 -.4500E+00 .0000E+00 .0000E+00
 ~~~~~~~~~~~~~~~~~~~~~~~~~~~~~~~~~~~~~~~~~~~~~~~~~~~~~~~~~~~~~~~~~~~~~~~~~~~~~~~~~~~~~~~
 46 -.3500E+02 -.3500E+02 -.3500E+02 -.3500E+02 -.8550E+01 -.4500E+00 .0000E+00 .0000E+00
 47 -.3500E+02 -.3500E+02 -.3500E+02 -.3500E+02 -.8550E+01 -.4500E+00 .0000E+00 .0000E+00
 48 -.3500E+02 -.3500E+02 -.3500E+02 -.3500E+02 -.8550E+01 -.4500E+00 .0000E+00 .0000E+00

           PO            PP          BMT(I)         BMP
  1    .00000E+00   -.76500E+00   .00000E+00   .94286E-01
  2    .00000E+00   -.76500E+00   .00000E+00   .94286E-01
  3    .00000E+00   -.76500E+00   .00000E+00   .94286E-01
  4    .00000E+00   -.76500E+00   .00000E+00   .94286E-01
  5    .00000E+00   -.76500E+00   .00000E+00   .94286E-01
  6    .00000E+00   -.76500E+00   .00000E+00   .94286E-01
  7    .00000E+00   -.76500E+00   .00000E+00   .94286E-01
  8    .00000E+00   -.76500E+00   .00000E+00   .94286E-01
  9    .00000E+00   -.76500E+00   .00000E+00   .94286E-01
 10    .00000E+00   -.76500E+00   .00000E+00   .94286E-01
 11    .00000E+00   -.76495E+00   .00000E+00   .94293E-01
 12    .00000E+00    .16213E-07   .00000E+00   .18857E+00
 ~~~~~~~~~~~~~~~~~~~~~~~~~~~~~~~~~~~~~~~~~~~~~~~~~~~~~~~~~
 46    .00000E+00   -.76500E+00   .00000E+00   .94286E-01
 47    .00000E+00   -.76500E+00   .00000E+00   .94286E-01
 48    .00000E+00   -.76500E+00   .00000E+00   .94286E-01

      MX          MY          P          XA(1)        XA(2)        XA(3)
-.48067E-03 .30000E+03 .10830E+03 -.25743E-05 .81884E-04 .35599E-03 ITERATION = 2
-.94976E-03 .35000E+03 .10830E+03 -.26101E-05 .95737E-04 .35590E-03 ITERATION = 1
-.14992E-02 .40000E+03 .10830E+03 -.26456E-05 .10959E-03 .35582E-03 ITERATION = 1
```

```
.34971E-02 .44974E+03 .10825E+03 -.26809E-05 .12345E-03 .35574E-03 ITERATION = 1
.73535E-02 .49933E+03 .10816E+03 -.27173E-05 .13747E-03 .35607E-03 ITERATION = 1
.12832E-01 .54892E+03 .10807E+03 -.27523E-05 .15173E-03 .35700E-03 ITERATION = 1
.11594E-01 .59854E+03 .10798E+03 -.27863E-05 .16623E-03 .35857E-03 ITERATION = 1
```

2.5.2 Long-Column Analysis

INPUT DATA OF THE PROGRAM "NEWMARK" FOR A LONG COLUMN PROBLEM

```
        3.463E+01            3
0.000E+00 0.000E+00 0.000E+00 0.000E+00
2.338E+01 2.338E+01 2.338E+01 2.338E+01
0.000E+00 0.000E+00
9.000E+01 2.700E+02 9.000E+01 2.700E+02
3.600E+01 3.600E+01 3.600E+01 3.600E+01
 Sample Calculation for Long Column
 3.600E+01  3.000E+04  1.500E+01 3.130E-01   48     4        0.30
-5.191E+01-1.000E+01  1 0.000E+00 5.000E+01 10 0.000E+00 5.000E+01  4
      1     0               90
  4   0 -12.6 0.156   3.6 0.156  -3.6 0.313  12.6
  6   0    36 0.380 -12.6 1.180   3.6 1.980     0 2.360-2.520 3.142 1.800
   0.000E+00             3.750E-02        90
```

OUTPUT DATA OF THE PROGRAM "NEWMARK" FOR A LONG COLUMN PROBLEM

```
1
  IND= 2, (SYMMETRY)
  LENGTH(FT)= 34.630   NEL= 3

  ELASTIC SUPPORT
   SPRING1=  .000E+00 SPRING2=  .000E+00

  ISHORT =2 (SHORTENING DEOS NOT INCLUDE GEOMETRICAL THAT OF IMPERFECTION)
  INITIAL OUT-OF-STRAIGHTNESS
   WL(I,J)=   .000E+00  .000E+00  .000E+00  .000E+00
   WL(I,J)=   .234E+02  .234E+02  .234E+02  .234E+02

  ARROW OF ASSUMED DEFLECTION CURVE
   ROW(IX)=  .0000E+00  .0000E+00

  ALPHI(I)=   .900E+02  .270E+03  .900E+02  .270E+03
  SYILD(I)=   .360E+02  .360E+02  .360E+02  .360E+02

Sample Calculation for Long Column
  YIELD ST =   36.00   E = 30000.00   D =15.000   T = .313   NTHTA = 48   NT = 4
POISSON RATIO = .30

  FORCE INCREMENT (MX, MY, P)
     FORCE     INCREMENT     TIMES
        .00        50.00        4
        .00        50.00       10
     -51.91       -10.00        1

  NCTRL = 1   NCTRL2 = 0

RESIDUAL STRESS
   ALPH = 90.000
    CIRCUM    .000-12.600   .156  3.600   .156-3.600    .313 12.600
    AXIAL     .000  36.000   .380-12.600  1.180  3.600  1.980   .000  2.360 -2.520  3.142  1.800

   EXTERNAL PRESSURE   =        .00000E+00
   INITIAL OUT-OF-ROUNDNESS=   .37500E-01   BETA = 90.00
```

```
DATA OF MEMBER
 AREA =  .14442E+02 I =  .38958E+03 S =  .51945E+02 Z =  .67527E+02  SHAPE FUNCTION =  .13000E+01
 R =  .51938E+01
 M(YIELD)=  .18700E+04 M(PLASTIC)=  .24310E+04 P(YIELD)=  .51991E+03 PHI(YIELD)=  .16000E-03

 DA =   .76421E-01   .75619E-01   .74818E-01   .74017E-01
 RI =   .75000E+01   .74218E+01   .73435E+01   .72653E+01   .71870E+01

 QCR =  .62486E+00   STRESS DUE TO PRESSURE =  .00000E+00   .00000E+00
 PCR =  .16699E+03

  RESTRAINT FACTOR
   ZO=  .281E+05 Z1  .000E+00 Z2=  .000E+00

P =  -.51910E+02
  XL    .00000E+00   .13852E+03   .27704E+03   .41556E+03
  YA    .00000E+00   .00000E+00   .00000E+00   .00000E+00
        .00000E+00   .52730+00    .91332E+01   .10546E+02
*******************************************************************************************
   BM    .00000E+00   .00000E+00   .00000E+00   .00000E+00
       -.12137E+04  -.14874E+04  -.16878E+04  -.17611E+04
  PHI    .22884E-05  -.34683E-05  -.60429E-06  -.35326E-05
       -.10528E-03  -.12737E-03  -.14556E-03  -.15180E-03
 ALPH  -.24164E-04  -.38091E-03  -.15057E-03  -.42173E-03
       -.78245E-02  -.17599E-01  -.20024E-01  -.20883E-01
       -.76880E-02  -.15431E-01  -.16891E-01  -.15879E-01
SLOPE         .74235E-03   .36144E-03   .21087E-03
              .48065E-01   .30466E-01   .10441E-01
   YB    .00000E+00   .10283E+00   .15290E+00   .18211E+00
         .00000E+00   .66579E+01   .10878E+02   .12324E+02
   EO    .00000E+00  -.17554E+00  -.25601E+00  -.27995E+00
*******************************************************************************************
   BM    .00000E+00   .53379E+01   .79368E+01   .94531E+01
       -.12137E+04  -.15593E+04  -.17783E+04  -.18534E+04
  PHI    .22884E-05  -.39294E-05  -.13151E-05  -.43079E-05
       -.10528E-03  -.13366E-03  -.15407E-03  -.16142E-03
 ALPH  -.36032E-04  -.44235E-03  -.24689E-03  -.52763E-03
       -.79932E-02  -.18423E-01  -.21191E-01  -.22190E-01
       -.76906E-02  -.15500E-01  -.17185E-01  -.16534E-01
SLOPE         .95306E-03   .51070E-03   .26382E-03
              .50709E-01   .32286E-01   .11095E-01
   YB    .00000E+00   .13202E+00   .20276E+00   .23930E+00
         .00000E+00   .70243E+01   .11497E+02   .13033E+02
   EO    .00000E+00  -.19372E+00  -.28229E+00  -.30768E+00
*******************************************************************************************
   BM    .00000E+00   .68530E+01   .10525E+02   .12422E+02
       -.12137E+04  -.15783E+04  -.18104E+04  -.18902E+04
  PHI    .22884E-05  -.40624E-05  -.15067E-05  -.44998E-05
       -.10528E-03  -.13598E-03  -.15757E-03  -.16591E-03
 ALPH  -.39531E-04  -.45992E-03  -.27276E-03  -.55421E-03
       -.80311E-02  -.18657E-01  -.21666E-01  -.22789E-01
       -.76827E-02  -.15545E-01  -.17505E-01  -.17090E-01
SLOPE         .10098E-02   .54987E-03   .27710E-03
              .51718E-01   .33061E-03   .11395E-01
   YB    .00000E+00   .13988E+00   .21604E+00   .25443E+00
         .00000E+00   .71640E+01   .11747E+02   .13322E+02
   EO    .00000E+00  -.20090E+00  -.29317E+00  -.31947E+00
*******************************************************************************************
   BM    .00000E+00   .72609E+01   .11215E+02   .13207E+02
       -.12137E+04  -.15855E+04  -.18233E+04  -.19052E+04
  PHI    .22884E-05  -.40988E-05  -.15525E-05  -.45352E-05
       -.10528E-03  -.13598E-03  -.15899E-03  -.16776E-03
 ALPH  -.40526E-04  -.46464E-03  -.27888E-03  -.55936E-03
       -.80451E-02  -.18747E-01  -.21859E-01  -.23036E-01
       -.76793E-02  -.15564E-01  -.17640E-01  -.17326E-01
SLOPE         .10232E-02   .55856E-03   .27968E-03
              .52124E-01   .33377E-01   .11518E-01
```

```
  YB    .00000E+00   .14173E+00   .21911E+00   .25785E+00
        .00000E+00   .72202E+01   .11844E+02   .13439E+02
  EO    .00000E+00  -.20384E+00  -.29764E+00  -.32431E+00
****  CONVERGENCE
*****************************************************************************************
```

REFERENCES

AISC (1993) Load and Resistance Factor Design Specification for Structural Steel Buildings, American Institute of Steel Construction, Chicago, IL.

AISC (1989) Specification for Structural Steel Buildings; Allowable Stress Design and Plastic Design, American Institute of Steel Construction.

API (1972) Specifications for Fabricated Structural Steel Pipe, American Petroleum Institute, API Specification 2B.

ASCE (1971) Plastic Design in Steel, A Guide and Commentary, ASCE Manuals and Report on Engineering Practice No. 41.

Chen, W. F. and Ross, D. A. (1977) Test of fabricated tubular columns, *Journal of the Structural Division,* 103 (ST3), 619–634.

Chen, W. F. and Atsuta, T. (1976) *Theory of Beam-Column, Vol. 1: In-Plane Behavior and Design,* McGraw-Hill, New York.

Marshall, P. W. (1970) Stability Problems in Offshore Structures, presentation at the Annual Technical Meeting of the Column Research Council, St. Louis.

Pan, R. B. (1978) Lower Bound Solutions for Pipeline Hydrostatic Collapse Pressure, Session 46 on Offshore Pipeline Design, Energy Technology Conference and Exhibition, Houston.

Santathadaporn, S. and Chen, W. F. (1972) Tangent Stiffness Method for Biaxial Bending, *Journal of the Structural Division,* 98(ST3), 153–163.

SSRC (1988) *Guide to Stability Design Criteria for Metal Structures,* Chapter 3, 4th Edition, The Structural Steel Research Council, editor T. V. Galambos, John Wiley & Sons, New York.

Timoshenko, S. P. and Gere, J. M. (1961) *Theory of Elastic Stability,* Chapter 5, McGraw-Hill, New York.

Toma, S. and Chen, W. F. (1979) Analysis of Fabricated Tubular Columns, *Journal of the Structural Division,* 105 (ST11).

Toma, S. and Chen, W. F. (1983) Design of Vertical Chords in Deepwater Platform, *Journal of the Structural Division,* 109 (ST11), 2343–2366.

Wagner, A. L., Mueller, W. H., and Erzurmlu, H. (1976) Design Interaction Curve for Tubular Steel Beam-Columns, OTC Paper No. 2684, Offshore Technology Conference.

3: Approximate Analysis of Beam-Columns

S. Toma
Department of Civil Engineering, Hokkai-Gakuen University, Sapporo, Japan

W. F. Chen
School of Civil Engineering, Purdue University, West Lafayette, Indiana

NOTATIONS

The following symbols are used in this chapter:

- A area of cross section
- a,b,c a constant for M-P-Φ curve expressions
- D diameter
- E Young's modulus
- E_{eff} effective Young's modulus
- f a constant for M-P-Φ curve expressions or parameter function for average flow moment
- g parameter function for effective stiffness
- h a constant for effective stiffness
- I moment of inertia of section
- K effective column length factor
- L column length
- M_{ext} external moment
- M_{int} internal moment
- M_{mc} average flow moment with axial load
- M_o applied end moment
- M_{pc} full plastic moment with axial load
- M_y yield moment
- M_{yc} initial yield moment with axial load
- m_i normalized moment due to axial load and initial imperfection
- m_{mc} normalized average flow moment
- m_{MQ} normalized moment due to lateral load and/or end moment
- m_{op} normalized moment due to axial load and deflection which is caused by lateral load and/or end moment
- m_{pc} normalized plastic moment
- m_Q normalized moment due to lateral load
- m_{yc} normalized yield moment

0-8493-8282-3/96/$0.00+$.50

N number of segments
P applied axial load
p normalized applied axial load
P_{BUCK} buckling load (ultimate strength of column)
P_{cr} elastic critical load (= Euler load)
Q applied lateral load at mid-span
q normalized applied lateral load at mid-span
r radius of gyration
t thickness
w_{1P} boundary value between elastic and primary yield range for deflection of pin-ended beam-column
w_{1F} boundary value between elastic and primary yield range for deflection of fix-ended beam-column
w_{2P} boundary value between primary and secondary yield range for deflection of pin-ended beam-column
w_{2F}: boundary value between primary and secondary yield range for deflection of fix-ended beam-column
w_i initial deflection
w_m deflection at mid-span due to axial load
$(w_m)_{EL}$ deflection at mid-span due to axial load in elastic range
$(w_m)_{EP}$ deflection at mid-span due to axial load in primary yield range
$(w_m)_{PL}$ deflection at mid-span due to axial load in secondary yield range
w_o deflection at mid-span due to lateral load and/or end moment
x distance from one end of a column
y_i deflection at station i
y_o deflection at distance x due to lateral load and/or end moment
y_w deflection at distance x due to axial load
ϵ axial strain
ϵ_o axial strain at centroid
ϵ_{oi} axial strain at station i
ϵ_y yielding axial strain
σ_y yield stress
Φ curvature
Φ_m curvature at mid-span
Φ_y yielding curvature
ϕ normalized curvature
θ_{y1},θ_{yz} angle to define yielded range
Δ total shortening
Δ_G shortening due to geometrical shape change
Δ_s: shortening due to axial strain
$(\Delta_s)_{EL}$ shortening due to axial strain by elastic analysis

3.1 INTRODUCTION

Behaviors of beam-columns will be described in this chapter for the entire range of monotonic axial loading up to ultimate load, including post-buckling unloading using approximate methods. The analytical method utilized in this chapter is the so-called "assumed deflection" method. This method is not only simple to use, but also gives a reasonably accurate solution for practical use. For the analysis of beam-columns in the elastic–plastic range, the exact solution is often intractable and recourse must be made to numerical procedures to obtain solutions. Often, numerical procedures require much computing time, as we saw in Chapter 2, and often result in an expensive solution. In this chapter, therefore, some reasonable simplifications are made to reduce the computing complexity so that the results can serve a practical use. The behavior of tubes subjected to cyclic loading sequences are not considered here and will be discussed in Chapter 4.

Here, the deflection shape of a beam-column, the moment-curvature relation of a cylindrical section and the equilibrium condition of a member must all be idealized to accomplish a solution. For example, the solution of the problem can be simplified drastically by establishing equilibrium only at mid-height of a column, by idealizing the moment-curvature relation as elastic–perfectly plastic (plastic hinge concept), and by using the elastic deflected shape of a column throughout the elastic–plastic range. The solution based upon these types of assumptions generally leads to analytical expressions for the behavior and strength of axially and/or laterally loaded columns of various cross sections. A comparison of results obtained by the approximate method with those by the "exact" theory, and also the results of tests for wide flange columns, indicates that these idealizations are reasonable (Chen and Atsuta, 1976).

The behavior of beam-columns can be described either in terms of an axial load-lateral deflection relation or axial load-axial shortening relation. A precise load-deflection relation is required for the development and study of load-shortening behavior. The four approximate methods are used in this chapter based on the assumption of different deflection shapes for the column and the simplification of moment-curvature relationships for the cross section. The four methods are: plastic hinge method, modified plastic hinge method, exact moment-curvature method, and average flow moment method. In these analyses, three types of deflected shapes for the column combined with three types of moment-curvature idealization for cylindrical cross sections, ranging from an almost exact to a rather crude approximation, are assumed. The basic features of these four methods are summarized in Table 3.1.

In the elastic range, all these methods use the same conventional elastic beam-column analysis (Timoshenko and Gere, 1961) that is described in Section 3.4. In the plastic range, only the plastic hinge method uses the two-straight-line deflected shape (see Sec. 3.5, Fig. 3.4), and the other analyses assume the deflection shape to be a sinusoidal function throughout the entire loading range (see Sec. 3.4, Fig. 3.3). A polynomial function is also used for deflection induced by lateral forces or end moments.

As for the M-P-Φ relation, only the exact moment-curvature method uses the exact M-P-Φ curves, and the others use the elastic–perfectly plastic type of bilinear relation where the average flow moment method modifies the full plastic moment to a somewhat averaged value (see Sec. 3.6, Fig. 3.6).

TABLE 3.1 Basic Features of the Four Analysis Methods

	1. Plastic hinge method	2. Modified plastic hinge method	3. Average flow moment method	4. Exact moment-curvature method
Deflected shape				
Elastic range:	Deflection: $W_m = (W_i + W_o)\dfrac{p/p_{cr}}{1 - p/p_{cr}}$ W_o = Deflection due to q and m_o			
	$\Delta = \dfrac{pL}{AE} + \left(\cos\dfrac{2W_i}{L} - \cos\dfrac{2(W_i + W_m)}{L}\right)L$	$\Delta = \dfrac{pL}{AE} + \dfrac{\pi^2 W_m^2}{4L}$		$\Delta = \dfrac{pL}{AE} + \Delta_G$ Δ_G = Geometrical shortening calculated by numerical integration

Primary yielded range:	Same as above		Same as above or below	$W_m = f_{Ep}(m_o, q, p, W_i)$ $\Delta = \Delta_S + \Delta_G$ Δ_S = Elastic-plastic axial strain shortening
Secondary yielded range	  W_i, W_m, Δ $W_m = \frac{M_{pc}}{p} - W_i$ $\Delta = \frac{pL}{AE} + \left(\cos\frac{2W_i}{L} - \cos\frac{2(W_i + W_m)}{L}\right)l$	$W_m = \frac{M_{pc}}{p} - W_i$ or $W_m = \frac{M_c}{p} - W_i$ $\Delta = \frac{pL}{AE} + \frac{\pi^2 W_m^2}{4L}$		 $W_m = f_p(m_o, q, p, W_i)$ $\Delta = \Delta_S + \Delta_G$ (f_{Ep}, f_p = cubic equation)
Stress-strain relationships	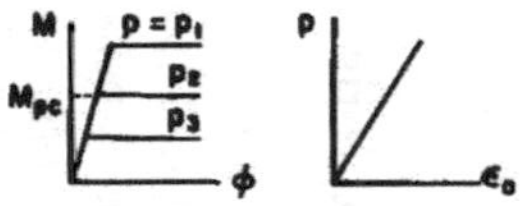 M, M_{pc}, $p = p_1$, p_2, p_3, ϕ; p, ϵ_o	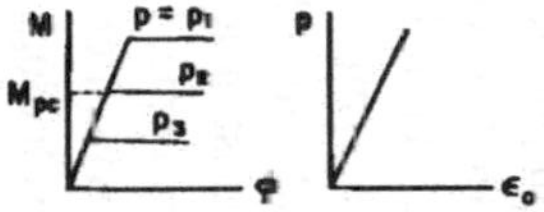 M, M_{pc}, $p = p_1$, p_2, p_3, ϕ; p, ϵ_o	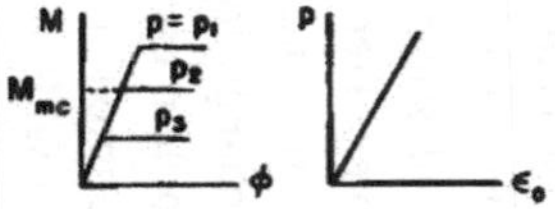 M, M_{mc}, $p = p_1$, p_2, p_3, ϕ; p, ϵ_o	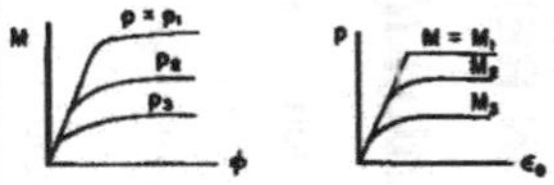 M, $p = p_1$, p_2, p_3, ϕ; p, $M = M_1$, M_2, M_3, ϵ_o

3.2 LOAD-DEFLECTION RELATION (P-W CURVE)

3.2.1 Assumed Deflection Method

The assumed deflection approach is used in this chapter to investigate parametrically the behavior of tubular beam-columns throughout the entire range of loading up to ultimate load and including post-buckling unloading. For an elastic beam-column, the deflection shapes are well known. For example, the deflected shape of an axially loaded elastic column is sinusoidal, and the deflections for beams subjected to lateral loads or end moments are polynomial functions. When the axial and lateral loads are applied simultaneously, the deflection becomes a combination of sinusoidal and polynomial functions. In the plastic range, however, the deflected shape due to axial load is generally very complex. Here, we shall assume that the deflected shape of a column be a certain function, and further that this general shape will not alter during loading but merely change its magnitude as the axial load increases. As a result, the beam-column problem can now be drastically reduced to a one-degree-of-freedom problem. This approach has been shown to give reasonable solutions for steel beam-columns with wide flange cross sections (Chen and Atsuta, 1976). Hence, the approach is worth pursuing for the present cylindrical cross section.

In this approach, we need to consider the equilibrium in terms of moment between external loads and internal resistance of a member only at one critical section of the column. For a symmetrically loaded column, this critical section is at the mid-length of the column. This simplifies significantly the beam-column analysis and is found to be most efficient for parametric studies of the behavior and strength of beam-column problems, including post-buckling, among many analysis methods available, most of which can only trace the behavior of the column up to ultimate load (buckling), excluding the post-buckling branch of the load-deflection curve.

3.2.2 Deflection Functions

It is obvious that a proper choice of deflection shape is one of the key factors in the assumed deflection method. The assumed deflection function should be as close to the actual deflected shape as possible. Its closeness will reflect the accuracy of the solution. In the present work, reasonable functions are the sinusoidal function, linear function, and polynomial function.

Since the sinusoidal function is the exact shape for an axially loaded column with pin-ends, it gives the exact solution for an axially loaded beam-column in the elastic range. Hence, this function is chosen for elastic as well as elastic–plastic analysis of beam-column up to ultimate load and is expected to give a good approximation when the axial load is predominant.

Near and beyond the ultimate load range, plastic hinges will form at critical sections, and a two-bar linkage type of mechanism will develop. This mechanism, which consists of two straight lines knuckled at the plastic hinge locations, mainly contributes the additional deflection of a beam-column in the fully plastic range.

Hence, the two-straight-line deflection shape will also be chosen in the present analysis, and its results will be compared with those of sinusoidal shape.

The polynomial function is the type of deflection shape for an elastic beam subjected to lateral loads or end moments. Hence, it is suitable for beam-columns with large lateral loads or end moments. Herein, this function is used only for the calculation of the initial shape of a beam-column resulting from the application of lateral loads or end moments but before the axial load is applied.

3.2.3 Moment-Thrust-Curvature Relationships (M-P-Φ Curve)

In order to satisfy the equilibrium condition at the critical section between internal and external moments, the moment-thrust-curvature (M-P-Φ) relationship must be known prior to the analysis. There are two ways to describe this relationship: exact and approximate. The exact M-P-Φ relation can be obtained either theoretically or experimentally. A numerical procedure for the M-P-Φ relation including the effect of residual stresses was reported (Fiala et al., 1972). A more efficient computer-based formulation based on the tangent stiffness method has been described in Chapter 2 (see Fig. 2.18).

The tangent stiffness method requires iterative procedures and often results in an increased cost of computing time. Here, in order to reduce the computing time, closed-form expressions are introduced to approximate the exact M-P-Φ relations. Details will be described in Section 3.7.2. These closed-form expressions, if necessary, can be further modified by simple manipulation to fit experimental data.

Experimental and theoretical M-P-Φ curves are compared in Fig. 3.1. Experimental curves are deduced from the member behavior tests reported by Sherman et al. (1979). There exists a significant discrepancy between the present theoretical prediction based on an assumed residual stress distribution and the tests, which may have a very different pattern and magnitude of residual stress distributions. Further, local buckling and cross sectional distortion, which are essential parts of an actual tube after extensive material yielding, are not considered in this chapter.

A major simplification of an actual M-P-Φ relation is idealized as described in Sec. 3.8 and in Fig. 3.11. The relation is either linearly elastic or perfectly plastic with a constant flow moment. This is probably a reasonable idealization for columns in the elastic range as well as in the post-buckling plastic range when considerable plastification in critical sections has been taken place. But appreciable errors must be expected for columns near the initial elastic–plastic range of loading where the peak or maximum load usually occurs. Nevertheless, this type of simplification is very appealing and worth pursuing. This method will be described in this chapter. The average flow moment M_{mc} (Chen and Atsuta, 1972) is also used instead of the usual full plastic moment M_{pc} as shown in Fig. 3.11 for the flow moment that closely approximates the behavior of an actual cylindrical member.

3.3 LOAD-SHORTENING RELATION (P-Δ CURVE)

The axial shortening Δ of a beam-column results from two sources: axial shortening due to lateral geometrical shape change of a column and axial shortening due to

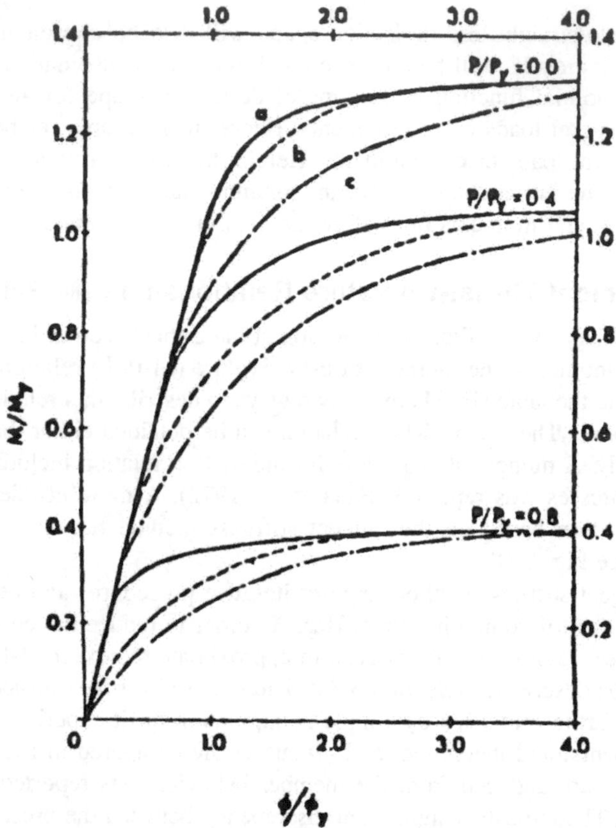

FIGURE 3.1 Comparison of M-P-Φ curves between theory and test (by Sherman et al., 1979): (a) no residual stress; (b) with assumed residual stress (same as Wagner's curve in Fig. 2.18); (c) test result.

axial compressive strain. The axial shortening due to column lateral movement can be calculated from the changes in lateral deflections of a column during loading. In this chapter, this can be done simply by computing directly from the assumed deflection configurations.

The exact axial shortening due to axial compressive strain must be obtained by an elastic–plastic analysis. In order to obtain this part of shortening due to axial strain, one must know the moment-thrust-axial strain (M-P-ϵ_o) relationship. Integrating this axial strain along the entire length of the column, we obtain the total axial shortening due to axial strain. The M-P-ϵ_o relationship can be obtained theoretically in a similar manner as that of the M-P-Φ relation using computer-based numerical technique such as the tangent stiffness method (Santathadaporn and Chen, 1972). Since closed-form expressions for M-P-Φ relation will be developed in this chapter, derivation of equations for exact M-P-ϵ_o relation will also be made (see Section 3.7.4). Simple expressions to approximate this behavior will also be attempted here. It is found that an elastic relation gives a good approximation for an elastic–plastic beam-column.

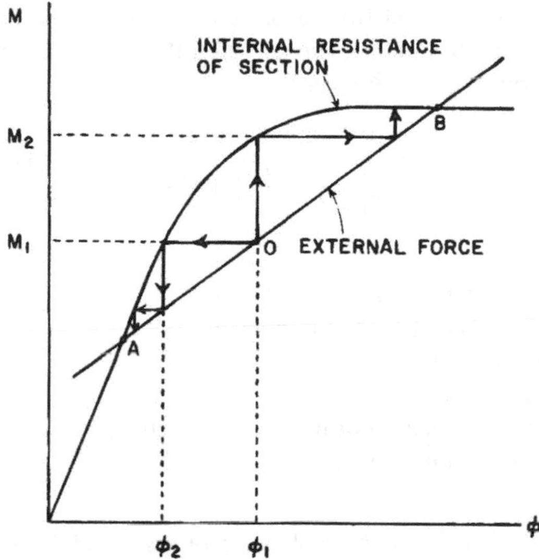

FIGURE 3.2 Interactive procedure diagram for equilibrium.

3.4 ELASTIC ANALYSIS

3.4.1 Basic Concept

In the elastic range, the deflection shape for a pin-ended column subjected to axial load is assumed to be a sinusoidal function that is known to be exact for a perfectly straight column. The term *elastic* implies that the internal resisting moment must remain in the straight-line portion of the M-P-Φ curve (Fig. 3.2). Therefore, the solution to the problem is now reduced to simply obtaining the intersection of two straight lines representing the moment equilibrium condition of the column. For an elastic–perfectly plastic type of idealization for the M-P-Φ relation, the present analysis is valid until the internal resisting moment reaches the flow moment. This elastic limit condition will be discussed further in each of the four methods to be described later.

3.4.2 Load-Deflection Relation (P-w Curve)

Pin-Ended Beam-Column

To start with, let us consider three components of a beam-column deflection: initial imperfection w_i (out-of-straightness), deflection due to end moment and/or lateral load w_o, and deflection amplified by axial force w_m (Fig. 3.3a). The curvatures corresponding to each stage of loading are also shown in the figure. Note that the initial imperfection w_i or curvature does not produce internal moment and should therefore be excluded from M-P-Φ calculations.

The deflection y_o resulted from equal end moments and lateral load at mid-span can be calculated from conventional beam theory. In the elastic range, it has the polynomial form

$$y_0 = \frac{M_o L^2}{2EI}\left(\frac{x}{L} - \frac{x^2}{L^2}\right) + \frac{QL^3}{16EI}\left(\frac{x}{L} - \frac{4x^3}{3L^3}\right) \tag{3.1}$$

where M_o = applied bending moment at ends
Q = applied lateral load at mid-span
L = column length
E = Young's modulus
I = moment of inertia of the column section
x = distance from one end

The additional deflection due to the application of axial load P at the ends is assumed to be a half-sine shape

$$y_w = w_m \sin\frac{\pi x}{L} \tag{3.2}$$

where w_m = deflection at mid-span.

The bending moment at mid-span induced by the external loads is

$$M_{ext} = M_{MQ} + P(w_i + w_o + w_m) \tag{3.3}$$

where M_{MQ} = bending moment due to M_o and/or Q (= M_o + QL/4)
w_i = initial imperfection at mid-span
w_o = deflection at mid-span (x = L/2) due to M_o and/or Q (= $M_oL^2/(8EI) + QL^3/(48EI)$)
P = axial load

or nondimensionalizing Eq. (3.3) by initial yield moment M_y, we have

$$m_{ext} = m_{MQ} + m_i + m_{op} + \frac{P}{M_y} w_m \tag{3.4}$$

where m_{ext} = M_{ext}/M_y
m_{MQ} = M_{MQ}/M_y = $(M_0 + QL/4)/M_y$
m_i = Pw_i/M_y
m_{op} = Pw_o/M_y

The internal resisting moment of a member in the elastic range has the value

$$m_{int} = a\phi_m = a(\phi_{MQ} + \phi_P) \tag{3.5}$$

where m_{int} = M_{int}/M_y
a = stiffness constant
ϕ_m = Φ_m/Φ_y, $\phi_{MQ} = m_{MQ}$, $\phi_p = \pi^2EI(L^2M_y)w_m$
Φ_y = initial yield curvature corresponding to M_y

The initial yield curvature is defined here as the curvature at which the extreme fiber stress is just equal to the yield stress under pure bending. In the present nondimensional form, the stiffness constant "a" has the value of unity (see Sec. 3.7.2). For a general moment-curvature relation including residual stresses, however, the initial stiffness constant may not be equal to unity.

The equilibrium condition requires that the external and internal bending moments be equal. By equating Eq. (3.4) to Eq. (3.5), we have

$$(w_m)_{EL} = \frac{m_{MQ}(1 - a) + m_i + m_{op}}{\left(a - \frac{P}{P_{cr}}\right)\frac{P_{cr}}{M_y}} \tag{3.6a}$$

where P_{cr} = Euler buckling load = π^2EI/L^2.

Note that if the constant "a" is equal to unity, the term m_{MQ} will vanish and Eq. (3.6a) becomes

$$(w_m)_{EL} = \frac{m_i + m_{op}}{\left(1 - \frac{P}{P_{cr}}\right)\frac{P_{cr}}{M_y}} \tag{3.6b}$$

This is the load-deflection relation for an elastic beam-column with pin-ended supports.

Fixed-Ended Beam-Column

Here, as in the pin-ended case, the deflection shape of a fixed-ended beam-column consists of the initial imperfection w_i, the deflection w_o due to lateral loads, and the additional deflection w_m due to axial load (Fig. 3.3b).

The deflection due to lateral load at mid-span for a fixed-ended beam is

$$y_o = \frac{QL^3}{16EI}\left(\frac{x^2}{L^2} - \frac{4}{3}\frac{x^3}{L^3}\right) \tag{3.7}$$

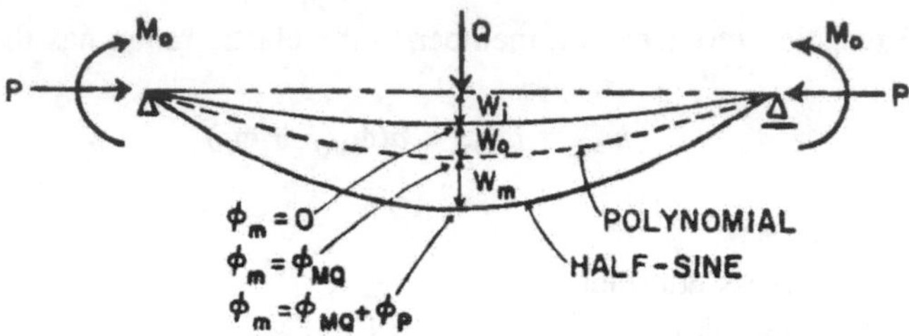

a) PIN-ENDS BEAM-COLUMN

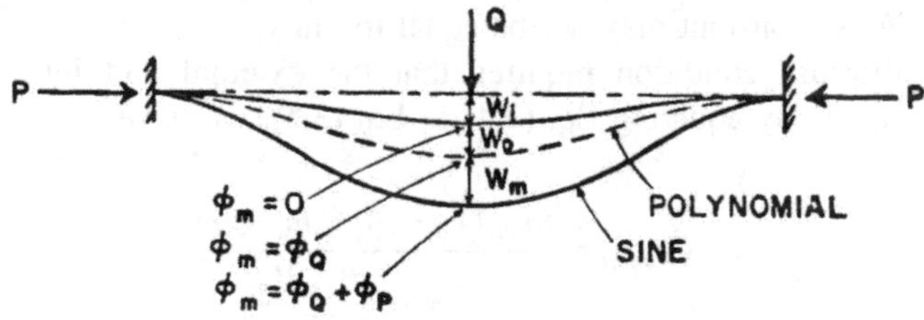

b) FIXED-ENDS BEAM-COLUMN

M
P
(Wi + Wo + Wm)
M
Q/2
P
Q/2

c) FREE BODY OF FIXED-ENDS BEAM-COLUMN

FIGURE 3.3 Deflection shape of a beam-column in elastic range.

and the additional deflection due to the axial load P is assumed to have the sinusoidal shape

$$y_m = \frac{w_m}{2}\left(1 - \cos\frac{2\pi x}{L}\right) \tag{3.8}$$

Hence, the external bending moment at the center of the beam-column is (Fig. 3.4c)

$$M_{ext} = \frac{QL}{4} + P(w_i + w_o + w_m) - M_{end} \tag{3.9}$$

in which M_{end} denotes the fixed-end moment. From the deflection functions Eq.

(3.7) and Eq. (3.8), it follows that the curvatures (or bending moments) at center and at the ends of the beam-column must have the same value. Normalizing by M_y, Eq. (3.9) can be rewritten in the form

$$m_{ext} = m_Q + m_i + m_{op} + \frac{Pw_m}{2M_y} \tag{3.10}$$

where $m_{ext} = M_{ext}/M_y$
$m_Q = M_Q/M_y = QL/(8M_y)$
$m_i = Pw_i/(2M_y)$
$m_{op} = Pw_o/(2M_y)$

and

$w_o = QL^3/192.$

By equating Eq. (3.10) to Eq. (3.5) and solving for w_m, we have

$$(w_m)_{EL} = 2\frac{m_Q(1-a) + m_i + m_{op}}{\left(a - \dfrac{P}{P_{cr}}\right)\dfrac{P_{cr}}{M_y}} \tag{3.11a}$$

where we have used $\phi_Q = m_Q$ instead of $\phi_{MQ} = m_{MQ}$ in Eq. (3.5) and $\phi_p = (2\pi^2EI/(L^2M_y))w_m$, and P_{cr} = Euler buckling load for a fixed-ended column $= 4\pi^2EI/L^2$. When $a = 1$, it reduces to

$$(w_m)_{EL} = 2\frac{m_i + m_{op}}{\left(1 - \dfrac{P}{P_{cr}}\right)\dfrac{P_{cr}}{M_y}} \tag{3.11b}$$

Note that the terms in the right-hand side of Eq. (3.11) are similar to those of Eq. (3.6) except that a factor of two is used in the nondimensionalizing of moments for the fixed-ended case.

3.4.3 Load-Shortening Relation (P-Δ Curve)

The axial shortening, Δ, of a beam-column consists of two parts: the one due to axial strain and the other due to lateral deflection.

Axial Shortening Due to Axial Strain

In the elastic range, axial shortening due to elastic axial compression strain is obtained by the equation

$$(\Delta_s)_{EL} = \frac{PL}{EA} \tag{3.12}$$

Axial Shortening Due to Lateral Deflection

Axial shortening caused by the lateral deflection of a beam-column depends on the particular shape of deflection assumed and the magnitude of its mid-span deflection,

a) PIN-ENDS BEAM-COLUMN

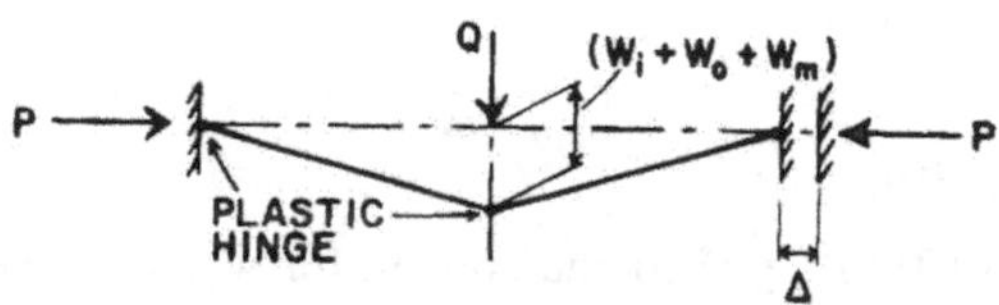

b) FIX-ENDS BEAM-COLUMN

FIGURE 3.4 Plastic hinge method.

w_m. Ignoring initial imperfection, the axial geometrical shortening for a sine-shape deflection has the simple form:

$$\Delta_G = \frac{\pi^2 w_m^2}{4L} \tag{3.13}$$

For a two-straight-line shape, it has the form:

$$\Delta_G = \left(1 - \cos \frac{2w_m}{L}\right)L \tag{3.14}$$

These relations will be used later for the calculation of axial geometrical shortening in connection with an elastic-plastic analysis.

3.5 PLASTIC HINGE METHOD

3.5.1 Basic Concept

Here, the moment-curvature relation is idealized as elastic–perfectly plastic with plastic hinge moment M_{pc} at which the curvature increases infinitely. The value M_{pc} is known as the full plastic moment including the effect of axial load P.

In the elastic range, the solutions presented in the previous section are applicable. In the elastic–plastic range, plastic hinges will be formed successively at critical sections of the frame. When a sufficient number of plastic hinges are formed, a collapse mechanism will be developed, and the segment between plastic hinges will now behave as a rigid body. The additional deflection beyond this stage of loading consists of two straight lines knuckled at the center (Fig. 3.4). This mechanism approach for deriving load-deflection relation is good when the deflection becomes large. However, in the elastic-plastic transition range, this approach may not be suitable.

3.5.2 Load-Deflection Relation (P-w Curve)

Pin-Ended Beam-Column

The column is capable of resisting external loads until the maximum moment at the center of a column is equal to the full plastic moment M_{pc}.

Solving the equilibrium equation for mid-span deflection, we have

$$(w_m)_{PL} = \frac{1}{P}\left(M_{pc} - M_o - \frac{QL}{4}\right) - w_i - w_o \tag{3.15a}$$

or in the nondimensionalized form

$$(w_m)_{PL} = \frac{M_y}{P}(m_{pc} - m_{MQ} - m_i - m_{op}) \tag{3.15b}$$

The definition for each term has been given in Section 3.4.2.

The full plastic moment M_{pc} of a cylindrical section with axial load is given in Eqs. (2.34) and (2.35), rewritten in the alternative form as follows:

$$m_{pc} = \begin{cases} 1.273(1 - 1.18\,p^2) & \text{for} \quad 0 \leq p \leq 0.65 \\ 1.82(1 - p) & \text{for} \quad 0.65 \leq p \leq 1.0 \end{cases} \tag{3.16}$$

where $m_{pc} = M_{pc}/M_y$ and $p = P/P_y$ (P_y = yield axial force = $\sigma_y A$).

The mid-span deflection $(w_m)_{PL}$ in the plastic range can be calculated from Eqs. (3.15) and (3.16). Equation (3.15) describes the post-peak branch of the load-deflection relation where for given loads the elastic deflection as calculated from Eq. (3.6) is smaller than that of the plastic one:

$$(w_m)_{EL} \leq (w_m)_{PL} \tag{3.17}$$

Fixed-Ended Beamed-Column

For a fixed-ended beam-column, a mechanism will be developed when plastic hinges are formed both at ends and at the center of a beam-column. In this case, the equilibrium at the center of the beam-column gives

$$(w_m)_{PL} = \frac{1}{P}\left(2M_{pc} - \frac{QL}{4}\right) - w_i - w_o \tag{3.18a}$$

or in the nondimensionalized form

$$(w_m)_{PL} = \frac{M_y}{P}(2m_{pc} - m_Q - m_i - m_{op}) \tag{3.18b}$$

Using Eq. (3.16), the mid-span deflection for a fixed-ended beam-column in the plastic range can be calculated. Equation (3.17) determines the range of validity of Eqs. (3.11) and (3.18).

3.5.3 Load-Shortening Relation (P-Δ Curve)

For an elastic–perfectly plastic M-P-Φ relation, we need to consider only elastic axial strain for the axial shortening of a beam-column throughout the entire range of loading, including post-peak unloading. Further, the axial shortening due to lateral deflection can be calculated directly from the two-straight-line deflection shape in the plastic as well as the elastic range. This is probably a good approximation, since the lateral deflection in the elastic range is generally small.

Based on these simplifications, the total axial shortening of the beam-column can be expressed in the form

$$\Delta = \frac{PL}{EA} + \left[\cos\frac{2w_i}{L} - \cos\frac{2(w_i + w_o + w_m)}{L}\right]L \tag{3.19}$$

The first term on the right-hand side of this equation is the contribution to axial shortening due to axial strain, and the second represents the contribution from lateral deflection.

3.6 MODIFIED PLASTIC HINGE METHOD

3.6.1 BASIC CONCEPT

The deflection shape in the plastic range is assumed here to be a sine or a cosine type of curve, as shown in Fig. 3.5. Here, we extend the usage of the shape of an elastic deflection into the plastic range. As in the plastic hinge method, the plastic hinges are assumed to develop at mid-span for the pin-ended case, and both at mid-span and ends for the fixed-ended case.

3.6.2 Load-Deflection Relation (P-w Curve)

For an elastic–perfectly plastic M-P-Φ relation, the maximum internal bending moment at the critical sections is always equal to the full plastic moment, M_{pc}. The

M₀ Q M₀
P (Wᵢ + W₀ + Wₘ) P
Δ Δ
PLASTIC HINGE Δ

a) PIN - ENDS BEAM - COLUMN

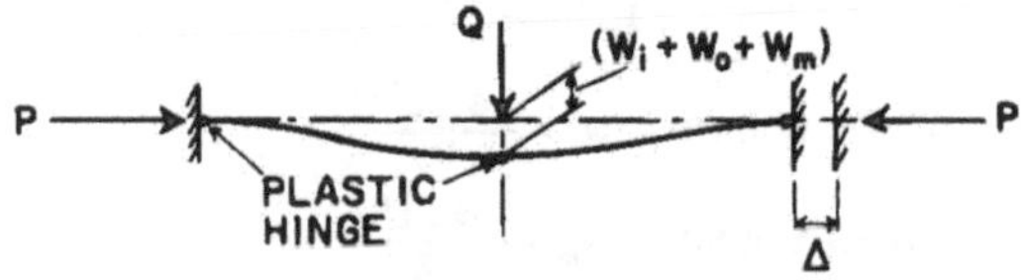

b) FIXED - ENDS BEAM - COLUMN

FIGURE 3.5 Modified plastic hinge method.

moment induced by external loads at the critical section is, of course, not affected by the particular type of deflection shape assumed. Therefore, the same load-deflection equations, i.e., Eqs. (3.15a,b) for a pin-ended beam-column and Eqs. (3.18a,b) for a fixed-ended beam-column as developed in the preceding section for the plastic hinge method, can be used here.

In these equations, M_{pc} is given by Eq. (3.16), and the condition that specifies the validity of the range of these equations is given by Eq. (3.17).

3.6.3 Load-Shortening Relation (P-Δ Curve)

Here, as in the plastic hinge method, only the elastic axial strain is needed to compute the axial strain shortening of a beam-column. The shortening due to lateral deflection can be calculated on the basis of the assumed deflection shape, as shown in Figs. 3.5a and b. Thus, the total axial shortening due to applied forces is obtained as follows:

$$\Delta = \frac{PL}{EA} + \frac{\pi^2(w_o + w_m)^2}{4L} \tag{3.20}$$

The first term of Eq. (3.20) is the axial shortening due to axial strain, and the second represents the geometrical shortening. It has the same form for both the pin-ended case and fixed-ended case.

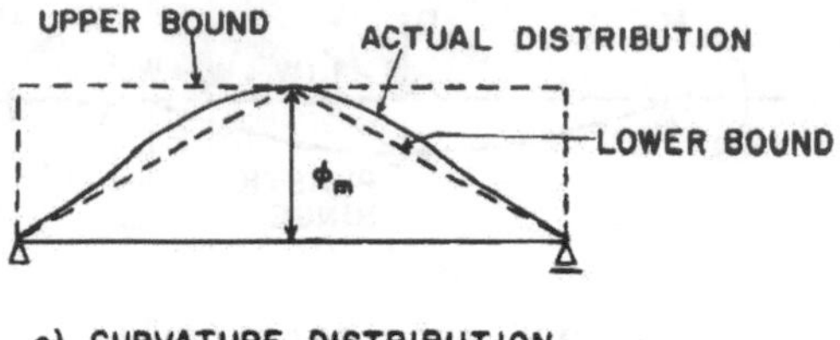

a) CURVATURE DISTRIBUTION

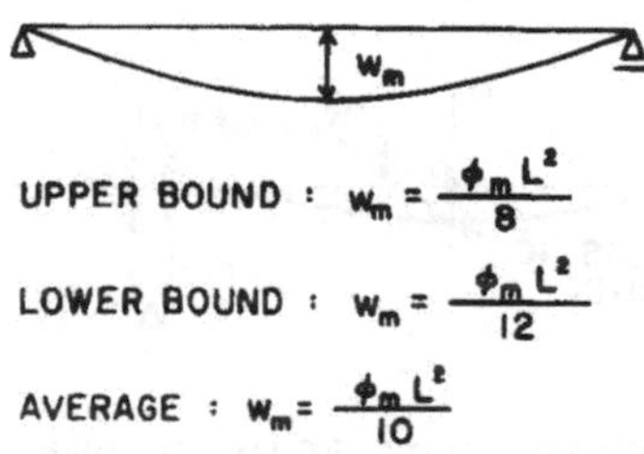

A HALF SINE DEFLECTION : $w_m = \frac{\phi_m L^2}{\pi^2}$

b) DEFLECTIONS

FIGURE 3.6 Relationship between curvature distribution and deflection at center.

3.7 EXACT MOMENT-CURVATURE METHOD

3.7.1 Basic Concept

The method of analysis described herein is intended to trace the load-deflection response of a beam-column from zero load upward as exactly as possible within the limitation of using an assumed deflection function. The present analysis is essentially the same as that of the elastic analysis in that a polynomial function is used for the deflection induced by lateral load or bending moment, and the sinusoidal function is used for the additional deflection amplified by the axial load. Here, unlike in the earlier analysis, we take an accurate account of the moment-curvature relation and moment-axial strain relation.

In order to save computing time, closed-form expressions are employed to approximate closely the exact M-P-Φ and M-P-ϵ_o relations. This approach is rigorous and can be used to assess the consequences of simplification made previously for moment-curvature relation and moment-axial strain relation. Thus, if these simplified relations are acceptable, analysis and design of tubular members can be based on a consideration of the elastic deformed shape with either elastic or plastic hinge condition at critical sections.

The general pattern of curvature distribution for an elastic–plastic pin-ended beam-column is shown as the solid line in Fig. 3.6a. Formal mathematical treatment of this problem by integrating the exact curvature distribution along the length will

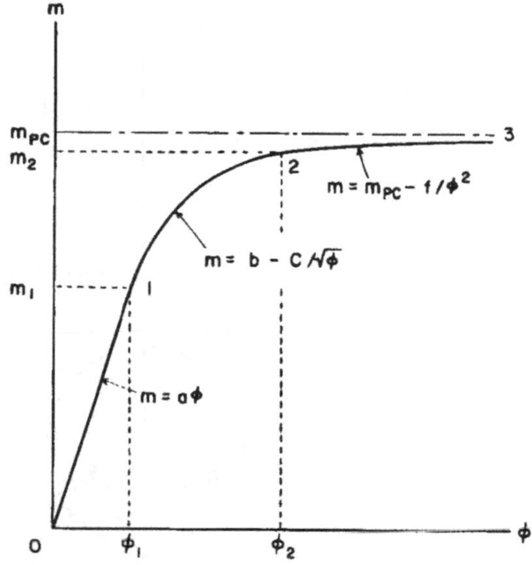

FIGURE 3.7 Closed-form expressions of M-P-Φ curves.

yield an exact equation for the bent shape. Taking the conservative triangular curvature distribution with the maximum curvature at center Φ_m and integrating it, we have the mid-span deflection $w_m = \Phi_m L^2/12$. This value may be considered as a lower bound. If we take the unconservative rectangular curvature distribution with the constant curvature Φ_m, the corresponding w_m will be ($\Phi_m L^2/8$). This may be considered as an upper bound. It is clear therefore, that a reasonable estimate of w_m, certainly suitable for design purposes, would be to take an average value of $\Phi_m L^2/10$. It is reported that this simplification results in a good agreement with the test results in the case of reinforced concrete columns (Cranston, 1967). Consider now the assumed deflection shape here, a half-sine curve for a pin-ended beam-column. The deflection at mid-span is $w_m = \Phi_m L^2/\pi^2$ (Fig. 3.6b), which is very close to the average deflection described above.

3.7.2 Closed-Form Expression of M-P-Φ

Closed-form M-P-Φ expressions for circular tube sections are developed by following the general procedure described in Chen and Atsuta (1976). The general M-P-Φ curve shown in Fig. 3.7 is divided into three parts: elastic (curve 0–1), primary yielded (curve 1–2), and secondary yielded (curve 2–3). The strain distributions over the cross section corresponding to each part of the curve are shown in Fig. 3.8. As seen in the primary yielded range, the strain at one extreme side of the section has exceeded the yield strain, but the other extreme still remains in the elastic range (Fig. 3.8b). In the secondary yielded range, the strains and stresses at both top and bottom regions of the section have been yielded (Fig. 3.8c).

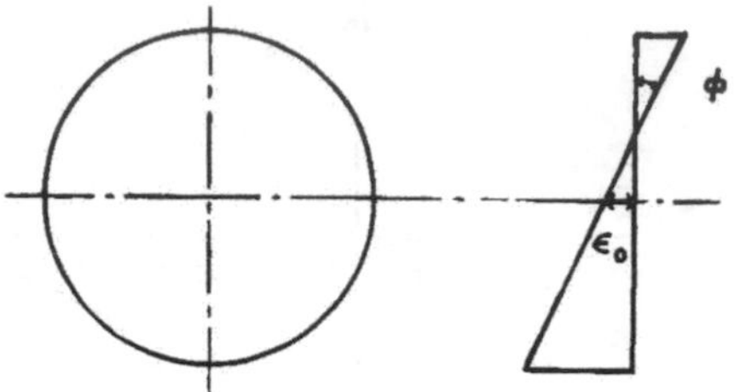

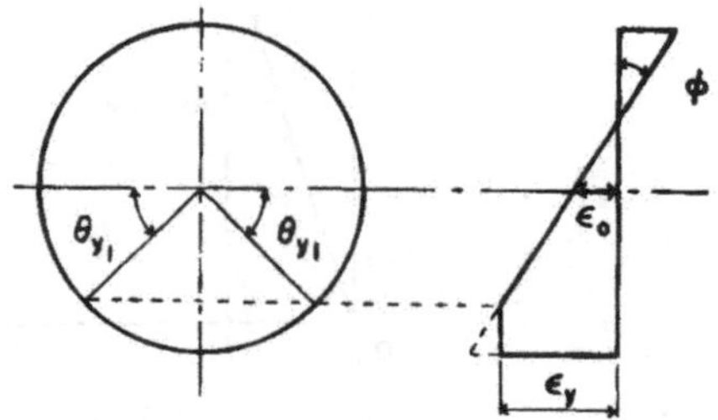

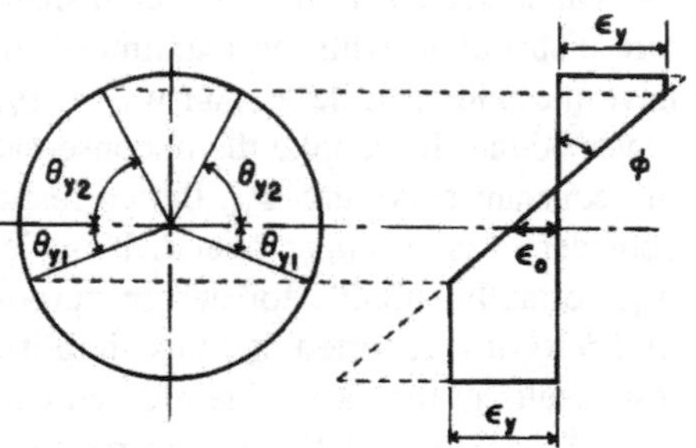

FIGURE 3.8 Axial strain of tubular section.

The nonlinear moment-curvature relation shown in Fig. 3.7 corresponds to a given axial load, P. This curve can be closely approximated by the expressions

$$m_{int} = \begin{cases} a\phi & \text{for} \quad \phi \leq \phi_1 \quad : \quad \text{elastic} \\ b - \dfrac{c}{\sqrt{\phi}} & \text{for} \quad \phi_1 < \phi \leq \phi_2 \quad : \quad \text{primary yielded} \\ m_{pc} - \dfrac{f}{\phi^2} & \text{for} \quad \phi_2 < \phi \quad : \quad \text{secondary yielded} \end{cases} \tag{3.21}$$

where the constants a, b, c, and f can be expressed in terms of the boundary values m_1, m_2, m_{pc}, ϕ_1, and ϕ_2 in the simple forms

$$a = \frac{m_1}{\phi_1}$$

$$b = \frac{m_2\sqrt{\phi_2} - m_1\sqrt{\phi_1}}{\sqrt{\phi_2} - \sqrt{\phi_1}} \tag{3.21a}$$

$$c = \frac{m_2 - m_1}{\frac{1}{\sqrt{\phi_1}} - \frac{1}{\sqrt{\phi_2}}}$$

$$f = (m_{pc} - m_2)\phi_2^2$$

in which the boundary values can be further expressed in terms of the normalized axial load p = P/P_y.

Without residual stress:

$$\begin{Bmatrix} m_1 = 1 - p \\ \phi_1 = 1 - p \end{Bmatrix} \tag{3.22}$$

$$\begin{cases} m_2 = 1 + 0.21\,p - 1.05\,p^2 \\ \phi_2 = \dfrac{1}{1 - 1.395\,p + 1.206\,p^2} & \text{for} \quad 0 \le p \le 0.4 \\ \left.\begin{aligned} m_2 &= 1.528(1 - p) \\ \phi_2 &= 2.625(1 - p) \end{aligned}\right\} & \text{for} \quad 0.4 < p \le 1.0 \end{cases} \tag{3.23}$$

With residual stresses:

$$\begin{cases} \left.\begin{aligned} m_1 &= 0.893 - 1.305\,p + 0.796\,p^2 \\ \phi_1 &= 0.893 - 1.305\,p + 0.796\,p^2 \end{aligned}\right\} & \text{for} \quad 0 \le p \le 0.6 \\ \left.\begin{aligned} m_1 &= 1.50 - 2.333\,p + 0.833\,p^2 \\ \phi_1 &= 1.50 - 2.333\,p + 0.833\,p^2 \end{aligned}\right\} & \text{for} \quad 0.6 < p \le 1.0 \end{cases} \tag{3.24}$$

$$\begin{cases} \left.\begin{aligned} m_2 &= 0.893 + 0.546\,p - 1.946\,p^2 \\ \phi_2 &= \frac{1}{1.12 - 2.491\,p + 4.308\,p^2} \end{aligned}\right\} & \text{for} \quad 0 \le p \le 0.6 \\ \left.\begin{aligned} m_2 &= 1.375 - 1.5\,p + 0.125\,p^2 \\ \phi_2 &= 1.0 + 0.875\,p - 1.875\,p^2 \end{aligned}\right\} & \text{for} \quad 0.6 < p \le 1.0 \end{cases} \tag{3.25}$$

and the full plastic moment is given by Eq. (3.16).

Figure 3.9 shows the comparison of the M-P-Φ relations between the closed-form expressions and the numerically computed exact solutions (Saleeb, 1981). It can be seen that the agreement is excellent. Although the theoretical M-P-Φ curve overestimates considerably the experimentally measured curves shown in Fig. 3.1 (curves b and c) due to inadequate information on residual stresses and stress-strain

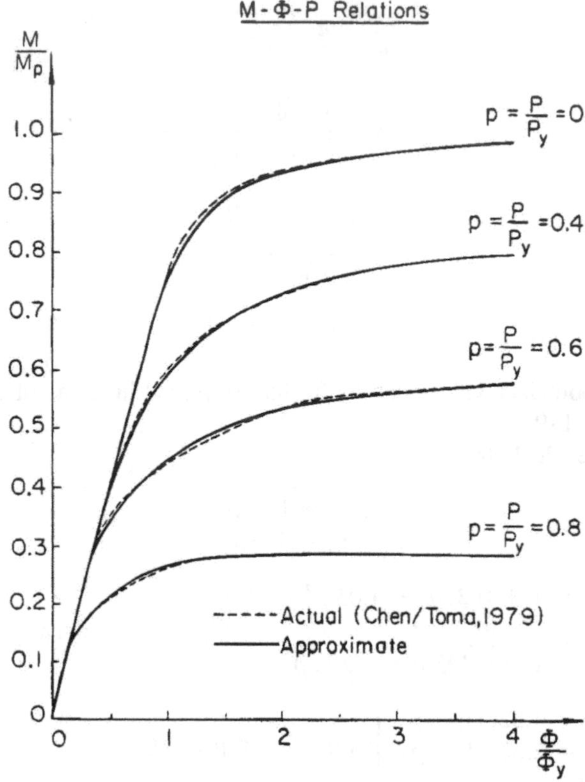

FIGURE 3.9 Comparison of M-P-Φ curves between closed-form expressions and numerical solution [6].

properties of the test specimen, nevertheless, the present approximate expressions can be modified, if necessary, by simply adjusting the boundary values for a better fit.

3.7.3 Load-Deflection Relation (P-w Curve)

Pin-Ended Beam-Column

Elastic Limit

The load-deflection relation in the elastic range is given by Eq. (3.6). Using the elastic boundary value shown in Fig. 3.7, the elastic deflection $(w_m)_{EL}$ must always be less than the deflection at the elastic limit so that Eq. (3.6) can be applied.

$$(w_m)_{EL} \leq w_{1P} \tag{3.26}$$

The limiting value of elastic deflection w_{1p} is given in Eq. (3.29).

Primary Yield Range

Equating the external bending moment, Eq. (3.4), to the internal resisting moment, the second equation of Eq. (3.21), we have

$$\left[m_{MQ} + m_i + m_{op} - b + \frac{P}{M_y}(w_m)_{EP}\right]^2 \times \left[m_{MQ} + \frac{P_{cr}}{M_y}(w_m)_{EP}\right] = c^2 \tag{3.27}$$

which is valid for

$$w_{1p} \leq (w_m)_{EP} \leq w_{2p} \tag{3.28}$$

where the limiting values of deflection are found such that the internal resisting moment is taken equal to the boundary values of Fig. 3.7 in the equilibrium condition.

$$w_{1p} = (\phi_1 - m_{MQ})\frac{M_y}{P_{cr}}$$

$$w_{2p} = (\phi_2 - m_{MQ})\frac{M_y}{P_{cr}} \tag{3.29}$$

Secondary Yield Range

Similarly, equating the external moment, Eq. (3.4), to the third equation of Eq. (3.21), we have

$$\left[m_{pc} - m_{MQ} - m_i - m_{op} - \frac{P}{M_y}(w_m)_{PL}\right] \times \left[m_{MQ} + \frac{P_{cr}}{M_y}(w_m)_{PL}\right]^2 = f \tag{3.30}$$

which is valid for

$$(w_m)_{PL} \geq w_{2p} \tag{3.31}$$

Fixed-Ended Beam-Column

Elastic Limit

The elastic solution is given by Eq. (3.11). The deflection at the elastic limit is denoted by w_{1F}, which is found to have the same form as that of w_{1P} (see Eq. (3.34)).

Primary Yield Range

Equating the external moment, Eq. (3.10), to the second equation of Eq. (3.21), we have

$$\left[m_Q + m_i + m_{op} - b + \frac{P}{M_y}\frac{(w_m)_{EP}}{2}\right]^2 \times \left[m_Q + \frac{P_{cr}}{M_y}\frac{(w_m)_{EP}}{2}\right]^2 = c^2 \tag{3.32}$$

which is valid for

$$w_{1F} \le \frac{(w_m)_{EP}}{2} \le w_{2F} \tag{3.33}$$

where

$$\begin{aligned} w_{1F} &= (\phi_1 - m_Q)\frac{M_y}{P_{cr}} \\ w_{2F} &= (\phi_2 - m_Q)\frac{M_y}{P_{cr}} \end{aligned} \tag{3.34}$$

Note that the Euler buckling load, P_{cr}, used in Eq. (3.34) must be calculated on the basis of effective column length for fixed-ended column, KL = L/2.

Secondary Yield Range

Equating Eq. (3.10) to the third equation of Eq. (3.21), the equilibrium condition becomes

$$\left[m_{pc} - m_Q - m_i - m_{op} - \frac{P}{M_y}\frac{(w_m)_{PL}}{2}\right] \times \left[m_Q + \frac{P_{cr}}{M_y}\frac{(w_m)_{PL}}{2}\right]^2 = f \tag{3.35}$$

which is valid for

$$\frac{(w_m)_{PL}}{2} \ge w_{2F} \tag{3.36}$$

Comparing the equations for fixed-ended beam-column with those of the pin-ended one, the expressions for both cases are almost identical except that the terms containing deflection w_m for the fixed-ended column have a factor of 1/2.

3.7.4 Load-Shortening Relation (P-Δ Curve)

Axial Shortening Due to Axial Strain

Here, as in load-deflection analysis, we attempt to trace the load-shortening response as exactly as possible. In general, the axial strain varies along the length of a beam-column except when the entire beam-column is in the elastic range. Therefore, the axial strain shortening of a beam-column must be calculated by numerical integration, that is, dividing the beam-column into a number of segments.

$$\Delta_s = \sum_{i=1}^{N} \left(\frac{L}{N}\right) \epsilon_{oi} \tag{3.37}$$

where N $=$ number of segments
ϵ_{oi} $=$ axial strain at station i

The moment-thrust-strain curve (M-P-ϵ_o) can also be divided into three parts in a similar manner to that of the M-P-Φ curve corresponding to the three strain distributions shown in Fig. 3.8. In the following derivation, we assume that the thickness of a cylindrical column cross section is thin in comparison with its diameter, and we use the following relation:

$$P = \int_{\text{elastic}} E\epsilon dA + \int_{\text{plastic}} \sigma_y dA \tag{3.38}$$

Elastic Range

In the elastic range (Fig. 3.8a), the load-shortening relation has the form

$$\epsilon_o = \frac{P}{2\pi Etr} \tag{3.39}$$

where ϵ_o $=$ axial strain at the centroid of the cylindrical section
t $=$ thickness of the cylindrical column
r $=$ radius of the tubular column

The elastic Φ-P-ϵ_o relation is applicable within the elastic deflection limit as given in Eq. (3.26). Expressing it in terms of strain, Eq. (3.26) is equivalent to the following limitations:

$$\begin{aligned} \epsilon_o - r\Phi &\geq -\epsilon_y \\ \epsilon_o + r\Phi &\leq \epsilon_y \end{aligned} \tag{3.40}$$

where ϵ_y $=$ yield strain $= \sigma_y/E$.

This condition, Eq. (3.40), limits the extreme strains not to exceed the yield strain ϵ_y.

Primary Yield Range

When one side of the cross section (Fig. 3.8b) is yielded, Eq. (3.38) leads to the following Φ-P-ϵ_o relation:

$$\epsilon_0 = \left[\frac{P}{Etr} + \epsilon_y(\pi - 2\theta_{y1}) - 2r\Phi\cos\theta_{y1}\right]\frac{1}{\pi + 2\theta_{y1}} \tag{3.41}$$

where

$$\theta_{y1} = \sin^{-1}\left(\frac{\epsilon_y + \epsilon_o}{r\Phi}\right) \tag{3.42}$$

Equation (3.41) is applicable with the axial strain in the region

$$\begin{aligned} \epsilon_o - r\Phi &\geq -\epsilon_y \\ \epsilon_o + r\Phi &\geq \epsilon_y \end{aligned} \tag{3.43}$$

It may be noted that Eq. (3.41) requires an iterative procedure for a solution. The convergence associated with such an iteration is found to be very fast.

Secondary Yield Range

Similarly, when both sides of the cross section (Fig. 3.8c) are yielded, Eq. (3.38) leads to

$$\epsilon_o = \left[\frac{P}{2Etr} - \epsilon_y(\theta_{y1} - \theta_{y2}) - r\Phi(\cos\theta_{y1} - \cos\theta_{y2})\right]\frac{1}{\theta_{y1} + \theta_{y2}} \tag{3.44}$$

where

$$\theta_{y2} = \sin^{-1}\left(\frac{\epsilon_y - \epsilon_o}{r\Phi}\right) \tag{3.45}$$

This equation is applicable in the region

$$\begin{aligned} \epsilon_o - r\Phi &\leq -\epsilon_y \\ \epsilon_o + r\Phi &\geq \epsilon_y \end{aligned} \tag{3.46}$$

Again, Eq. (3.44) requires an iterative procedure for a numerical solution.

Once the curvature-thrust-strain (Φ-P-ϵ_o) relations become known, the moment-thrust-strain (M-P-ϵ_o) relations can be obtained using the moment-thrust-curvature

(M-P-Φ) relations. The results of M-P-ϵ_o relations computed from Eqs. (3.39), (3.41), and (3.44) are compared with those obtained numerically by the tangent stiffness method in Fig. 3.10 (Toma and Chen, 1979). It can be seen that the agreement is good.

Axial Shortening Due to Lateral Deflection

The geometrical shortening can be calculated exactly from the assumed deflection functions. For an axially loaded column, this geometrical shortening is given by the second term of Eq. (3.20). When the lateral load or bending moment is also applied, the deflection shape is a combination of polynomial and sinusoidal functions.

To calculate this geometrical shortening exactly from the combined deflection functions, we divide a beam-column into a number of segments and summarize the shortening for each segment along the length of a column and obtain the total shortening:

$$\Delta_G = \Sigma \left[\sqrt{\left(\frac{L}{N}\right)^2 + (y_i - y_{i-1})^2} - \frac{L}{N} \right] \tag{3.47}$$

where N = number of segments, and y_i, y_{i-1} = total deflection at the (i)th and (i − 1)th station, respectively.

The total deflection at a station is calculated by summing up the deflections due to the lateral load, bending moment, and axial force. Hence, the total deflection at distance x from the end is obtained by adding Eqs. (3.1) and (3.2) for the pin-ended beam-column and Eqs. (3.7) and (3.8) for the fixed-ended beam-column.

3.8 AVERAGE FLOW MOMENT METHOD

3.8.1 Basic Concept

The bilinear moment-curvature relation has been used in both the plastic hinge method and the modified plastic hinge method. The solutions based on these methods indicate that this simplified relation is a reasonable idealization for predicting the response of beam-columns in the initial elastic region and in the large post-buckling region, but it grossly overestimates the intermediate transition region, which is probably the most important for analysis of offshore structures.

The above is, of course, an extremely simplified approach to the problem; nevertheless, the general behavior illustrated in the load-deflection and load-shortening diagram is valid. In order to improve this prediction near the peak load, but still using the bilinear moment-curvature simplification, the concept of an average flow moment may be applied. According to this concept, the moment-curvature relation for a constant thrust, as shown in Fig. 3.11, is assumed to be linear up to a certain average flow moment, M_{mc}. From here on the section acts as a plastic hinge and flows plastically at this average constant moment M_{mc}. The average

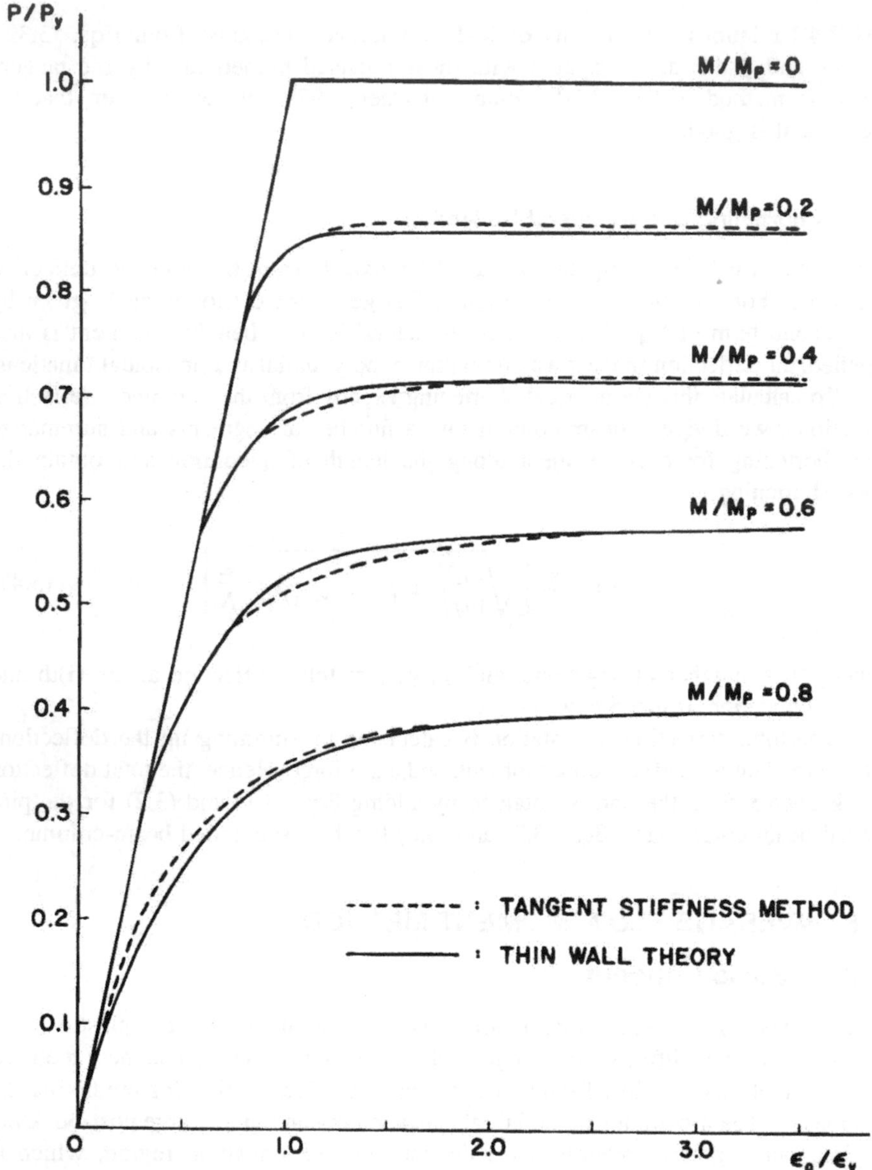

FIGURE 3.10 Comparison of M-P-Φ curves between closed-form expressions and numerical solution.

moment M_{mc} must lie between the initial yield moment M_{yc} and the full plastic moment M_{pc} (Fig. 3.11).

The average flow moment will be a function of thrust P, length L, and the boundary conditions of a beam-column. For example, if $P = 0$, it is a beam problem and the full plastic moment M_{pc} will govern the ultimate state, i.e., $M_{mc} = M_{pc}$. If

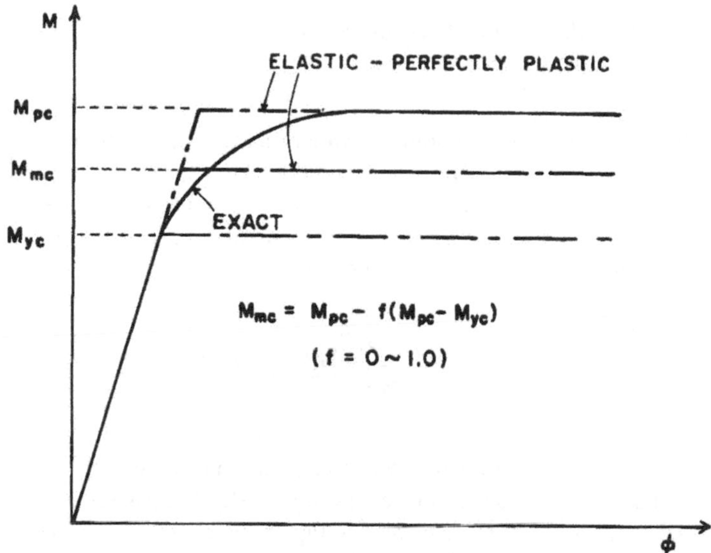

FIGURE 3.11 Average flow moment.

$P \approx P_y$, it is an axially loaded short-column problem and the initial yield moment will be the governing one, i.e., $M_{mc} = M_{yc}$. Similar reasoning can be advanced for the other two factors: L/r (slenderness ratio) and boundary condition. The application of this concept to beam-columns of wide-flange cross section has resulted in an accurate prediction of the maximum load-carrying capacity of all beam-columns studied (Chen and Atsuta, 1972). In the following section an attempt is made to apply the same approach to the beam-column problem of cylindrical cross sections.

3.8.2 Average Flow Moment

Regarding the elastic–perfectly plastic moment-curvature relation (Fig. 3.11), it is obvious that the elastic–fully plastic moment-curvature relation will give an upper bound solution (Fig. 3.11), and the elastic–initial yield moment-curvature relation will result in a lower bound. Therefore, the true response of a beam-column must lie between these two extreme idealizations. Hence, if the elastic–average flow moment-curvature relation is introduced, a plastic hinge type of analysis based on a consideration of an elastic, deformed shape corresponding to an average flow moment condition being presented at the critical cross sections will be much improved.

The normalized average flow moment $m_{mc} = M_{mc}/M_y$ can be written (Chen and Atsuta, 1972) in the form

$$m_{mc} = m_{pc} - f(m_{pc} - m_{yc}) \tag{3.48}$$

where $m_{pc} = M_{pc}/M_y$ = normalized full plastic moment

M_{yc} = M_{yc}/M_y = normalized initial yield moment
f = parameter function

It can be seen that when the parameter function f takes the extreme values, 0 and 1, the corresponding average flow moment is reduced to m_{pc} and m_{yc}, respectively. As mentioned earlier, the parameter function may be a function of thrust $p = P/P_y$, slenderness ratio L/r, and the boundary conditions of a beam-column. Herein, the parameter function is assumed to have the form

$$f = f_1\left(\frac{P}{P_{BUCK}}\right)f_2\left(\frac{L}{r}\right)f_3(B.C.) \tag{3.49}$$

where $f_1(P/P_{BUCK})$ = the parameter depending on axial force
f_2 (L/r) = the parameter depending on slenderness ratio
f_3 (B. C.) = the parameter depending on boundary conditions

For a tubular beam-column, the parameters, f_1, f_2, and f_3 are assumed to have the simple forms:

$$f_1 = \left(\frac{P}{P_{BUCK}}\right)^n, \quad \left(n = \frac{4}{1 + m_o + q}\right) \tag{3.50}$$

$$f_2 = \frac{1}{70}\frac{L}{r} \tag{3.51}$$

$$f_3 = \begin{cases} 1.0 & \text{for pin-ends} \\ 0.5 & \text{for fixed-ends} \end{cases} \tag{3.52}$$

where P_{BUCK} = the ultimate strength (buckling load) of a column by the plastic hinge type of analysis using the full plastic moment M_{pc}
m_o = M_o/M_y = normalized applied moment at ends
q = Q/Q_y = normalized lateral load at center
r = radius of gyration

P_{BUCK} is the maximum load of a beam-column when the full plastic moment, M_{pc}, is attained at critical sections. This value can be determined simply by equating the lateral deflection from the elastic branch of the load-deflection curve to that of the plastic branch and solving for the maximum axial force. For example, for the case of a pin-ended beam-column, equate Eq. (3.6) to Eq. (3.15) and solve for P. For the fixed-ended case, equate Eq. (3.11) to Eq. (3.18) and solve for P.

3.8.3 Load-Deflection Relation (P-w Curve)

Since the present analysis is identical to that of the modified plastic hinge method (except that we use the average flow moment), it follows that we can use the same equations by simply substituting the full plastic moment, M_{pc}, by the average flow moment, M_{mc}.

For a pin-ended beam-column, we have

$$(w_m)_{PL} = \frac{1}{P}\left(M_{mc} - M_o - \frac{QL}{4}\right) - w_i - w_o \tag{3.53a}$$

or

$$(w_m)_{PL} = \frac{M_y}{P}(m_{mc} - m_{MQ} - m_i - m_{op}) \tag{3.53b}$$

and for a fixed-ended beam-column, we have

$$(w_m)_{PL} = \frac{1}{P}\left(2M_{mc} - \frac{QL}{4}\right) - w_i - w_o \tag{3.54a}$$

or

$$(w_m)_{PL} = \frac{M_y}{P}(2m_{mc} - m_Q - m_i - m_{op}) \tag{3.54b}$$

which is valid for

$$(w_m)_{EL} \leq (w_m)_{PL} \tag{3.17}$$

3.8.4 Load-Shortening Relation (P-Δ Curve)

To account for the effect of the development of plastic strain near critical sections along the length of a column throughout most of the loading range on the column axial stiffness, the concept of effective axial stiffness is introduced here for the calculation of axial shortening due to axial strain. The effective axial stiffness may take the form

$$E_{\text{eff}} = hE + gE \tag{3.55}$$

where h = a constant (= 0.5)

g = parameter function depending on axial force

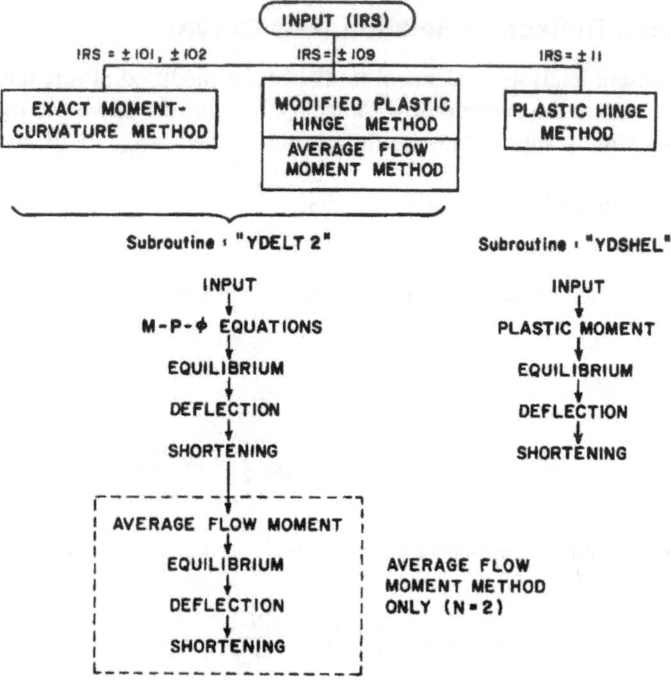

FIGURE 3.12 Brief chart of computer program.

The following function for the parameter function g was found suitable for tubular columns:

$$g = 0.5 \frac{P}{P_{BUCK}} \tag{3.56}$$

P_{BUCK} has been defined in Section 3.8.2.

3.9 COMPUTER IMPLEMENTATION

In order to provide numerical results, a computer program covering each of the four methods of analysis has been developed and named "ADMCOL." The program is based on the equations formulated in Sections 3.4 to 3.8. Figure 3.12 shows a brief summary of this program. Since a theoretically rigorous solution for M-P-Φ relation requires considerable computing time, the approximate closed-form expressions for M-P-Φ curves described in Section 3.7.2 are incorporated in the program.

3.10 USER'S MANUAL FOR THE PROGRAM ADMCOL

All inputs are written in free format.

```
Line 1. Control data for output form.
```

```
IPLOT = 1: Output data are real value.
      = 2: Output data are normalized by corresponding yield value.
Note: 1) When IRS = ±11, IPLOT is set as 1.
      2) Line 1 is required at only the beginning of a run.
When the other case follows next, repeat Lines 2 through 4.

Line 2. Control data for analytical method.
      IRS = ±11: Plastic Hinge Method
           ±101: Exact Moment-Curvature Method (without
                 residual stresses)
           ±102: Exact Moment-Curvature Method (with residual stresses)
           ±109: Modified Plastic Hinge Method and Average Flow Moment
                 Method
Note: Positive sign indicates pin-ended supports.
      Negative sign indicates fixed-ended supports.

Line 3. Member data
When IRS = ±11,
      SY = Yield stress (ksi)
       E = Young's modulus (ksi)
       D = Diameter of a member (in.)
       T = Thickness of a member (in.)
      AL = Length of a member (ft.)
When IRS = ±102 or ±109,
      AL = Length of a member (ft.)
       E = Young's modulus (ksi)
      SY = Yield stress (ksi)
       D = Diameter of a member (in.)
       T = Thickness of a member (in.)
      wi = Initial imperfection (=wi *L (in.))
Note: Input the data in sequence.

Line 4. Load data
When IRS = ±11,
      Yi = Initial imperfection (=Yi *L (in.))
      SQ = Starting value of lateral load at mid-span (=Q/Qy)
      DQ = Increment of lateral load
      NQ = Number of repetition for lateral load
      SP = Starting value of axial load at ends (kips)
      DP = Increment of axial load (kips)
      NP = Number of repetitions for axial load
    EMOS = Moment applied at ends (=Mo/My)
When IRS = ±102 or ±109,
      PP = Starting value of axial load at ends (kips)
   PDELT = Increment of axial load (kips)
      NP = Number of repetitions for axial load
    EMOS = Moment applied at ends (=Mo/My)
      SQ = Lateral load at mid-span (=Q/Qy)
Note: 1) Input the data in sequence.
      2) EMOS is ignored for fix-ends.
```

3.11 SAMPLE CALCULATIONS

Lists of the input data and the corresponding output for a sample problem are shown in the following.

INPUT DATA OF THE PROGRAM "ADMCOL" FOR A SAMPLE PROBLEM

```
1
11
 3.600E+01 3.000E+04 4.500E+00 9.375E-02 1.558E+01
 1.000E-03 0.000E+00 3.000E-01 2 1.000E+00 1.000E+00 50 0.000E+00
```

OUTPUT DATA OF THE PROGRAM "ADMCOL" FOR A SAMPLE PROBLEM

```
IRS= 11
INPUT DATA
  YIELD STRESS= .3600E+02 YOUNG'S MODULUS= .3000E+05 PIPE ... D= .4500E+01 T= .9375E-01
   L= .1870E+03
  INITIAL DEFLECTION= .1000E-02

  LATERAL LOAD . . . SQ= .0000E+00 DQ= .3000E+00 NQ= 2  AXIAL LOAD . . . SP= .1000E+01
  DP= .1000E+01 NP=50
  END-MOMENTS= .0000E+00

  I= .3151E+01 A= .1298E+01 PCR= .2669E+02 PY= .4672E+02 MY= .5041E+02

Q= .0000E+00
N= 1
       P           YELP        DELP        YPLP        DPLP
   .1000E+01   .7277E-02   .4824E-02   .6396E+02   .5096E+02
   .2000E+01   .1514E-01   .9660E-02   .3183E+02   .1132E+02
   .3000E+01   .2368E-01   .1450E-01   .2110E+02   .4927E+01
   .4000E+01   .3296E-01   .1934E-01   .1572E+02   .2745E+01
   .5000E+01   .4310E-01   .2420E-01   .1248E+02   .1747E+01
   .6000E+01   .5422E-01   .2905E-01   .1030E+02   .1209E+01
   .7000E+01   .6646E-01   .3392E-01   .8738E+01   .8874E+01
   .8000E+01   .8002E-01   .3880E-01   .7558E+01   .6808E+00
   .9000E+01   .9511E-01   .4369E-01   .6632E+01   .5409E+00
   .1000E+02   .1120E+00   .4860E-01   .5884E+01   .4423E+00
   .1100E+02   .1311E+00   .5352E-01   .5266E+01   .3708E+00
   .1200E+02   .1527E+00   .5848E-01   .4745E+01   .3176E+00
   .1300E+02   .1775E+00   .6347E-01   .4299E+01   .2774E+00
   .1400E+02   .2062E+00   .6850E+01   .3911E+01   .2466E+00
   .1500E+02   .2399E+00   .7360E-01   .3571E+01   .2228E+00
   .1600E+02   .2798E+00   .7878E-01   .3269E+01   .2043E+00
   .1700E+02   .3280E+00   .8409E-01   .2998E+01   .1898E+00
   .1800E+02   .3872E+00   .8958E-01   .2754E+01   .1786E+00
   .1900E+02   .4619E+00   .9536E-01   .2532E+01   .1699E+00
   .2000E+02   .5589E+00   .1016E+00   .2328E+01   .1633E+00
   .2100E+02   .6899E+00   .1087E+00   .2140E+01   .1584E+00
   .2200E+02   .8769E+00   .1174E+00   .1967E+01   .1549E+00
   .2300E+02   .1165E+01   .1296E+00   .1805E+01   .1525E+00
   .2400E+02   .1668E+01   .1517E+00   .1654E+01   .1511E+00
N= 2
       P           YELP        DELP        YPLP        DPLP
   .1000E+01   .7277E-02   .4824E-02   .6396E+02   .5096E+02
   .2000E+01   .1514E-01   .9660E-02   .3183E+02   .1132E+02
   .3000E+01   .2368E-01   .1450E-01   .2110E+02   .4926E+01
```

```
Q=  .3000E+00
N= 1
       P           YELP        DELP        YPLP        DPLP
   .1000E+01   .4915E+00   .9344E-02   .4883E+02   .2777E+02
   .2000E+01   .5189E+00   .1455E-01   .2427E+02   .6521E+01
   .3000E+01   .5487E+00   .1982E-01   .1606E+02   .2859E+01
   .4000E+01   .5812E+00   .2514E-01   .1194E+02   .1598E+01
   .5000E+01   .6166E+00   .3054E-01   .9450E+01   .1020E+01
   .6000E+01   .6554E+00   .3602E-01   .7780E+01   .7087E+00
   .7000E+01   .6982E+00   .4162E-01   .6578E+01   .5234E+00
```

```
  .8000E+01  .7456E+00  .4734E-01  .5667E+01  .4050E+00
  .9000E+01  .7983E+00  .5322E-01  .4951E+01  .3255E+00
  .1000E+02  .8573E+00  .5931E-01  .4371E+01  .2701E+00
  .1100E+02  .9239E+00  .6564E-01  .3891E+01  .2304E+00
  .1200E+02  .9995E+00  .7230E-01  .3484E+01  .2015E+00
  .1300E+02  .1086E+01  .7939E-01  .3135E+01  .1802E+00
  .1400E+02  .1186E+01  .8703E-01  .2831E+01  .1643E+00
  .1500E+02  .1304E+01  .9543E-01  .2563E+01  .1526E+00
  .1600E+02  .1443E+01  .1049E+00  .2324E+01  .1439E+00
  .1700E+02  .1612E+01  .1159E+00  .2109E+01  .1376E+00
  .1800E+02  .1819E+01  .1291E+00  .1914E+01  .1333E+00
  .1900E+02  .2079E+01  .1458E+00  .1736E+01  .1304E+00
N= 2
     P         YELP       DELP       YPLP       DPLP
  .1000E+01  .4915E+00  .9344E-02  .4883E+02  .2777E+02
  .2000E+01  .5189E+00  .1455E-01  .2426E+02  .6516E+01
  .3000E+01  .5487E+00  .1982E-01  .1603E+02  .2850E+01
  .4000E+01  .5812E+00  .2514E-01  .1189E+02  .1585E+01
  .5000E+01  .6166E+00  .3054E-01  .9368E+01  .1003E+01
  .6000E+01  .6554E+00  .3602E-01  .7657E+01  .6878E+00
  .7000E+01  .6982E+00  .4162E-01  .6403E+01  .4984E+00
  .8000E+01  .7456E+00  .4734E-01  .5431E+01  .3760E+00
  .9000E+01  .7983E+00  .5322E-01  .4644E+01  .2927E+00
  .1000E+02  .8573E+00  .5931E-01  .3984E+01  .2338E+00
  .1100E+02  .9239E+00  .6564E-01  .3413E+01  .1912E+00
  .1200E+02  .9995E+00  .7230E-01  .2909E+01  .1598E+00
  .1300E+02  .1086E+01  .7939E-01  .2453E+01  .1366E+00
  .1400E+02  .1186E+01  .8703E-01  .2036E+01  .1197E+00
  .1500E+02  .1304E+01  .9543E-01  .1650E+01  .1078E+00
  .1600E+02  .1443E+01  .1049E+00  .1289E+01  .9975E-01
```

3.12 NUMERICAL RESULTS

3.12.1 Plastic Hinge and Modified Plastic Hinge Methods

Typical numerical results of the plastic hinge and the modified plastic hinge methods are shown in Fig. 3.13. For beam-columns with a bilinear type of moment-curvature relation, it can be seen that the maximum or peak load corresponds to the attainment of plastic hinge moment at the center of the pin-ended case or plastic hinge moments at the center and two ends of the fixed-ended case. There is a sharp drop of load corresponding to the formation of mechanism. This is, of course, an extremely simplified approach to the problem; nevertheless, the general behavior illustrated in the load-deflection or load-shortening diagrams of Fig. 3.13 is valid.

3.12.2 Exact Moment-Curvature Method

Numerical results for pin-ended columns using the exact moment-curvature method are shown in Figs. 3.14 and 3.15 for two column lengths, L/r = 80 and 120. Also shown are the results of the modified plastic hinge method. Initial imperfection is assumed to be 0.001L in both analyses. It can be seen from the figures that the ultimate strengths of beam-columns by the exact moment-curvature method are significantly lower than those obtained by the modified plastic hinge method. This is because the nonlinear part of the M-P-Φ curve results in a smooth transition curve near the peak portion of a P-w curve. It is clear that the more pronounced the nonlinearity of the moment-curvature relation, the smoother, and thus the less

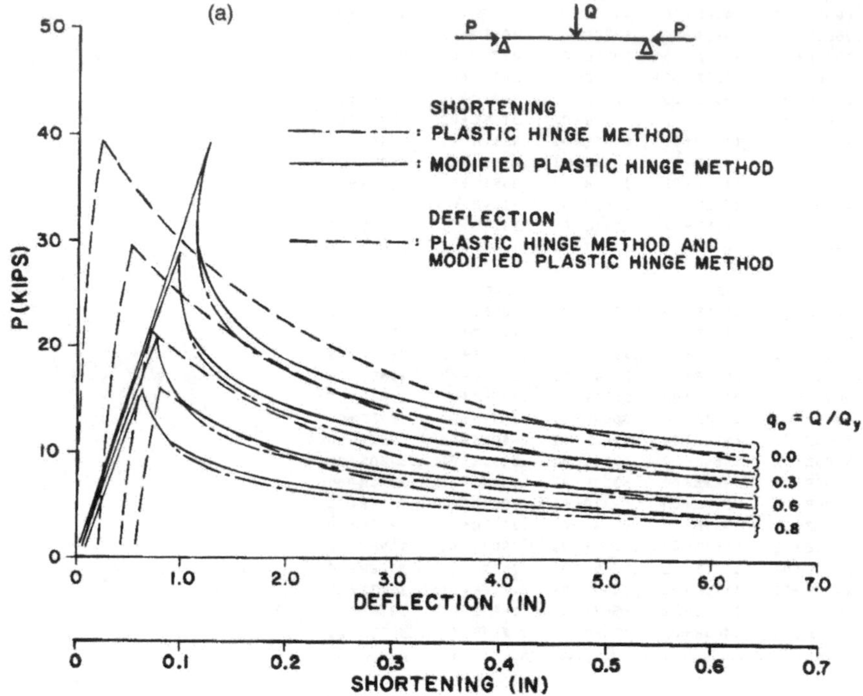

FIGURE 3.13 (a) Comparison between plastic hinge method and modified plastic hinge method, pin-ends, L/r = 80. (b) Comparison between plastic hinge method and modified plastic hinge method, pin-ends, L/r = 120.

peak, the load-deflection curve is likely to be. Thus, if the bilinear moment-curvature relation is accepted as a simplification for practical analysis, the ultimate strength of a tubular member may be considerably overestimated.

The results of beam-column tests using small-size specimens (Wagner et al., 1977) are shown in Fig. 3.16. The analytical results obtained by Wagner et al. are also shown in the figure. Wagner's analysis is based on a finite difference approach in which the column is divided into a number of segments, and the equilibrium is satisfied at each station (each end of the segment) by an iterative procedure. This method, however, traces the load-deflection curve from zero only up to the ultimate strength, and the post-peak response cannot be obtained. As seen in Fig. 3.16, these analytical results give somewhat higher values than those of the test results. On the other hand, the present exact moment-curvature method results in somewhat lower values for the case L/r = 90 (Fig. 3.16a). For the case L/r = 60, the curve obtained by the exact moment-curvature method agrees very well with the test results (Fig. 3.16b).

Strength of Cylindrical Members

Figure 3.17 shows the comparison of strengths of an axially loaded column with 0.1% out-of-straightness as predicted by the exact moment-curvature method and

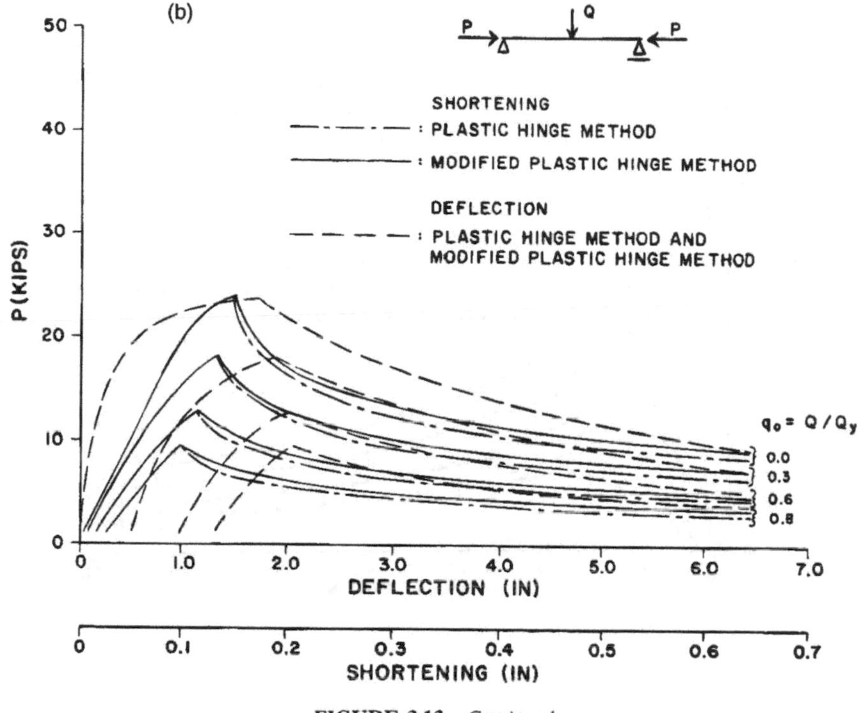

FIGURE 3.13 *Continued*

by Newmark's method (Toma and Chen, 1979). The theoretical strength as well as the CRC and AISC design curves are all shown in the figure for reference. Note that in Newmark's method, the beam-column is divided into a number of segments, and the equilibrium at each station between segments is satisfied by an iterative procedure. It can be seen that there is a very good agreement for the column strength as predicted by both methods, even though the exact moment-curvature method described here is much more simplified than that of Newmark's method.

Figure 3.18 shows the comparison of the interaction curves for a pin-ended beam-column with end moment. The solid line and the dotted line indicate the results obtained by the exact moment-curvature method and by Newmark's method, respectively. Initial imperfection and residual stress are not considered in these calculations. It can be seen from the figure that when the column is short, the exact moment-curvature method compares well with Newmark's method. When the column length becomes larger, the exact moment-curvature method gives somewhat more conservative results.

Effect of Initial Imperfection

The effect of initial imperfection (out-of-straightness) on the tubular column is studied here with four initial imperfections: w_i = 0.0001L, 0.001L, 0.005L, and

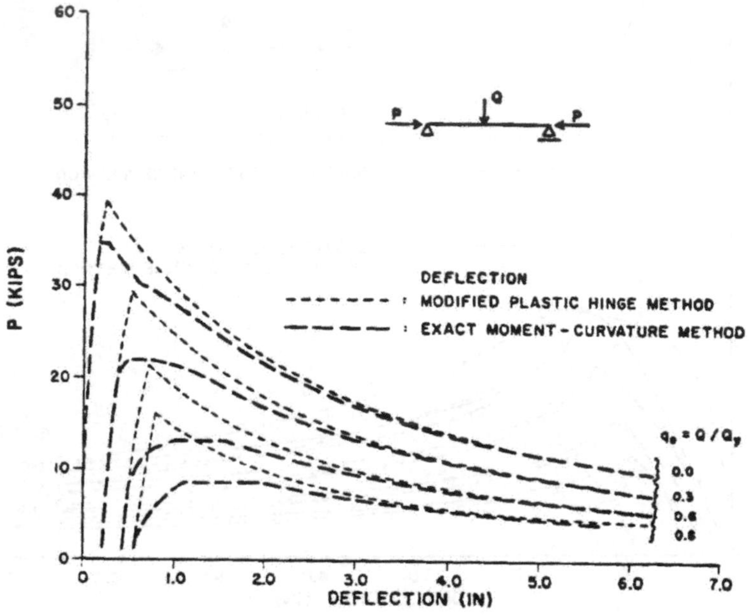

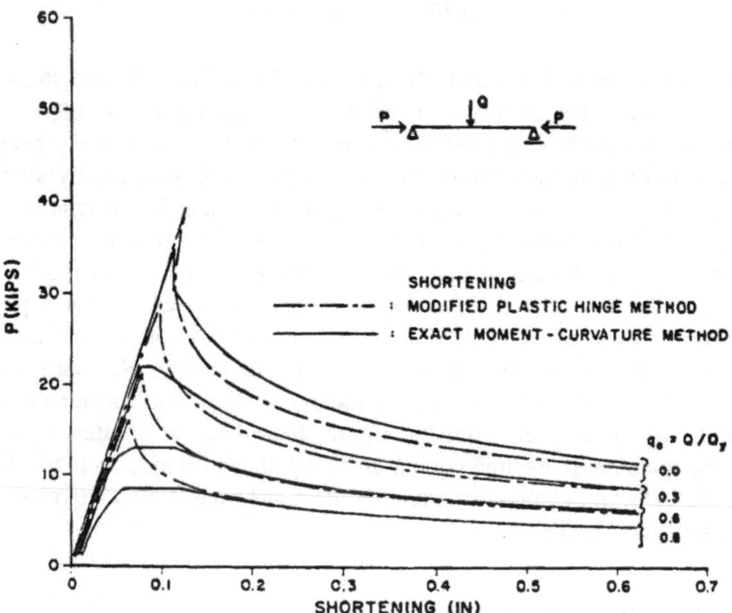

FIGURE 3.14 Comparison between exact moment-curvature method and modified plastic hinge method, pin-ends, L/r = 80.

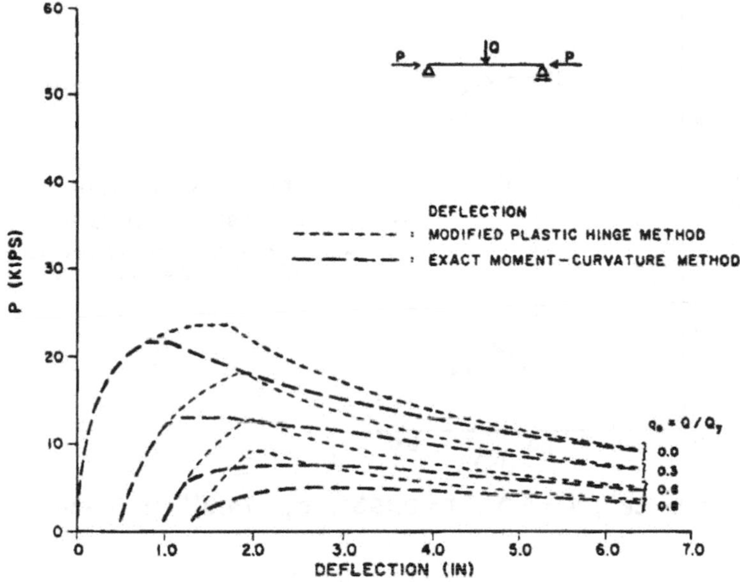

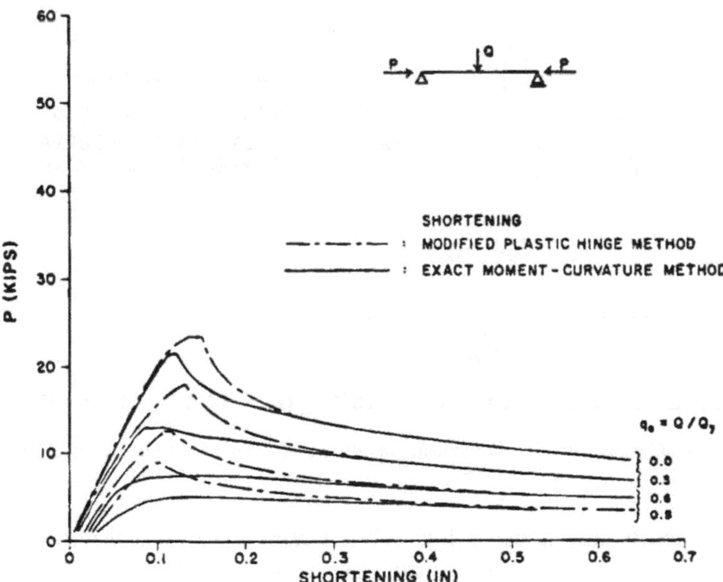

FIGURE 3.15 Comparison between exact moment-curvature method and modified plastic hinge method, pin-ends, L/r = 120.

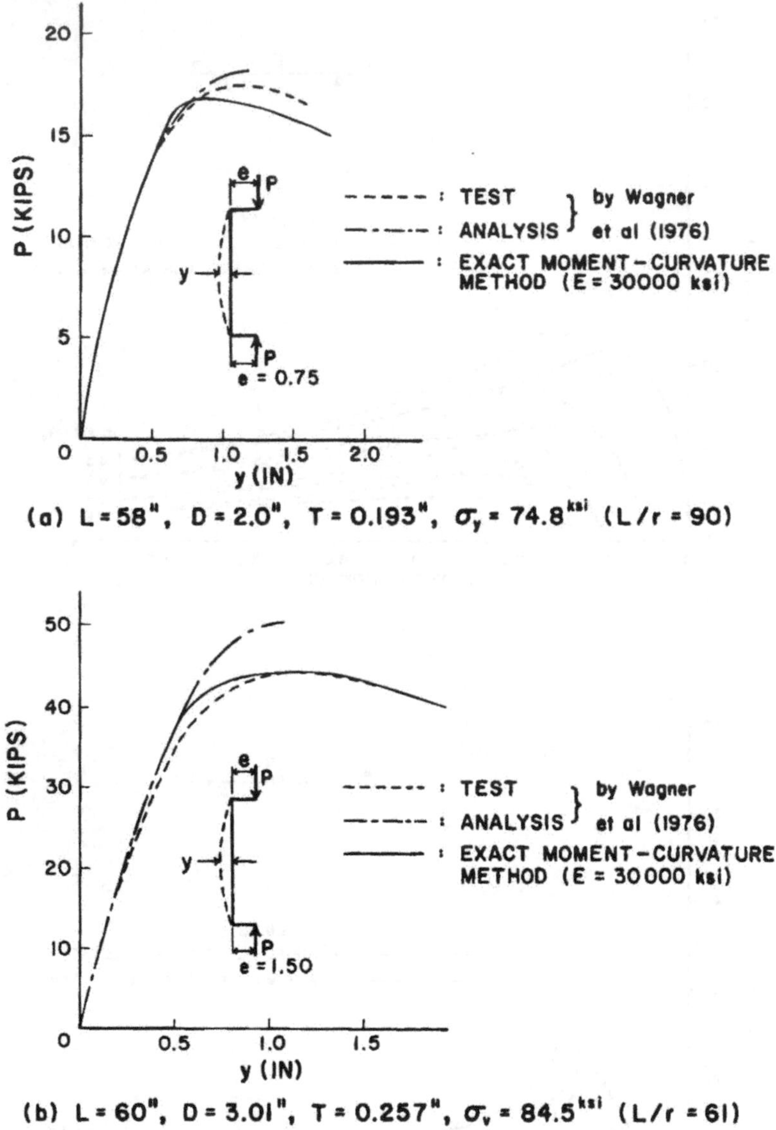

FIGURE 3.16 Comparison of exact moment-curvature method with test results, small-scale test.

0.01L. The results are shown in Fig. 3.19. It can be seen that the increase in the initial imperfection reduces the ultimate strength significantly. The column with L/r = 80 and w_i = 0.001L, which are typical in a practical situation, has an ultimate strength about 80% less than that of a perfect case.

For beam-columns with small imperfection (less than 0.001L), the load-deflection or load-shortening curve shows an apparent sharp peak. However, when the imperfec-

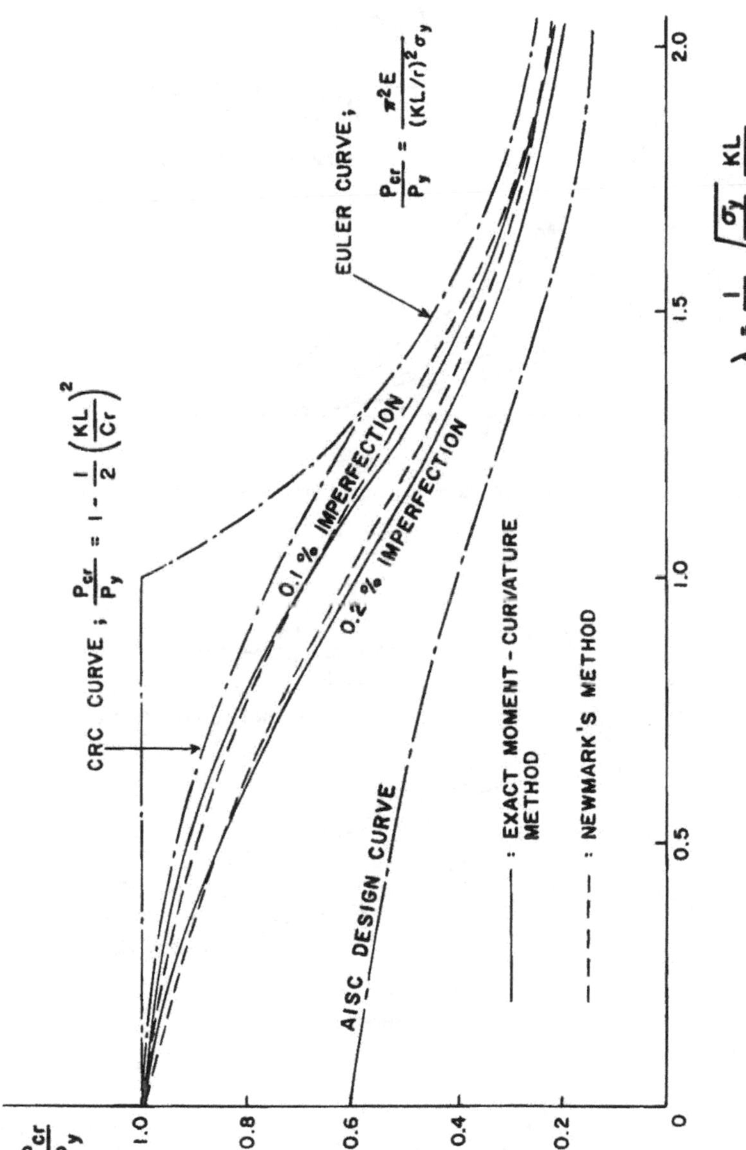

FIGURE 3.17 Comparison of column strength between exact moment-curvature method and Newmark's method.

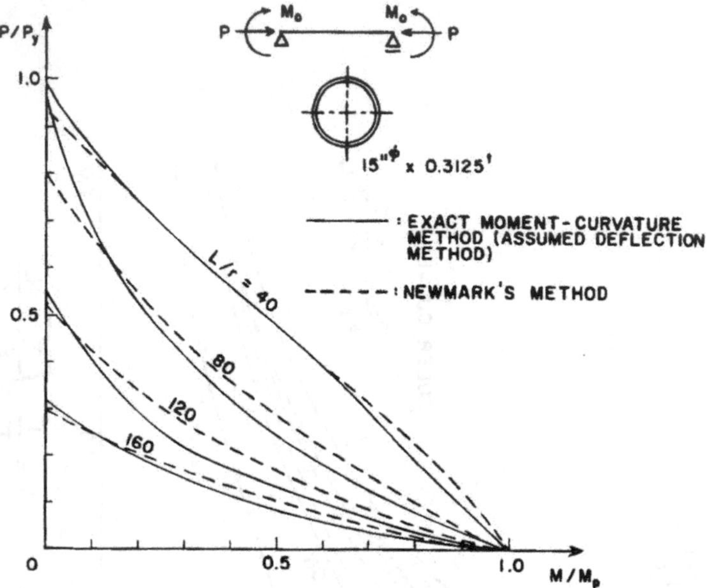

FIGURE 3.18 Comparison of interaction curves for a pin-ended beam-column.

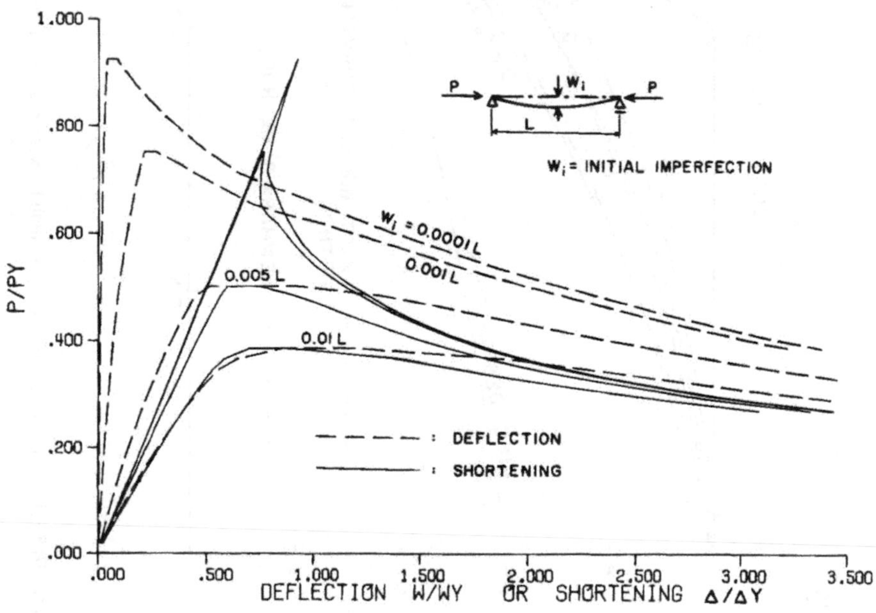

FIGURE 3.19 Effect of initial imperfection, L/r = 80.

tion becomes large (0.01L), the curves become flatter. After reaching the ultimate strength, the peak load remains almost constant in the post-peak range. It is found that this observation is also truc for beam-columns with other slenderness ratios.

Effect of Lateral Load and End Moment

The effects of lateral load at mid-span and bending moment at ends are investigated here. The out-of-straightness is taken here to be 0.1% of the column length. Figure 3.20 shows the comparisons of results for various combinations of lateral load and end moments applied to pin-ended and fixed-ended beam-columns. The external loads are seen to reduce the ultimate strength and post-peak behavior of a beam-column significantly. The lateral load and the end moment show similar effects on the behavior and strength of a beam-column.

Comparing the fixed-ended beam-column (Fig. 3.20c) with the pin-ended beam-column (Fig. 3.20a), the normalized load-deflection relationship for both cases is seen to be identical. However, there is a small difference in their load-shortening behavior. This difference results from the fact that they have different actual lengths, although their effective column length is the same.

Figure 3.21 summarizes the results for the case of a longer beam-column, KL/r = 120. The general behavior is seen to be similar to that of a shorter column. However, the curves are much flatter beyond the peak value, and the maximum strength throughout the entire range of loading combinations is significantly lower than that of a shorter column. It has been confirmed that for cases with other slenderness ratios, similar characteristics to the cases shown are observed.

Components for Axial Shortening

The total axial shortening consists of two parts: the effect of geometrical shape change and the effect of axial strain. Typical contributions from each of these effects is shown in Fig. 3.22. The effect due to lateral geometrical change is indicated by the dotted line, which is calculated directly from the corresponding load-deflection curve (dashed line). The total axial shortening for an elastic–plastic beam column is indicated by the solid curve. Therefore, the difference between the total and the geometrical shortening is the contribution due to axial strain. If an elastic axial stiffness is used for an axial shortening calculation throughout the entire loading range, the corresponding total axial shortening curve in the post-peak range lies below the more exact curve based on the closed-form expressions for elastic–plastic M-P-ϵ_o curves, which take a more accurate account of the moment-axial strain relation in the elastic–plastic range. The curve based on an elastic axial stiffness is shown by the dashed-dotted line in the figure. It can be seen that the difference between these two approaches is not large.

It should be noted that up to the ultimate load, the geometrical change hardly contributes to the axial shortening, and the main contribution is from axial strain. After buckling, the contribution due to geometrical shortening

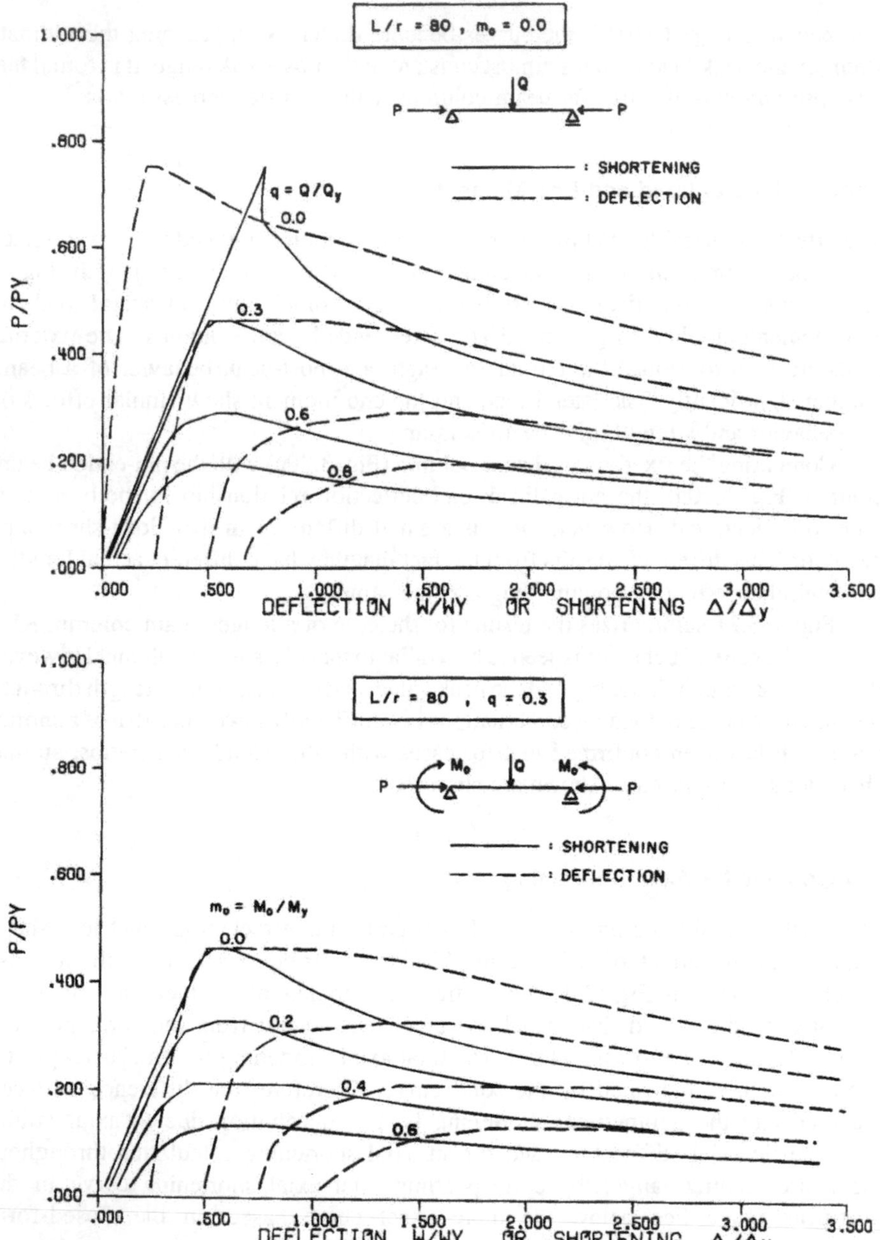

FIGURE 3.20 (a) Effect of lateral load, L/r = 80, m_o = 0.0, pin-ends. (b) Effect of end-moment, L/r = 80, q = 0.3, pin-ends. (c) Effect of lateral load, KL/r = 80, fixed-ends.

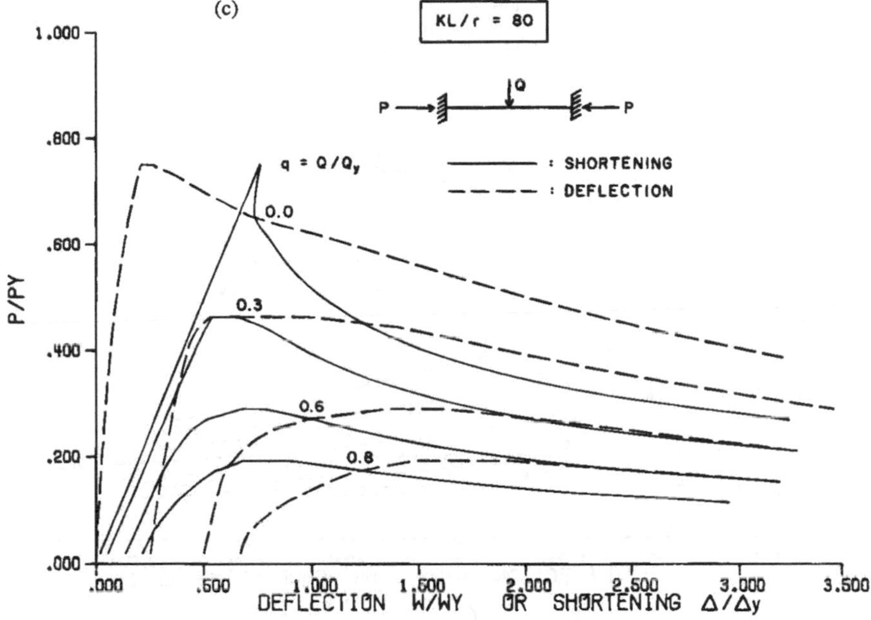

FIGURE 3.20 *Continued*

increases rapidly, while the percentage of contribution from axial strain decreases steadily as the axial load decreases. It is of interest that in the case of a relatively shorter column with a small initial imperfection, the decrease of axial shortening as the result of axial load drop beyond the peak value may be greater than that of the increase of geometrical shortening. If this happens, the load-shortening curve will show a shape concave to the left immediately after the peak value. Note that in the present analysis, material strain hardening, Bauschinger effect and local buckling effect are not considered. It seems likely that these, together with differences in the cross sectional size and the assumed pattern of residual stresses, could well be responsible, since such factors would be expected to become increasingly significant as the column approaches the pure axial load case.

3.12.3 Average Flow Moment Method

Typical numerical results using an average flow moment are shown in Fig. 3.23 for L/r = 80 and 120, respectively. Also shown are the results by the exact moment-curvature method. It can be seen that for a more slender column (say, L/r over 90), the average flow moment method agrees well with the exact moment-curvature method (Fig. 3.23b). However, when the beam-column is relatively short while lateral load is large, the average flow moment method shows some errors. It should be noted that where a bilinear moment-curvature relation is used, the load-deflection response of a beam-column will always show a sharp peak at the ultimate strength. This is unavoidable, for this type of idealized moment-curvature relation.

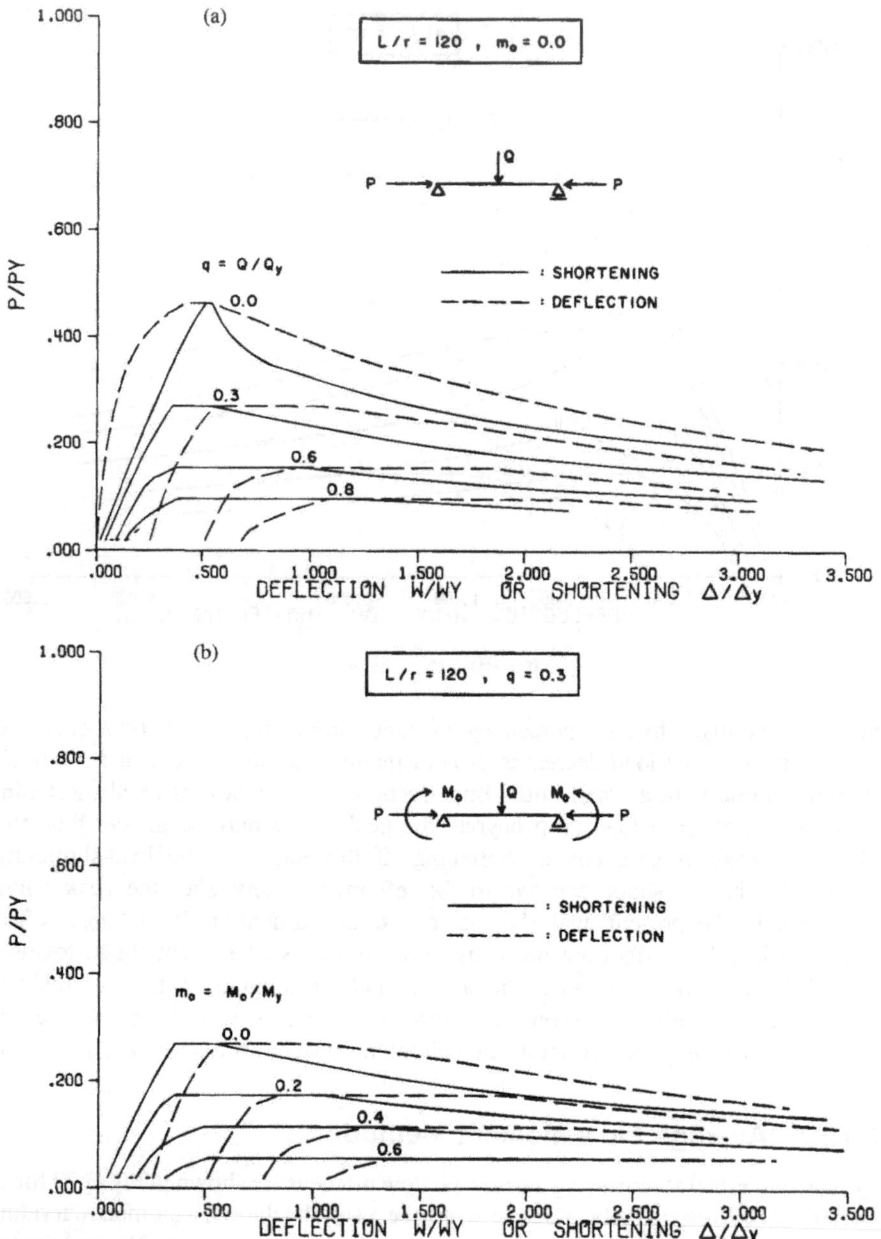

FIGURE 3.21 (a) Effect of lateral load, L/r = 120, m_o = 0.0, pin-ends. (b) Effect of end-moment, L/r = 120, q = 0.3, pin-ends. (c) Effect of lateral load, KL/r = 120, fixed-ends.

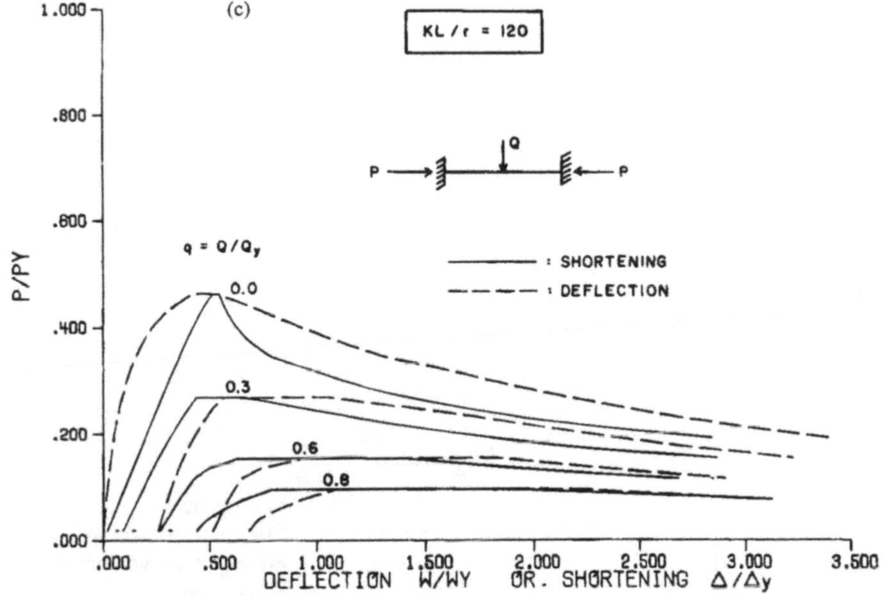

FIGURE 3.21 *Continued*

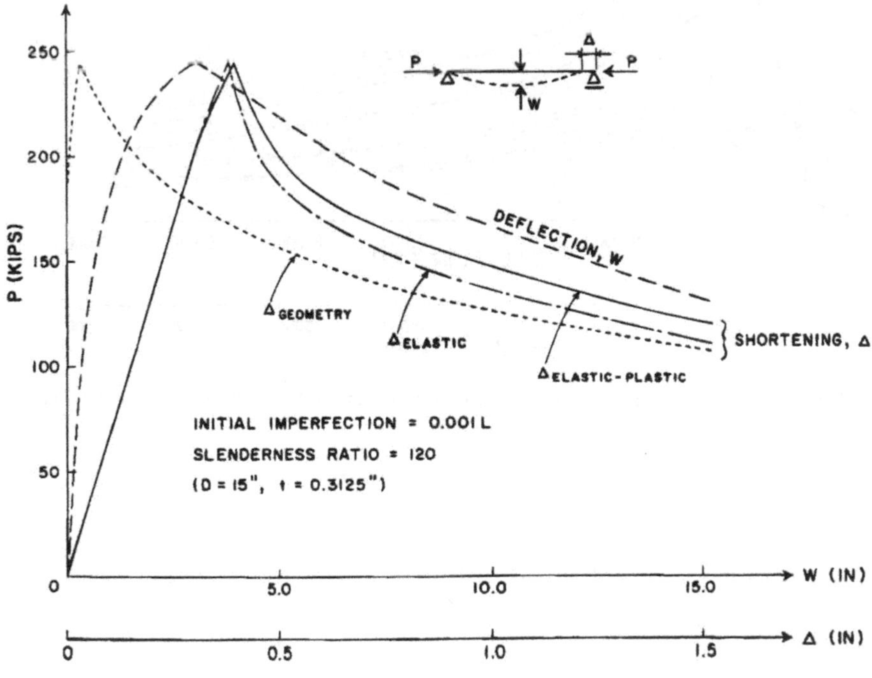

FIGURE 3.22 Components of shortening.

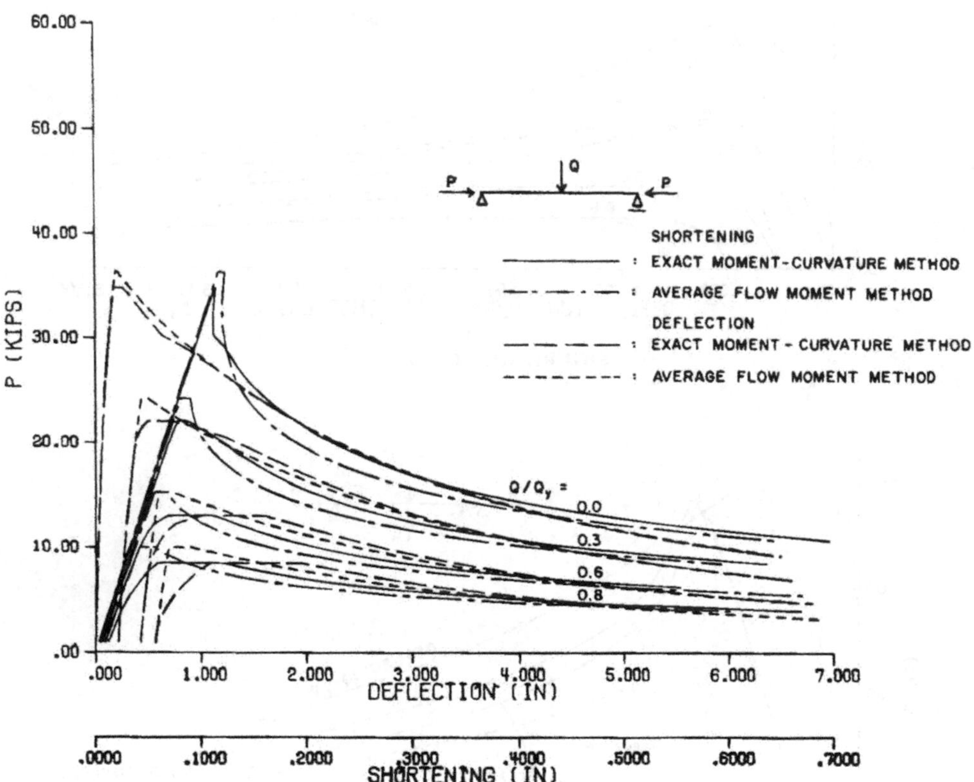

FIGURE 3.23a Comparison between exact moment-curvature method and average flow moment method, pin-ends, L/r=80.

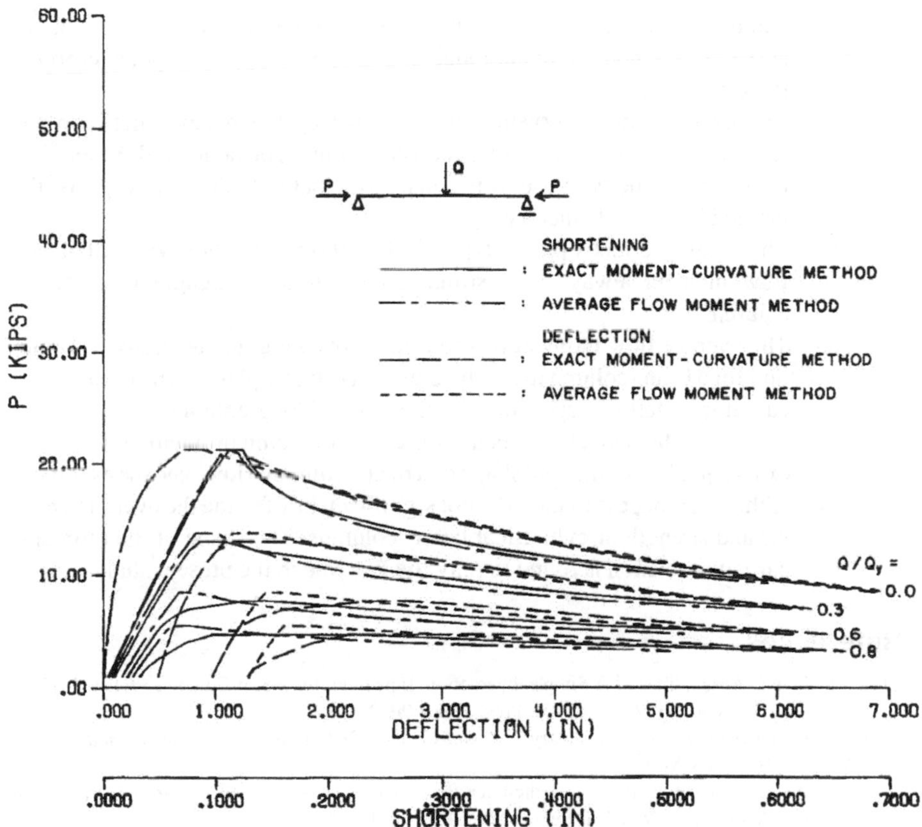

FIGURE 3.23b Comparison between exact moment-curvature method and average flow moment method, pin-ends, L/r=120.

3.12.4 Conclusions

1. The exact moment-curvature method is most suitable among the assumed deflection methods studied in this chapter. This method enables designers to assess in a simple manner the behavior of beam-columns up to ultimate load, including post-buckling behavior.
2. The general validity of the exact moment-curvature method has been demonstrated by comparisons with other analytical procedures as well as available test data.
3. Although solving a cubic equation is required in the exact moment-curvature method, computational time required by a computer is found to be very reasonable.
4. The exact moment-curvature method strikes the balance between the requirement of realistic representation of column behavior and the requirement for simplicity in use. It is considered that in both these respects, the method is most satisfactory.
5. The elastic–perfectly plastic type of M-P-Φ simplification with full plastic flow moment always overestimates the ultimate strength of a beam-column.
6. The average flow moment method improves the ultimate strength prediction for a beam-column using the elastic–perfectly plastic type of moment-curvature relation, especially for the case of long columns.
7. Formal mathematical treatment of the beam-column problem will yield an exact equation for the bent shape of a beam-column. However, the exactness of this bent shape is found to be not significant in affecting the overall behavior and strength of cylindrical beam-columns. The shape of the moment-curvature relation is found to play the key role in the present study.

References

Chen, W. F. and Atsuta, T. (1972) Simple Interaction Equations for Beam-Columns, *Journal of the Structural Division, ASCE,* 98, (ST7), Proc. Paper 9020, 1413–26.

Chen, W. F. and Atsuta, T. (1976) *Theory of Beam-Column,* Vol. 1: *In-Plane Behavior and Design,* McGraw-Hill, New York.

Cranston, W. B. (1967) A Computer Method for the Analysis of Restrained Columns, Cement and Concrete Association, Technical Report TRA 402, 20, London.

Fiala, D. W. and Erzurumlu, H. (1972) Moment-Thrust-Curvature of Tubular Members by Iteration, OTC Paper No. 1668, Offshore Technology Conference, Houston.

Saleeb, A. F. and Chen, W. F. (1981) Elastic-Plastic Large Displacement Analysis of Pipes, *Journal of the Structural Division, ASCE,* 107 (ST4), Proc. Paper 16199.

Santathadaporn, S. and Chen, W. F. (1972) Tangent Stiffness Method for Biaxial Bending, *Journal of the Structural Division, ASCE,* 98 (ST1), Proc. Paper 8637, 153–63.

Sherman, D. R., Erzurumlu, H. and Mueller, W. H., (1979) Behavioral Study of Circular Tubular Beam-Columns, *Journal of the Structural Division, ASCE,* (ST7), 1055–68.

Timoshenko, S. P. and Gere, J. M. (1961) *Theory of Elastic Stability,* Chapter 5, McGraw-Hill, New York.

Toma, S. and Chen, W. F. (1979) Analysis of Fabricated Tubular Columns, *Journal of the Structural Division, ASCE,* 105, (ST11), Proc. Paper 14994, 2343–66.

Wagner, A. L., Mueller, W. H. and Erzurumlu, H. (1977) Ultimate Strength of Tubular Beam-Columns, *Journal of the Structural Division, ASCE,* (ST1), Proceeding Paper 12670, 9–22.

4: Cyclic Behavior and Modeling

S. Toma
Department of Civil Engineering, Hokkai-Gakuen University, Sapporo, Japan

W. F. Chen
School of Civil Engineering, Purdue University, West Lafayette, Indiana

NOTATIONS

The following symbols are used in this chapter:

- A area or an integration constant
- a distance from the end of column to the lateral load or a constant for M-P-Φ curve expressions
- B an integration constant
- b a constant for M-P-Φ curve expressions
- C an integration constant
- c a constant for M-P-Φ curve expressions
- D diameter of the tube or an integration constant
- E Young's modulus or an integration constant
- F an integration constant
- f a constant for M-P-Φ curve expressions
- G an integration constant
- H an integration constant
- I moment of inertia of the section
- K effective column length factor
- k $\sqrt{P/EI}$
- L column length
- M bending moment
- M_c bending moment at midspan
- M_E bending moment at ends
- M_{pc} plastic moment with axial force
- M_x bending moment with respect to x axis
- M_y yield moment or bending moment with respect to y axis
- m normalized bending moment ($=M/M_y$)
- m_{pc} normalized plastic moment with axial force ($=M_{pc}/M_y$)
- m_s normalized bending moment at turning point
- P axial force at centroid
- p normalized axial force at centroid ($=P/P_y$)
- P_y yield axial force
- Q lateral load
- Q_y yield lateral load

0-8493-8282-3/96/$0.00+$.50

r radius of the tube or radius of gyration
t thickness of the tube
X, Y coordinates of an element
x distance from one end of the column
y deflection
y_1 deflection on left side to the lateral load
y_2 deflection on right side to the lateral load
σ stress
σ_c stress at current state
σ_y yield stress
ϵ strain
ϵ_o axial strain at centroid
ϵ_{y1} upper yield strain
ϵ_{y2} lower yield strain
Φ curvature
Φ_x curvature with respect to x axis
Φ_y yield curvature or curvature with respect to y axis
ϕ normalized curvature ($=\Phi/\Phi_y$)
ϕ_o normalized curvature at $m=0$
ϕ_1 normalized boundary curvature between elastic and primary yielded ranges on initial M-P-Φ curve
ϕ_1' normalized boundary curvature between elastic and primary yielded ranges on unloaded M-P-Φ curve
ϕ_2 normalized boundary curvature between primary and secondary yielded ranges on initial M-P-Φ curve
ϕ_2' normalized boundary curvature between primary and secondary yielded ranges on unloaded M-P-Φ curve
ϕ_s normalized curvature at turning point
ϕ_s' normalized symmetric curvature of ϕ_s
θ_c slope of the column at midspan
θ_E slope of the column at ends
θ_{y1}, θ_{y2} angles to define yielded range
Δ shortening
Δ_y yield shortening

4.1 INTRODUCTION

If the loads are cyclic in nature, such as from a severe earthquake or wave force, some members of the structure may suffer considerable inelastic cyclic deformation, and analytical predictions based on the monotonic behavior of members are not reliable. In this chapter, the cyclic inelastic behavior of cylindrical members is presented.

The main element required to maintain the overall integrity of a braced frame subjected to lateral forces is the brace. The principal objective of this chapter is to develop an analytical model that defines the nonlinear behavior of such braces

TABLE 4.1 Summary of Column Analyses

End Condition	Pin-Ends	Fixed-Ends
1. Loads	Cyclic axial compression-tension	Cyclic axial compression-tension
2. Method of Analysis	Newmark's method	Hinge-by-hinge method
3. M-P-Φ Curve	Exact, Doubly symmetric, Symmetric and elastic–perfectly plastic approximations for unloading	Elastic–perfectly plastic
4. P-M-ϵ_0 Curve	Exact, Symmetric approximation for unloading	Elastic–perfectly plastic
5. Computer Programs	Exact solution: BMCYCL Approximate solution: APCYCL	FIXCYCL

alternatively subjected to compressive loads, causing inelastic buckling, followed by tensile loads.

The hysteresis behavior of bracing members is quite complex because it is influenced by both buckling and yielding. Theoretical studies of the hysteresis behavior of members have assumed perfect members with simple end conditions, using the drastically simplified concept of plastic hinges (Higginbotham et al., 1976; Jain et al., 1978). It is evident that members in structures do not resist loads in this idealized manner. A real bracing member has geometric and material imperfections such as out-of-straightness and residual stresses and always has some flexural rigidity due to connections at its ends that provide a secondary load-resisting mechanism. The early analytical models for inelastic behavior of members have not been substantiated by experiments. In this chapter, the effort is made to develop a refined analytical model that can describe adequately the hysteretic characteristics of axially loaded steel cylindrical bracing members subjected to cyclic loading and to compare these theoretical predictions with available experimental results.

The analyses described in this chapter can be categorized into the following three parts:

1. M-P-Φ and P-M-ϵ_0 relations for cyclic loading will be studied by a numerical exact method (called tangent stiffness method) and also by the closed-form approximations. The doubly symmetric, symmetric, and elastic–perfectly plastic curves for unloading M-P-Φ are used for the approximate solutions.
2. The pin-ended column is analyzed by Newmark's method using the exact and approximate M-P-Φ curves.
3. The fix-ended column is analyzed by the hinge-by-hinge method using an elastic–perfectly plastic type of M-P-Φ curve.

The key features considered in the cyclic analysis of pin-ended and fixed-ended columns are summarized in Table 4.1.

4.2 CYCLIC BEHAVIOR OF SHORT TUBE

4.2.1 Basic Concept

The basic relation required in any beam-column analysis is the moment-curvature relation of the cross section. In particular, we need the moment-thrust-curvature (M-P-Φ) relation for load-deflection (P-w) analysis and the moment-thrust-axial strain (M-P-ϵ_o) relation for load-shortening (P-Δ) analysis. To obtain the exact moment-curvature relations, a computer program has been developed in Chapter 2, using the tangent stiffness method, in which the loads are increased monotonically. Therefore, it is now necessary to develop M-P-Φ and M-P-ϵ_o relations for cyclic loading to analyze the cyclic behavior of beam-columns.

In considering the history of loading, stress-strain characteristics of materials must be prescribed. For a tubular member, a bilinear stress-strain curve is assumed herein so that elastic–perfectly plastic can be taken into account in addition to elastic unloading. In the analysis only normal stress and strain are treated, and longitudinal residual stresses of the type reported by Chen et al. (1977) are used.

It is commonly recognized that the calculations of M-P-Φ or M-P-ϵ_o result in the most time-consuming part of a beam-column analysis because this moment-curvature subroutine must be used for every stage of loading at every station along the length of a beam-column in an iterative and incremental procedure. Therefore, for practical purposes, closed-form expressions of M-P-Φ and M-P-ϵ_o will be derived to approximate the actual behavior of the cross section under cyclic loading.

A three-regime type of expression proposed by Chen (1971) has been used successfully in Chapter 3 for describing the post-buckling behavior of cylindrical beam-columns under monotonic loading. A similar attempt will be made here for the cylindrical segment under cyclic, proportional loading. This will reduce significantly the computational time for the study of the behavior of individual members as well as the case of braced frames.

Other key features of this section for the development of moment-curvature relation under cyclic loading are:

1. Material: elastic–perfectly plastic type
2. Method of Analysis: tangent stiffness method
3. Residual Stresses: longitudinal residual stress
4. Loading: Case 1 − P = constant and M = cyclic
Case 2 − P and M = cyclic proportionally, (M = αP)
5. Results: M-P-Φ
M-P-ϵ_o curves
6. Developed Computer Program: MPCYCL

4.2.2 Stress-Strain Relation

The stress-strain relationship of the material is assumed to be elastic–perfectly plastic, and strain hardening is not considered. Figure 4.1 shows the stress-strain

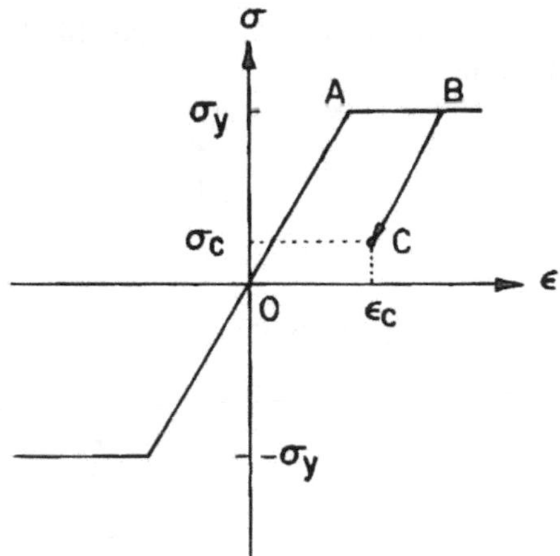

a) Initial Stress-Strain Relation

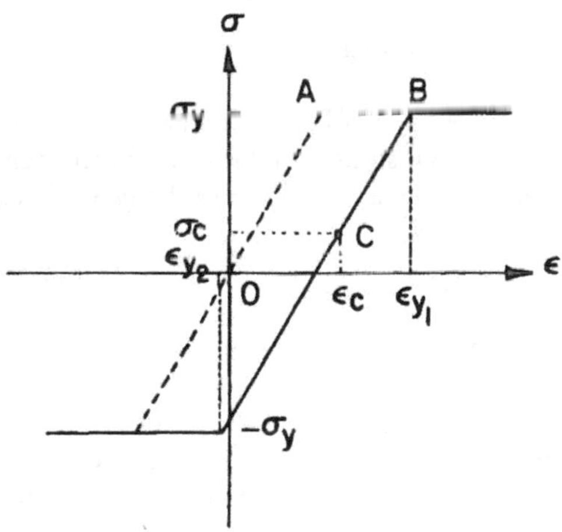

b) Current Stress-Strain Relation

FIGURE 4.1 Stress-strain relationship for cyclic loading.

curve for cyclic loading. The material is initially stressed up to the yield point "B" from the origin "O" and then unloaded to the point "C" (Fig. 4.1a). The Young's moduli for loading and unloading are, of course, the same, but the current stress-strain relation is now represented by the solid curve shown in Fig. 4.1b. The new curve is identical to the initial curve except that the position of the curve is shifted in such a way that it now passes through the current stress and strain state (the point "C"). Accordingly, the updated stress-strain relationship can be written in the form:

$$\sigma = \begin{cases} \sigma_c + E(\epsilon - \epsilon_c) & \text{for} \quad \epsilon_{y2} \leq \epsilon \leq \epsilon_{y1} \\ \sigma_y & \text{for} \quad \epsilon_{y1} < \epsilon \\ -\sigma_y & \text{for} \quad \epsilon < \epsilon_{y2} \end{cases} \tag{4.1}$$

The boundary values are

$$\epsilon_{y1} = \epsilon_c + \frac{\sigma_y - \sigma_c}{\epsilon}$$
$$\epsilon_{y2} = \epsilon_c - \frac{\sigma_y + \sigma_c}{\epsilon} \tag{4.2}$$

in which

σ_c, ϵ_c = current stress and strain
σ_y = yield stress
E = Young's modulus

The axial strain ϵ in an element of a cross section can be expressed in terms of the axial strain ϵ_o at the centroid and bending curvatures Φ_x and Φ_y of the cross section, which are known as the generalized strains in a beam-column analysis.

$$\epsilon = \Phi_x Y - \Phi_y X + \epsilon_o \tag{4.3}$$

in which

X, Y = coordinates of a element
Φ_x, Φ_y = curvatures with respect to the X and Y axis, respectively
ϵ_o = axial strain at the centroid of the section

Integrating the stress given by Eqs. (4.1) to (4.3) for the entire cross section, one can relate the generalized stresses, i.e., bending moments and axial force of the section, in terms of the generalized strains, ϵ_o, Φ_x, and Φ_y. That is,

$$P = \int_A \sigma \, dA$$
$$M_x = \int_A \sigma Y \, dA \tag{4.4}$$
$$M_y = \int_A \sigma X \, dA$$

in which

P = axial force at centroid
M_x, M_y = bending moment with respect to x and y axis, respectively

4.2.3 Exact M-P-Φ and P-M-ϵ_o Relations

The exact moment-thrust-curvature (M-P-Φ) relation has been studied extensively under monotonic loading using the tangent stiffness method described in Chapters 2 and 3. A brief outline of the procedure is as follows: The section is first divided into numerous finite elements, and the stress in each element is then calculated in accordance with the usual assumption of linear strain distribution. This is followed by an integration over the entire section to obtain the M-P-Φ relation.

The same approach is extended here to the cyclic loading case. The only difference between the monotonic and cyclic loading cases in developing the generalized stress-strain relations is that the stress-strain curve in the cyclic loading case changes according to the current stress state, rather than only a fixed curve in the case of monotonic loading (see Fig. 4.1).

The thrust-moment-axial strain (P-M-ϵ_o) relations are obtained based on the same assumptions and procedures as in the M-P-Φ relations; that is, the tangent stiffness method is used.

4.2.4 Computer Implementation

A computer program has been developed for the numerical solution of the exact cyclic M-P-Φ and P-M-ϵ_o curves based on the theory described above. The program is named MPCYCL, which is written in FORTRAN, and its brief flow chart is given in Fig. 4.2. The program is stored in the attached floppy disk and the sample problem is given in the next section.

4.2.5 User's Manual for the Program MPCYCL

Input manual of the program MPCYCL for the exact M-P-Φ and P-M-ϵ_o curves is given below. All data are in free format unless otherwise noted and should be written in sequence.

```
Line 1:   Title
     Column 1-68.  Title of a job
     Column 72.    NMP=Controller for Plotting type
                   NMP=0 . . . No plotting
                      =1 . . . Plotting of Moment-
                               Curvature
                      =2 . . . Plotting of Axial Force
                               -Axial Strain
```

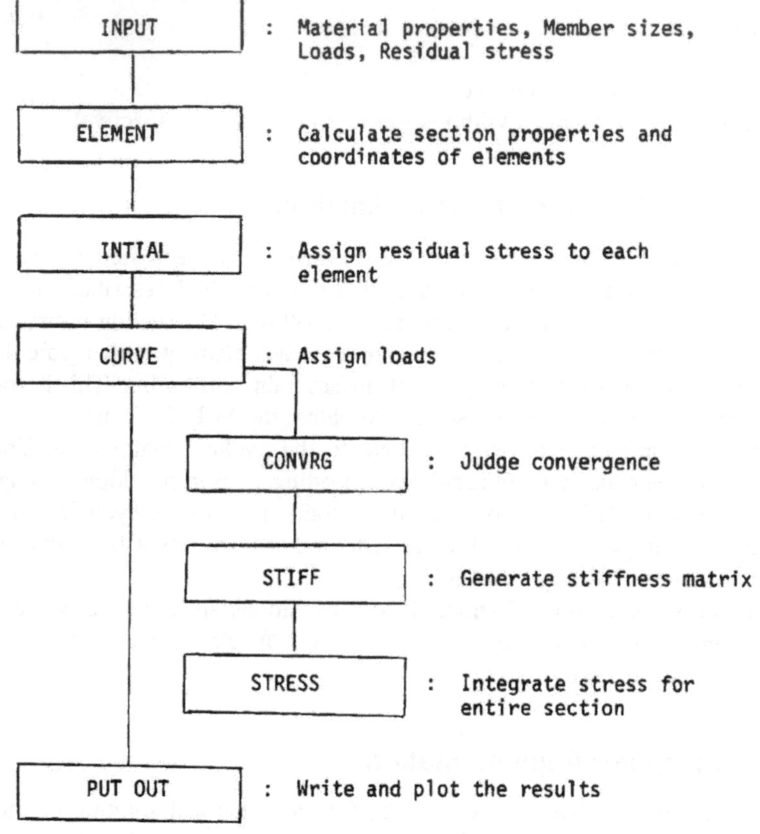

FIGURE 4.2 Flow chart of MPCYCL.

```
                        = 3 . . . Plotting of Moment-
                                  Curvature and Axial Force
                                  -Axial Strain

Line 2:  Material and Member Information
     SY = Yield stress (ksi)
     E = Young's modulus (ksi)
     D = Diameter (in.)
     T = Thickness (in.)
     NTH = Number of segments
     NOUT = Controller of kind of plotted lines
                = 1 Solid line
                = 2 Dotted line
                = 3 Dashed-dotted line

Line 3:  Load (1)
     QF = Shear force (kips)
     NM = Number of input for each force
```

```
Line 4:  Load (2)
     MX(i)  = bending moment about x-axis (i = 1,2, . . .
              NM) (kip-in.)
     DMX    = Load increment between MX(i) and MX(i + 1)
              (kip-in.)

Line 5:  Load (3)
     MY(i)  = Bending moment about Y-axis (i = 1,2, . . .
              NM) (kip-in.)
     DMY    = Load increment between MY(i) and MY(i + 1)
              (kip-in.)

Line 6:  Load (4)
     P(i)   = Axial force (i = 1,2,
     . . . NM) (kips)
     DP     = Load increment
              between P(i) and
              P(i + 1)
```

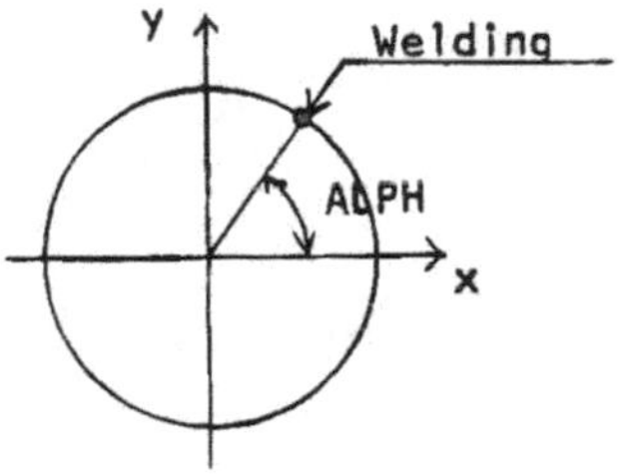

```
Line 7:  Residual Stress
     ALPH   = Angle of the location
              of welding seam
              from X - axis
              (degree)
     NRT    = Number of input

     RLA(i) = Circumferential location of the point
              (radian, i = 1,2, . . . NRT, = 0 to π . . .
              Symmetric with respect to weld seam)
     FLA(i) = Residual stress at the point (i = 1,2, . . .
              NRT) (ksi)
Remarks:
(1) To continue the calculations, repeat Lines 2 through 7.
(2) All Lines except Line 1 are in free format.
(3) The program in the attached floppy disk is not capable of
 plotting.
```

Following the input manual described above, the input data and the corresponding output data for a sample calculation is shown next.

INPUT DATA OF THE PROGRAM "MPCYCL" FOR A SAMPLE PROBLEM

```
Sample Calculation of MPCYCL
36 30000 15 0.3125 48 3
0 5
 0.000E+00  1.950E+03 -1.950E+03 1.950E+03 -1.950E+03 5.000E+01
 0.000E+00  0.000E+00  0.000E+00 0.000E+00  0.000E+00 1.000E+00
```

```
 0.000E+00  1.038E+02 -1.038E+02 1.038E+02 -1.038E+02 1.000E+01
0 6
 0.000E+00  3.500E+01
 3.800E-01 -1.260E+01
 1.180E+00  3.600E+00
 1.980E+00  0.000E+00
 2.360E+00 -2.520E+00
 3.141E+00  1.800E+00
```

....

OUTPUT DATA OF THE PROGRAM "MPCYCL" FOR A SAMPLE PROBLEM

```
Sample Calculation of MPCYCL              NMP = 0
 SY= 36.00  E= 30000.00  D=15.000  T= .3125  NTH= 48   NOUT = 3

 QF    = .000E+00
 MX(I) = .000E+00   .195E+04 -.195E+04 .195E+04  -.195E+04  .500E+02
 MY(I) = .000E+00   .000E+00  .000E+00 .000E+00   .000E+00  .100E+01
 P(I)  = .000E+00   .104E+03 -.104E+03 .104E+03  -.104E+03  .100E+02

 ALPH= .00
 RLA(I)= .000E+00   .380E+00  .118E+01 .198E+01  .236E+01 .314E+01
 FLA(I) = .350E+02 -.126E+02  .360E+01 .000E+00 -.252E+01 .180E+01

DATA OF MEMBER
 AREA= .14419E+02 I= .38900E+03 S= .51867E+02 Z= .67424E+02 SHAPE FUNCTION= .1300E+01
RGI= .51940E+01 DA= .30040E+00
 M(YIELD)= .18672E+04 M(PLASTIC)= .24272E+04 P(YIELD)= .51910E+03 PHI(YIELD)= .16000E-03
EY= .12000E-02

COORDINATES
     XO          YO          XO          YO          XO          YO          XO          YO
 .7324E+01  .4800E+0  .7199E+01  .1432E+01  .6950E+01  .2359E+01  .6583E+01  .3246E+01
       XO          YO
      .6103E+01  .4078E+01
 .5518E+01  .4839E+01  .4839E+01  .5518E+01  .4078E+01  .6103E+01  .3246E+01  .6583E+01
      .2359E+01  .6950E+01
 .1432E+01  .7199E+01  .4800E+00  .7324E+01 -.4800E+00  .7324E+01 -.1432E+01  .7199E+01
     -.2359E+01  .6950E+01
-.3246E+01  .6583E+01 -.4078E+01  .6103E+01 -.4839E+01  .5518E+01 -.5518E+01  .4839E+01
     -.6103E+01  .4078E+01
-.6583E+01  .3246E+01 -.6950E+01  .2359E+01 -.7199E+01  .1432E+01 -.7324E+01  .4800E+00
     -.7324E+01 -.4800E+00
-.7199E+01 -.1432E+01 -.6950E+01 -.2359E+01 -.6583E+01  -3246E+01 -.6103E+01 -.4078E+01
     -.5518E+01 -.4839E+01
-.4839E+01 -.5518E+01 -.4078E+01 -.6103E+01 -.3246E+01 -.6583E+01 -.2359E+01 -.6950E+01
     -.1432E+01 -.7199E+01
-.4800E+00 -.7324E+01  .4800E+00 -.7324E+01  .1432E+01 -.7199E+01  .2359E+01 -.6950E+01
      .3246E+01 -.6583E+01
 .4078E+01 -.6103E+01  .4839E+01 -.5518E+01  .5518E+01 -.4839E+01  .6103E+01 -.4078E+01
      .6583E+01 -.3246E+01
 .6950E+01 -.2359E+01  .7199E+01 -.1432E+01  .7324E+01 -.4800E+00

RESIDUAL STRESS
 RSTA(I)
 26.3796  9.9827 -6.4142 -11.4394  -8.7887 -6.1380 -3.4873   -.8366  1.8142  2.8921
  2.3030  1.7140  1.1249    .5359   -.0532  -.7465 -1.6146  -2.4827 -2.6010 -1.8769
 -1.1529  -.4288   .2952   1.0193   1.0193   .2952  -.4288  -1.1529 -1.8769 -2.6010
 -2.4827 -1.6146  -.7465   -.0532    .5359  1.1249  1.7140   2.3030  2.8921  1.8142
  -.8366 -3.4873 -6.1380  -8.7887 -11.4394 -6.4142  9.9827  26.3796

A(I)
 26.8015  10.4046 -5.9923 -11.0175  -8.3668 -5.7160  -3.0653   -.4146  2.2361  3.3140
  2.7250   2.1359  1.5469    .9578    .3688  -.3246  -1.1927  -2.0607 -2.1790 -1.4550
  -.7309   -.0069   .7172   1.4413   1.4413   .7172   -.0069   -.7309 -1.4550 -2.1790
 -2.0607  -1.1927  -.3246    .3688    .9578  1.5469   2.1359   2.7250  3.3140  2.2361
  -.4146  -3.0653 -5.7160  -8.3668 -11.0175 -5.9923  10.4046  26.8015
```

	MX	MY	P	PHIX	PHIY	EO	ITERATION
1	.50000E+02	-.16779E-04	.26615E+01	.42913E-05	.52001E-05	.61527E-05	2
2	.10000E+03	-.86931E-05	.53231E+01	.85825E-05	.52001E-05	.12305E-04	1
3	.15000E+03	-.28504E-06	.79846E+01	.12874E-04	.52001E-05	.18458E-04	1
4	.20000E+03	.23382E-04	.10646E+02	.17165E-04	.52001E-05	.24611E-04	1
5	.25000E+03	-.10172E-04	.13308E+02	.21456E-04	.52001E-05	.30763E-04	1
6	.30000E+03	.13495E-04	.15969E+02	.25748E-04	.52001E-05	.36916E-04	1
7	.35000E+03	-.13073E-05	.18631E+02	.30039E-04	.52001E-05	.43069E-04	1
8	.40000E+03	-.43432E-05	.21292E+02	.34330E-04	.52001E-05	.49221E-04	1
9	.45000E+03	-.73792E-05	.23954E+02	.38621E-04	.52001E-05	.55374E-04	1

```
10   .50000E+03   .12473E-04   .26615E+02   .42913E-04  .52001E-05   .61527E-04   1
11   .55000E+03  -.54994E-05   .29277E+02   .47204E-04  .52001E-05   .67679E-04   1
12   .60000E+03   .14030E-04   .31938E+02   .51495E-04  .52001E-05   .73832E-04   1
13   .65000E+03   .19511E-06   .34600E+02   .55786E-04  .52001E-05   .79985E-04   1
14   .70000E+03  -.37173E-04   .37262E+02   .60078E-04  .52001E-05   .86137E-04   1
15   .75000E+03   .20181E-04   .39923E+02   .64369E-04  .52001E-05   .92290E-04   1
16   .80000E+03  -.13372E-04   .42585E+02   .68660E-04  .52001E-05   .98443E-04   1
17   .85000E+03  -.20223E-04   .45246E+02   .72951E-04  .52001E-05   .10460E-03   1
18   .90000E+03   .22517E-04   .47908E+02   .77243E-04  .52001E-05   .11075E-03   1
19   .95000E+03   .80373E-05   .50569E+02   .81534E-04  .52001E-05   .11690E-03   1
20   .10000E+04  -.26281E-05   .53231E+02   .85825E-04  .52001E-05   .12305E-03   1
21   .10500E+04  -.56640E-05   .55892E+02   .90116E-04  .52001E-05   .12921E-03   1
22   .11000E+04   .18003E-04   .58554E+02   .94408E-04  .52001E-05   .13536E-03   1
23   .11500E+04   .11797E-04   .61215E+02   .98699E-04  .52001E-05   .14151E-03   1
24   .12000E+04  -.48459E-04   .63877E+02   .10299E-03  .52001E-05   .14766E-03   1
25   .12500E+04   .16524E-04   .66538E+02   .10728E-03  .52001E-05   .15382E-03   1
~~~~~~~~~~~~~~~~~~~~~~~~~~~~~~~~~~~~~~~~~~~~~~~~~~~~~~~~~~~~~~~~~~~~~~~~~~~~~~~~~~~~~~~
258 -.12000E+04  -.41572E-06  -.63877E+02  -.86961E-04  .96573E-05  -.94990E-04   1
259 -.12493E+04  -.91462E-05  -.66445E+02  -.91252E-04  .96573E-05  -.10114E-03   1
260 -.13000E+04   .53570E-05  -.69200E+02  -.96302E-04  .96573E-05  -.11010E-03   2
261 -.13500E+04  -.14704E-04  -.71862E+02  -.10171E-03  .96573E-05  -.12038E-03   2
262 -.14000E+04   .78646E-06  -.74523E+02  -.10783E-03  .96573E-05  -.13340E-03   2
263 -.14500E+04  -.17264E-04  -.77185E+02  -.11404E-03  .96573E-05  -.14679E-03   2
264 -.15000E+04   .20117E-05  -.79846E+02  -.12126E-03  .96573E-05  -.16416E-03   2
265 -.15500E+04   .44749E-04  -.82508E+02  -.12853E-03  .96573E-05  -.18174E-03   2
266 -.16000E+04  -.80225E-05  -.85169E+02  -.13670E-03  .96573E-05  -.20304E-03   2
267 -.16500E+04  -.11619E-05  -.87831E+02  -.14537E-03  .96573E-05  -.22639E-03   2
268 -.16996E+04   .15615E-05  -.90414E+02  -.15403E-03  .96573E-05  -.24975E-03   2
269 -.17500E+04  -.40517E-04  -.93154E+02  -.16468E-03  .96573E-05  -.28166E-03   2
270 -.18000E+04  -.34608E-04  -.95815E+02  -.17522E-03  .96573E-05  -.31309E-03   2
271 -.18500E+04  -.44838E-06  -.98477E+02  -.18648E-03  .96573E-05  -.34781E-03   3
272 -.19000E+04  -.60901E-04  -.10114E+03  -.19956E-03  .96573E-05  -.39074E-03   2
273 -.19500E+04  -.53272E-05  -.10380E+03  -.21264E-03  .96573E-05  -.43367E-03   2
```

4.2.6 Numerical Results

Typical numerical results are given in Fig. 4.3, which shows the M-Φ relations for proportional loading cases without and with the effects of residual stress. It can be seen from the figures that the transition from elastic to plastic state is gradual and smooth. Figure 4.3a shows that if the influence of residual stress is neglected, the M-Φ relations are symmetric and their traces are the same under the second cycle of loading. However, this is not the case when the effect of residual stress is considered (Fig. 4.3b). The loops with residual stress are shifted slightly, and the traces under the second cycle are somewhat different from the first cycle of loading.

Figure 4.4 shows typical results of P-ϵ_o curves for the same proportional loading cases, with and without the effect of residual stress. Without the effect of residual stress, the P-ϵ_o curves are symmetric and trace the same loop for the second cycle of loading. With the effect of residual stress, however, the curves are not symmetric and are shifted slightly for the second cycle of loading, as in the M-P-Φ case.

4.2.7 Closed-Form Expressions of M-P-Φ Curves

Analytical expressions of M-P-Φ curves for monotonic loading have been described in Chapter 3. Rewriting again, they are given in the forms:

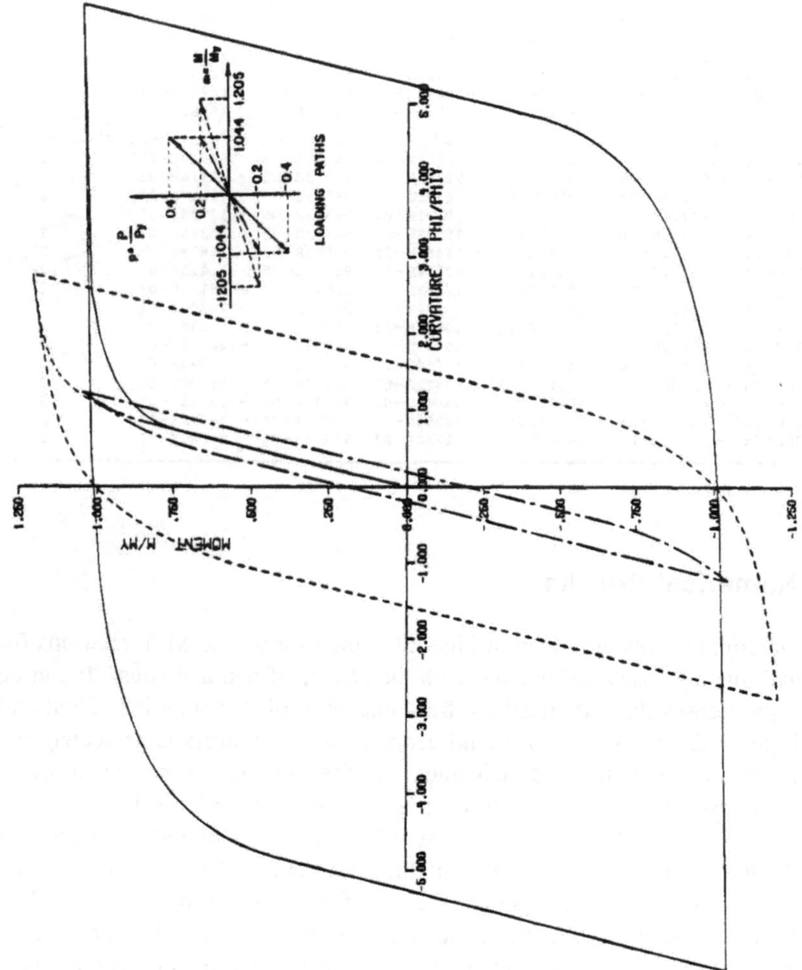

FIGURE 4.3a M-Φ relations for proportional loading without residual stress.

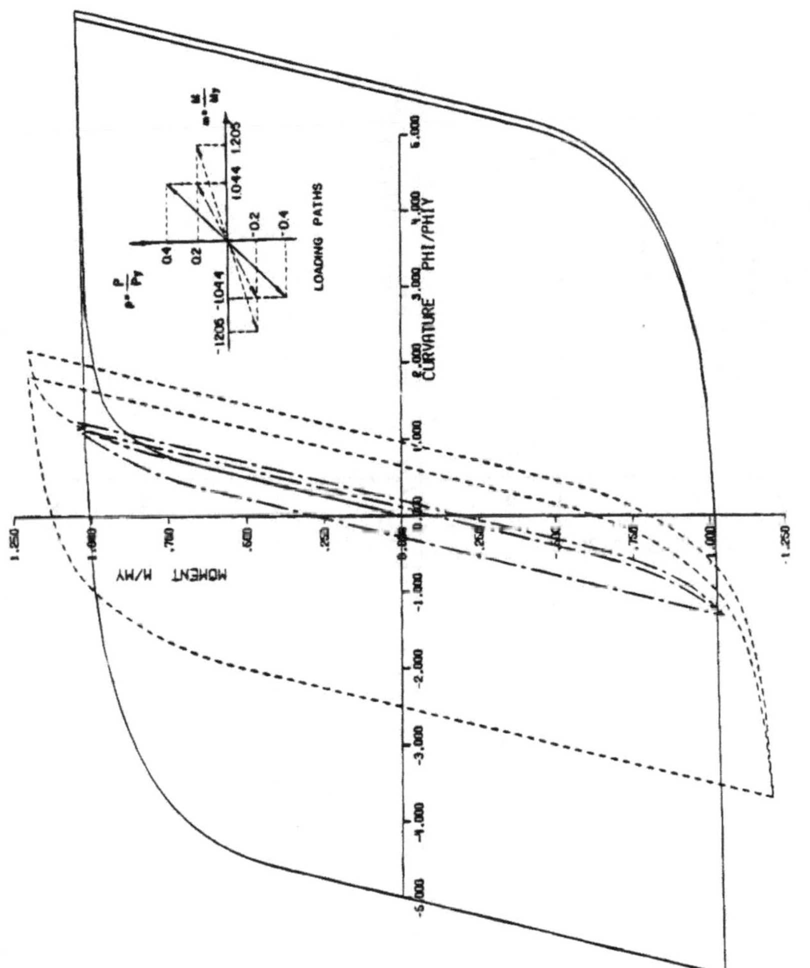

FIGURE 4.3b M-Φ relations for proportional loading with residual stress.

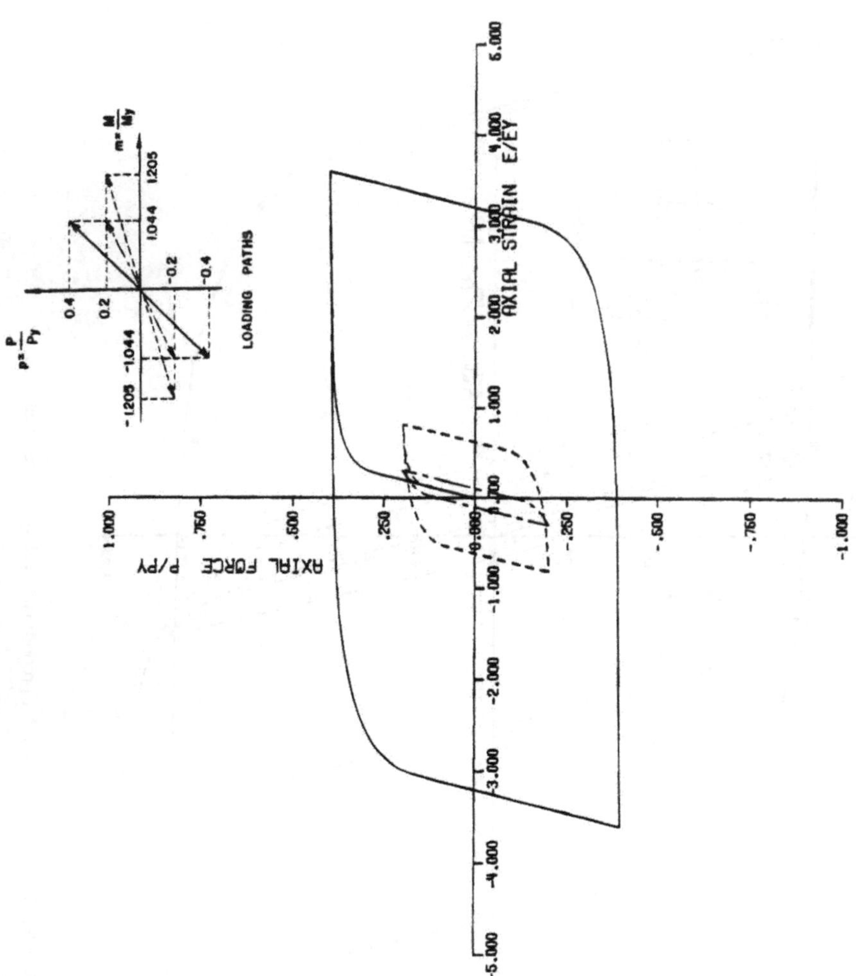

FIGURE 4.4a P-ϵ_o relations for proportional loading without residual stress.

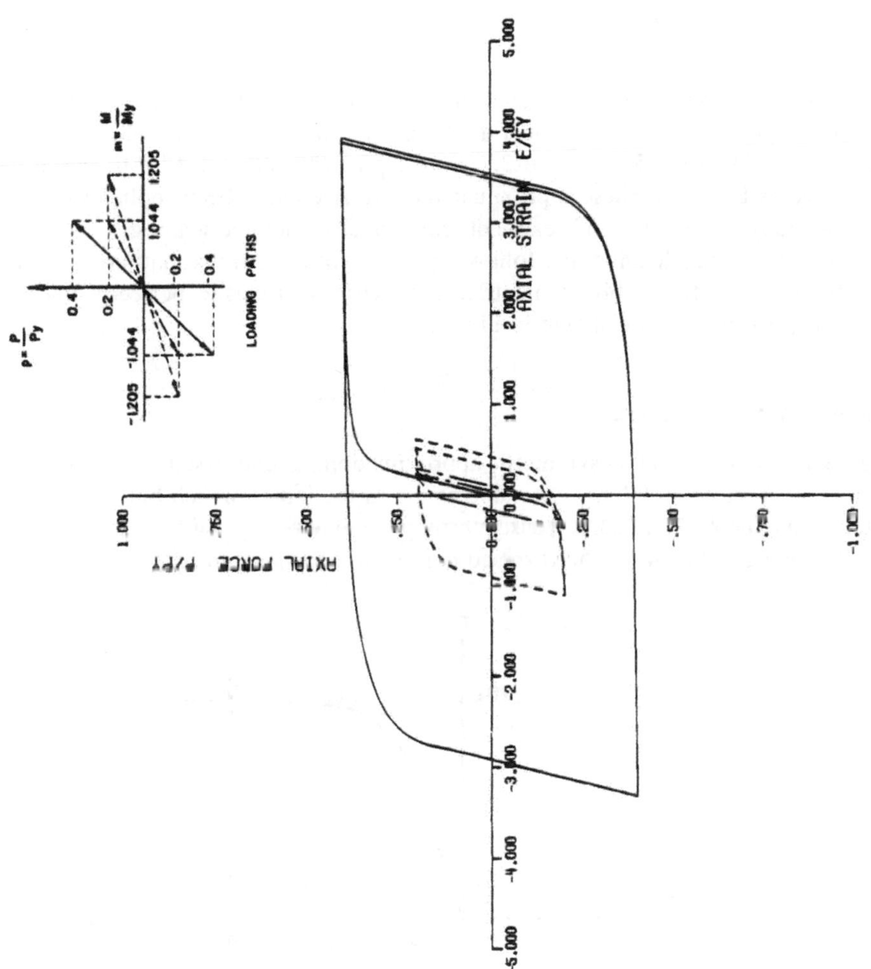

FIGURE 4.4b P-ϵ_0 relations for proportional loading with residual stress.

$$m = \begin{cases} a\phi & \text{for} \quad 0 \leq \phi \leq \phi_1\text{: elastic} \\ b - \dfrac{c}{\sqrt{\phi}} & \text{for} \quad \phi_1 < \phi \leq \phi_2\text{: primary yielded} \\ m_{pc} - \dfrac{f}{\phi^2} & \text{for} \quad \phi_2 < \phi\text{: secondary yielded} \end{cases} \tag{4.5}$$

in which m_{pc} = normalized fully plastic moment considering the effect of axial force.

The constants a, b, c, and f and the boundary curvatures ϕ_1 and ϕ_2 are given by Eqs. (3.21) to (3.25). Figure 3.7 shows schematically the M-P-Φ curve as expressed by Eq. (4.5). These approximated expressions have been applied successfully to study the behavior of beam-columns under monotonic loading.

As for reversed loading, the following three types of approximations of M-P-Φ curves are made: doubly symmetric, symmetric, and elastic–perfectly plastic. These approximations are shown in Fig. 4.5.

Doubly Symmetric Curve

Figure 4.5a shows a doubly symmetric approximation. The reversed loading branch of M-P-Φ from point "S" has the same shape as the initial branch but doubled in size. Thus, the reversed curve passes through symmetrical point "S."

Referring to Fig. 4.5a, the reversed curve can be expressed by

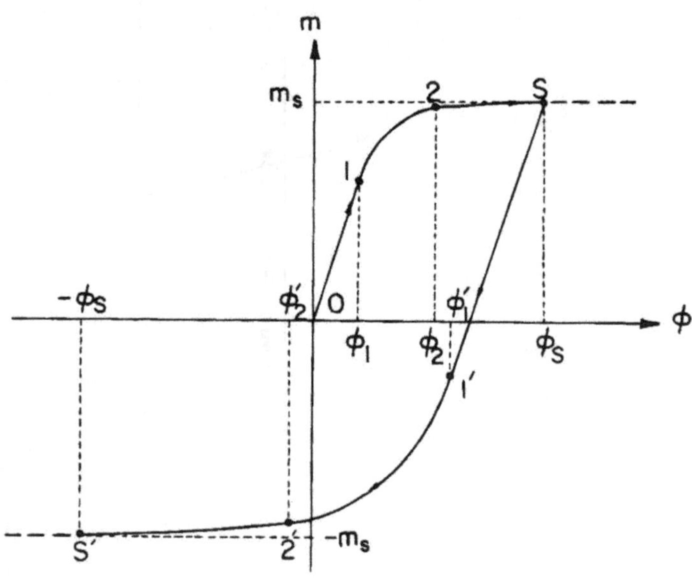

FIGURE 4.5 (a) Doubly symmetric approximation of cyclic M-P-Φ curves. (b) Symmetric approximation of cyclic M-P-Φ curves. (c) Elastic-plastic approximation of cyclic M-P-Φ curves.

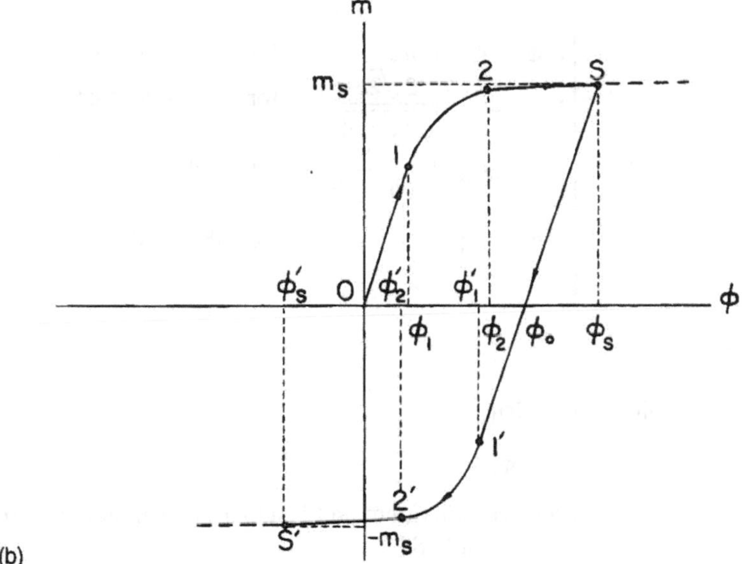

(b)

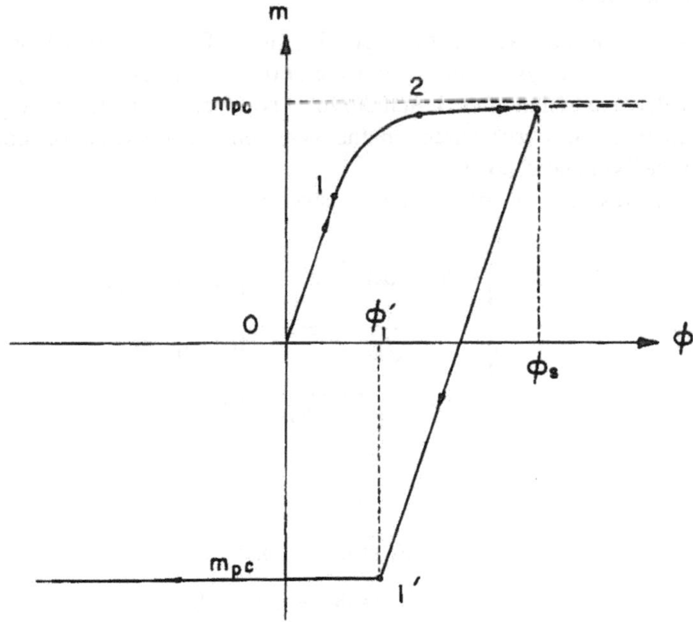

(c)

FIGURE 4.5 *Continued*

$$m = \begin{cases} a(\phi - \phi_s) + m_s & \text{for} \quad \phi_1' \le \phi \le \phi_s \\ m_s - 2b + \dfrac{2\sqrt{2}c}{\sqrt{\phi_s - \phi}} & \text{for} \quad \phi_2' \le \phi < \phi_1' \\ m_s - 2m_{pc} + \dfrac{8f}{(\phi - \phi_s)^2} & \text{for} \quad -\phi_s \le \phi < \phi_2' \\ -m_{pc} + \dfrac{f}{\phi^2} & \text{for} \quad \phi < -\phi_s \end{cases} \tag{4.6}$$

in which

$$\begin{aligned} \phi_1' &= \phi_s - 2\phi_1 \\ \phi_2' &= \phi_s - 2\phi_2 \end{aligned} \tag{4.7}$$

ϕ_s, m_s = normalized curvature and bending moment at reversed loading point "S"

Symmetric Curve

Figure 4.5b shows the symmetric approximation. The unloading branch up to the zero moment is always elastic. From the zero moment point ϕ_o on, the reversed curve is identical to that of the initial curve. Therefore, unlike the doubly symmetric approximation, the curve based on the symmetric approximation does not pass through the symmetric point "S."

The expressions for the symmetric curve are given by

$$m = \begin{cases} a(\phi - \phi_0) & \text{for} \quad \phi_1' \le \phi \le \phi_s \\ -b + \dfrac{c}{\sqrt{\phi_0 - \phi}} & \text{for} \quad \phi_2' \le \phi < \phi_1' \\ -m_{pc} + \dfrac{f}{(\phi - \phi_0)^2} & \text{for} \quad \phi < \phi_2' \end{cases} \tag{4.8}$$

in which

$$\begin{aligned} \phi_0 &= \phi_s - am_s \\ \phi_1' &= \phi_s - am_s - \phi_1 \\ \phi_2' &= \phi_s - am_s - \phi_2 \end{aligned} \tag{4.9}$$

Elastic–Perfectly Plastic Curve

To simplify drastically the cyclic M-P-Φ curves, the elastic–perfectly plastic approximation is studied (Fig. 4.5c). In this approximation, the unloading branch is assumed to be either elastic or perfectly plastic, although the initial loading branch is described by the three-regime expressions.

The expressions of the elastic–perfectly plastic approximation are

$$m = \begin{cases} q(\phi - \phi_0) & \text{for} \quad \phi_1' \le \phi \le \phi_s \\ -m_{pc} & \text{for} \quad \phi < \phi_1' \end{cases} \tag{4.10}$$

in which

$$\phi_0 = \phi_s - am_s \tag{4.11}$$

$$\phi_1' = \phi_s - a(m_s + m_{pc})$$

The comparison of M-P-Φ curves between the doubly symmetric approximation and exact numerical solutions is shown in Fig. 4.6. A good agreement between the two curves is observed.

4.2.8 Closed-Form Expressions of P-M-ϵ_O Curves

Analytical expressions of P-M-ϵ_O curves for monotonic loading has been described in Section 3.7.4. These expressions are summarized again here:

Elastic Range

$$\epsilon_0 = \frac{P}{2\pi Etr} \quad \text{for} \quad \begin{cases} \epsilon_0 - r\Phi \ge -\epsilon_y \\ \epsilon_0 + r\phi \le \epsilon_y \end{cases} \tag{4.12}$$

in which

ϵ_O = axial strain at centroid of the section
t = wall thickness of the tube
r = radius of the tube
P = axial force
ϵ_y = yield strain

Primary Yielded Range

$$\epsilon_o = \left[\frac{P}{Etr} + \epsilon_y(\pi - 2\theta_{y1}) - 2r\Phi\cos\theta_{y1}\right] \times \frac{1}{\pi + 2\theta_{y1}} \tag{4.13}$$

$$\text{for} \quad \begin{cases} \epsilon_o - r\Phi \ge \epsilon_y \\ \epsilon_o + r\Phi \ge \epsilon_y \end{cases}$$

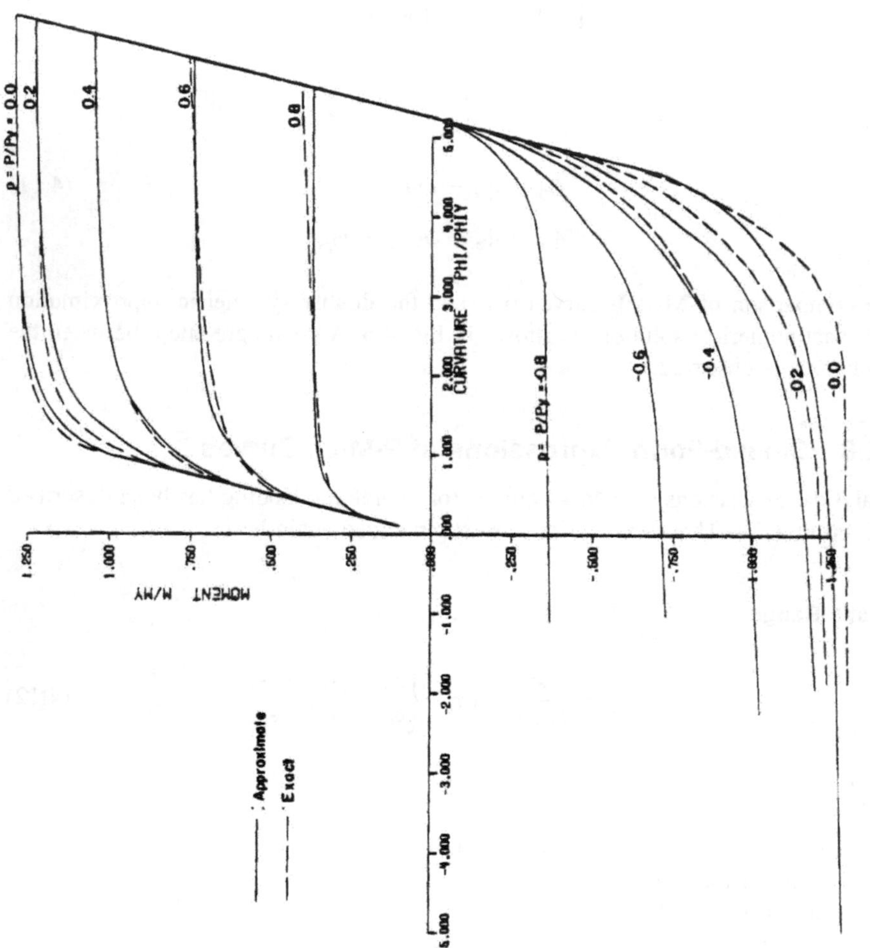

FIGURE 4.6 Comparison of M-P-Φ curves between exact and approximate solutions.

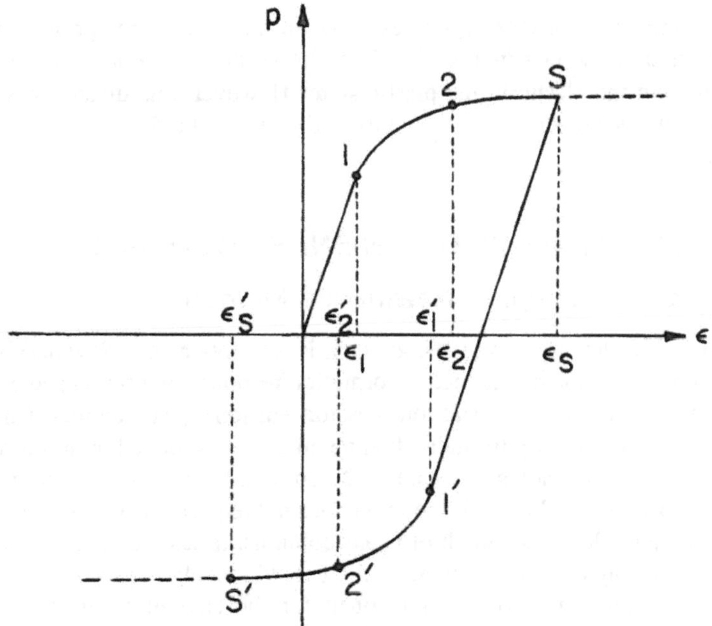

FIGURE 4.7 Symmetric approximation of cyclic P-M-ϵ_O curves.

in which

$$\theta_{y1} = \sin^{-1} \frac{\epsilon_y + \epsilon_o}{r\Phi} \tag{4.14}$$

Secondary Yielded Range

$$\epsilon_o = \left[\frac{P}{2Etr} - r\Phi(\cos \theta_{y1}) - \cos \theta_{y2} - \epsilon_y(\theta_{y1} - \theta_{y2}) \right] \times \frac{1}{\theta_{y1} + \theta_{y2}} \tag{4.15}$$

$$\text{for} \quad \begin{cases} \epsilon_o - r\Phi \geq -\epsilon_y \\ \epsilon_o + r\Phi \geq \epsilon_y \end{cases}$$

in which

$$\theta_{y2} = \sin^{-1} \frac{\epsilon_y - \epsilon_o}{r\Phi} \tag{4.16}$$

These expressions have been applied successfully for the case of monotonic loading. The last two equations are nonlinear for ϵ_O and require an iterative procedure for a solution. Further, it is too complicated to derive the doubly symmetric approximation for the unloading branch. Also, the elastic–perfectly plastic approximation is not suitable here because the axial strain becomes indeterminate. Therefore, in the following, only the symmetric approximation is used in all the analyses (Fig. 4.7).

The comparison of P-M-ϵ_O curves between the symmetric approximation and exact solutions is shown in Fig. 4.8. The discrepancy is found largest near the transition from the elastic to the plastic state. However, this discrepancy will not affect the behavior of the member significantly, as will be discussed in the following sections.

4.3 CYCLIC ANALYSIS OF PIN-ENDED COLUMNS

4.3.1 Basic Concept — Newmark's Method

On the basis of our previous investigations, it was found that Newmark's method is most suitable for tracing the load-deformation response of a pin-ended cylindrical member subjected to a cyclic axial compression-tension type of loading. This method has been used successfully for individual beam-columns loaded up to ultimate load and post-buckling unloading in Chapters 2 and 3 and the computer programs have been developed. Since Newmark's method does not require one to assume a distribution of curvature along the length of beam-columns, it has the capability to handle any type of loading as required in the cyclic behavior study of pipes. In this section, new computer programs will be developed for the case of cyclic loading using Newmark's numerical procedure, taking accurate account of the moment-curvature relation.

In outline form, the key features considered in this part of the study for inelastic cyclic behavior of cylindrical members are:

1. Load: cyclic axial compression-tension with a constant lateral load or cyclic lateral upward-downward load with a constant axial load
2. End Conditions: ideally pinned
3. Method of Analysis: Newmark's method
4. M-P-Φ Curves: exact solution, doubly symmetric, symmetric and elastic–perfectly plastic type of approximations for unloading
5. P-M-ϵ_O Curves: exact solution and symmetric type of approximation for unloading
6. Results: P-w (load-deflection) curves and P-Δ (load-shortening) curves
7. Developed Computer Programs: Exact Solution — BMCYCL
   Approximate Solution — APCYCL

4.3.2 Computer Implementation

Newmark's Method Using Exact Generalized Stress-Strain Relationships

The exact M-P-Φ and P-M-ϵ_O relationships have been discussed previously based on the assumption that the material is elastic-perfectly plastic (Fig. 4.1).

The computer program BMCYCL for the cyclic solution of pin-ended columns has been developed using the exact M-P-Φ and P-M-ϵ_O relations. A brief flow chart of the program is given in Fig. 4.9. In this program four subroutines for Newmark's

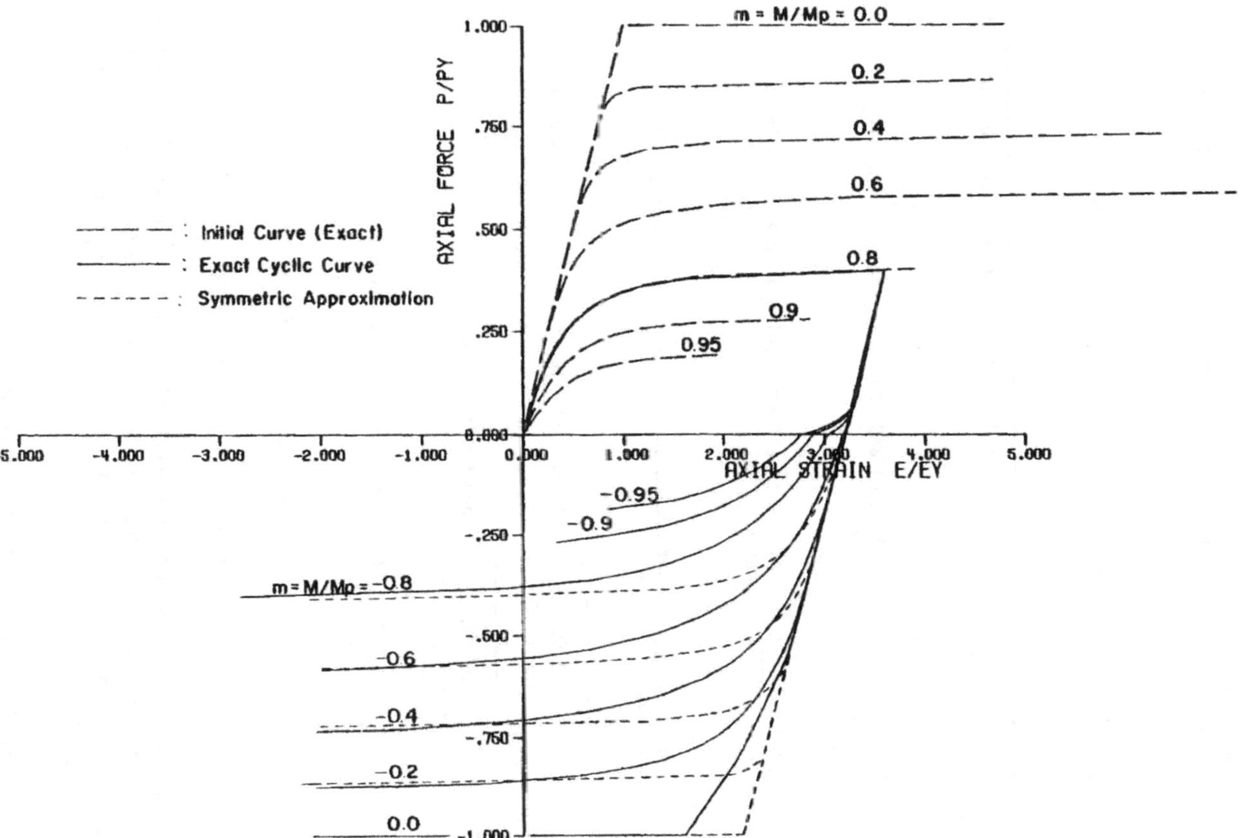

FIGURE 4.8 Comparison of M-P-ϵ_o curves between exact and approximate solutions.

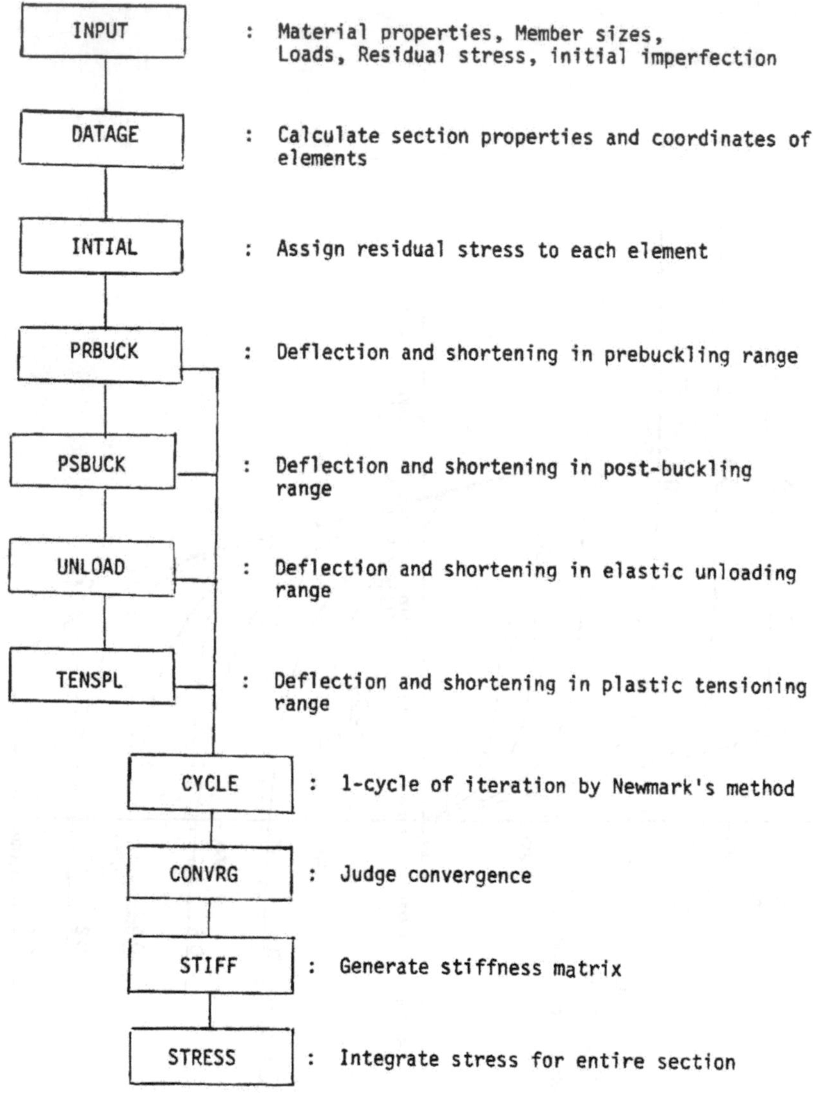

FIGURE 4.9 Flow chart of BMCYCL.

iterative procedures are included. These correspond to four different stages of loading: (1) initial compressive loading (pre-buckling); (2) post-buckling branch; (3) elastic unloading and tensioning; and (4) tensioning in the plastic range (see Sec. 4.3.5, Fig. 4.11).

Newmark's Method Using Approximate Generalized Stress-Strain Relationships

It was found in the previous work that the calculation of curvature from forces is the most time-consuming part of a numerical iteration, and analytical expressions

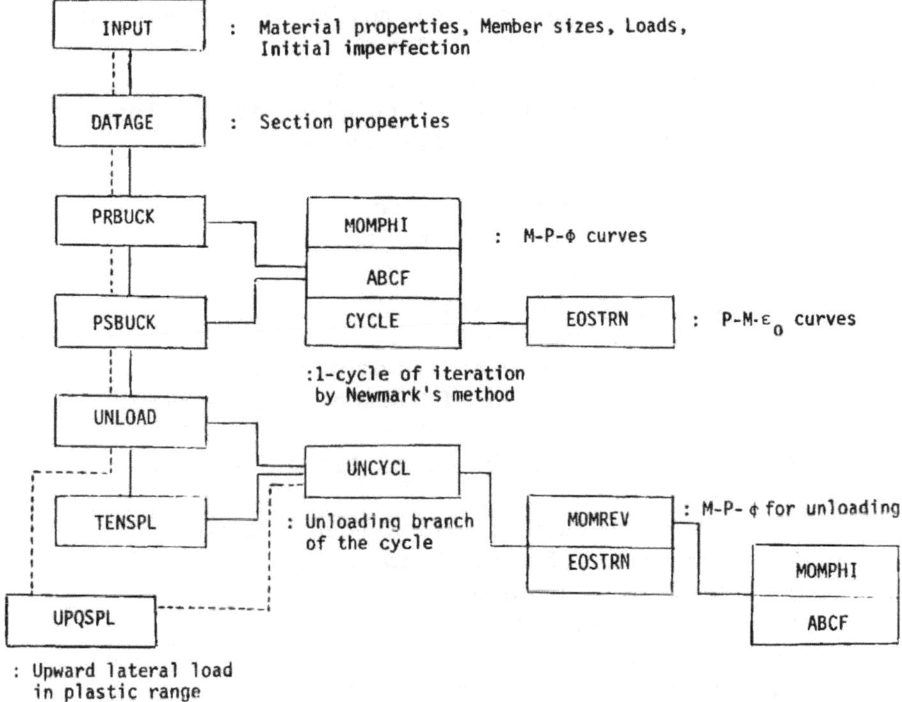

FIGURE 4.10 Flow chart of APCYCL.

approximating generalized stress-strain curves can significantly save computing time. Herein, an alternative computer program APCYCL has also been developed for the cyclic analysis of beam-columns using several analytical expressions of M-P-Φ and P-M-ϵ_O. A brief flow chart of the program is given in Fig. 4.10. The program is essentially similar to the previous one (BMCYCL), except that the calculation of curvature and axial strain part is modified. Further, this program has more flexibility in loading conditions, i.e., the lateral load at mid-length of the column can vary with the axial load.

4.3.3 User's Manual for the Program BMCYCL

Input manual of the program BMCYCL for pin-ended columns is given below. The program uses the exact M-P-Φ and P-M-ϵ_O relations. All data are written in free format and are given in sequence.

```
Line 1:  Title of a Job (Column 1−72)
Line 2:  Member Data
      IND = 1 . . . Unsymmetric column, 1-cycle of loading
          = 2 . . . Symmetric column, 1-cycle of loading
          = 3 . . . Unsymmetric column, Buckling Behavior
```

= 4 . . . Symmetric column, Buckling behavior
= 5 . . . Symmetric column with lateral load increased, 1-cycle of loading
= 6 . . . Symmetric column with lateral load increased, Buckling behavior

AL = Length of a column (ft.)
D = Diameter of a column (in.)
T = Thickness of tube (in.)
SY = Yield Stress (ksi)
E = Young's modulus (ksi)
NEL = Number of segments
NTH = Number of elements of the section
SPRG = Spring stiffness of end-restraints (kip-in./rad)

Note: For a symmetric column, total length is 2*AL.

Line 3: Loading Data

PP = Starting value of axial load (kips)
PDELT = Increment of axial load (kips)
NP = Number of increments
EMOS = End-Moment (=M/M_y)
QS = Lateral load at midspan (=Q/Q_y)
YLMT = Maximum limiting deflection (in.)

Notes: 1) P=PP+NP*PDELT > $P_{ultimate}$
2) Loading turns back at Deflection = YLMT for cyclic behavior.
3) When IND = 5 or 6, lateral and axial loads are interchanged.

Line 4: Initial Imperfection

NCW = 1; Initial imperfection is a half-sine.
= 2; Initial imperfections are to be input at each station.
WL(i) = Initial imperfections normalized by the member length (=W(i)/L)

Notes: 1) When NCW = 1, input the amplitude only.
2) When NCW = 2, input (NEL + 1) deflections.

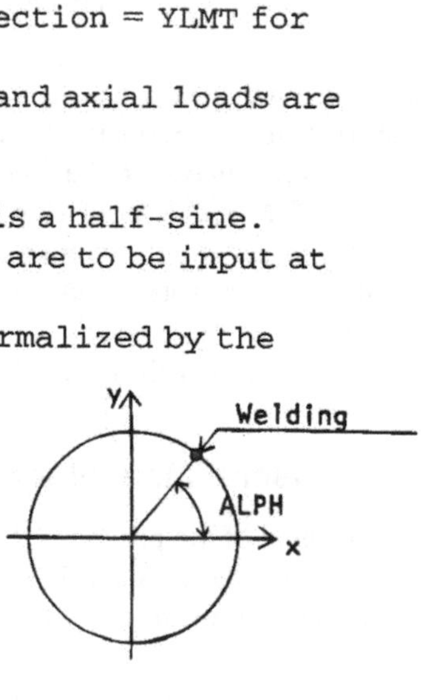

Line 5: Longitudinal Residual Stress

ALPH = Angle of welding seam location (degrees)
NRT = Number of input
RLA(i) = Position of the point (radians), i = 0 to NRT
FLA(i) = Residual stresses (ksi), i = 0 to NRT

Remark: Repeat from Line 1 to continue another job.

Following the input manual described above, the input and the corresponding output data for a sample problem shown in the next section.

INPUT DATA OF THE PROGRAM "BMCYCL" FOR A SAMPLE PROBLEM

```
COMPARISON WITH THE PROGRAM 'CYCLIC'
2 5.1935 4.5 0.09375 36 30000 4 48 0
31 2 10 0 0 6.5
1
0.001 0.0407 0.08814 0.1152 0.1246
0 6
0           3.600E+01
3.800E-01  -1.260E+01
1.180E+00   3.600E+00
1.980E+00   0
2.360E+00  -2.520E+00
3.142E+00   1.800E+00
```

OUTPUT DATA OF THE PROGRAM "BMCYCL" FOR A SAMPLE PROBLEM

```
COMPARISON WITH THE PROGRAM 'CYCLIC'
 IND= 2 AL= .5194E+01 D= .4500E+01 T= .9375E-01 SY= .3600E+02 E= .3000E+05 NEL= 4 NTH= 48
 SPRG= .0000E+00

 P= .3100E+02 PDELT= .2000E+01 NP= 10 EMOS= .0000E+00 QS= .0000E+00 Y(LIMIT)= .6500E+01

 INITIAL OUT-OF-STRAIGHTNESS     NCW= 1
  WL(I)= .000E+00 .477E-01 .881E-01 .115E+00 .125E+00

 ALPH= .00
 RLA(I)= .000E+00  .380E+00 .118E+01 .198E+01  .236E+01 .314E+01
 FLA(I)= .360E+02 -.126E+02 .360E+01 .000E+00 -.252E+01 .180E+01

 SI= .3151E+01 SA= .1298E+01 PHIY= .5333E-03 PES= .1285E+01 PY= .4672E+02 EMY= .5041E+02

 WOQM= .0000E+00 SRT= .7999E+02 WYILD= .8395E+00 DYILD= .1496E+00

COORDINATES
      X0          Y0          X0          Y0          X0          Y0          X0          Y0
  .2197E+01  .1440E+00  .2160E+01  .4296E+00  .2085E+01  .7078E+00  .1975E+01  .9739E+00
             X0          Y0
        .1831E+01  .1223E+01
  .1655E+01  .1452E+01  .1452E+01  .1655E+01  .1223E+01  .1831E+01  .9739E+00  .1975E+01
        .7078E+00  .2085E+01
  .4296E+00  .2160E+01  .1440E+00  .2197E+01 -.1440E+00  .2197E+01 -.4296E+00  .2160E+01
       -.7078E+00  .2085E+01
 -.9739E+00  .1975E+01 -.1223E+01  .1831E+01 -.1452E+01  .1655E+01 -.1655E+01  .1452E+01
       -.1831E+01  .1223E+01
 -.1975E+01  .9739E+00 -.2085E+01  .7078E+00 -.2160E+01  .4296E+00 -.2197E+01  .1440E+00
       -.2197E+01 -.1440E+00
 -.2160E+01 -.4296E+00 -.2085E+01 -.7078E+00 -.1975E+01 -.9739E+00 -.1831E+01 -.1223E+01
       -.1655E+01 -.1452E+01
 -.1452E+01 -.1655E+01 -.1223E+01 -.1831E+01 -.9739E+00 -.1975E+01 -.7078E+00 -.2085E+01
       -.4296E+00 -.2160E+01
 -.1440E+00 -.2197E+01  .1440E+00 -.2197E+01  .4296E+00 -.2160E+01  .7078E+00 -.2085E+01
        .9739E+00 -.1975E+01
  .1223E+01 -.1831E+01  .1452E+01 -.1655E+01  .1655E+01 -.1452E+01  .1831E+01 -.1223E+01
        .1975E+01 -.9739E+00
  .2085E+01 -.7078E+00  .2160E+01 -.4296E+00  .2197E+01 -.1440E+00

RESIDUAL STRESS
RSTA(I)
  27.1476   10.4062  -6.3351 -11.4992  -8.8484  -6.1977  -3.5470   -.8963   1.7544   2.8324
   2.2433    1.6543   1.0652    .4782   -.1129   -.8063  -1.6743  -2.5424  -2.6611  -1.9380
  -1.2149    -.4918    .2314    .9545    .9545    .2314   -.4918  -1.2149  -1.9380  -2.6611
  -2.5424   -1.6743   -.8063   -.1129    .4762   1.0652   1.6543   2.2433   2.8324   1.7544
   -.8963   -3.5470  -6.1977  -8.8484 -11.4992  -6.3351  10.4062  27.1476
```

```
A(I)
 27.6293   10.8879 -5.8535 -11.0175  -8.3668 -5.7160   -3.0653    -.4146  2.2361  3.3140
  2.7250    2.1359  1.5469    .9578    .3688  -.3246   -1.1927   -2.0607 -2.1795 -1.4563
  -.7332    -.0101   .7131   1.4362   1.4362   .7131    -.0101    -.7332 -1.4563 -2.1795
 -2.0607   -1.1927  -.3246    .3688    .9578  1.5469    2.1359    2.7250  3.3140  2.2361
  -.4146   -3.0653 -5.7160  -8.3668 -11.0175 -5.8535   10.8879   27.6293

P = -.3100E+02
 X(I)     .0000E+00   .1558E+02   .3116E+02   .4674E+02   .6232E+02
   YA     .0000E+00   .5090E-01   .9405E-01   .1229E+00   .1330E+00
   YB     .0000E+00   .5169E-01   .9552E-01   .1248E+00   .1351E+00
   YB     .0000E+00   .5211E-01   .9630E-01   .1259E+00   .1362E+00
   YB     .0000E+00   .5211E-01   .9630E-01   .1259E+00   .1362E+00
*****************************************************************************************
     BM     .0000E+00   .3081E+01   .5693E+01   .7440E+01   .8053E+01
    PHI   -.3490E-10  -.3309E-04  -.6100E-04  -.8001E-04  -.8667E-04
     EO   -.7971E-03  -.7973E-03  -.7979E-03  -.7990E-03  -.7994E-03
   ALPH   -.8927E-04  -.5088E-03  -.9389E-03  -.1231E-02  -.1333E-02
 ALPHEO   -.1242E-01  -.1242E-01  -.1243E-01  -.1245E-01  -.1245E-01
  SLOPE         .3345E-02   .2836E-02   .1897E-02   .6665E-03   .0000E+00
 SLOPEO     .0000E+00  -.1251E-01  -.2500E-01  -.3747E-01  -.4992E-01

CONVERGENCE
P = -.3300E+02
X(I)   .0000E+00  .1558E+02  .3116E+02  .4674E+02  .6232E+02
  YA   .0000E+00  .5819E-01  .1075E+00  .1405E+00  .1521E+00
  YB   .0000E+00  .6013E-01  .1111E+00  .1453E+00  .1573E+00
  YB   .0000E+00  .6126E-01  .1132E+00  .1480E+00  .1603E+00

---POST-BUCKLING---
 P = -.3500E+02
  YA   .0000E+00  .1388E+00  .2627E+00  .3569E+00  .3934E+00
  YB   .0000E+00  .1745E+00  .3314E+00  .4522E+00  .4998E+00
 P = -.3202E+02
  YA   .0000E+00  .1388E+00  .2627E+00  .3569E+00  .3934E+00
  YB   .0000E+00  .1083E+00  .2008E+00  .2635E+00  .2858E+00
 P = -.3351E+02
  YA   .0000E+00  .1388E+00  .2627E+00  .3569E+00  .3934E+00
  YB   .0000E+00  .1279E+00  .2393E+00  .3188E+00  .3497E+00
 P = -.3425E+02
  YA   .0000E+00  .1388E+00  .2627E+00  .3569E+00  .3934E+00
  YB   .0000E+00  .1483E+00  .2796E+00  .3771E+00  .4153E+00
 P = -.3388E+02
  YA   .0000E+00  .1388E+00  .2627E+00  .3569E+00  .3934E+00
  YB   .0000E+00  .1373E+00  .2579E+00  .3457E+00  .3801E+00
 P = -.3388E+02
  YA   .0000E+00  .1388E+00  .2627E+00  .3569E+00  .3934E+00
  YB   .0000E+00  .1373E+00  .2579E+00  .3457E+00  .3801E+00
*****************************************************************************************
    BM     .0000E+00   .6318E+01   .1189E+02   .1600E+02   .1755E+02
   PHI   -.2057E-10  -.6930E-04  -.1328E+03  -.2216E-03  -.2952E-03
    EO   -.8781E-03  -.8793E-03  -.8870E-03  -.9487E-03  -.1027E-02
  ALPH   -.1837E-03  -.1072E-02  -.2102E-02  -.4408E-02
ALPHEO   -.1368E-01  -.1371E-01  -.1389E-01  -.1480E-01  -.1580E-01
 SLOPE        .8812E-02  .7740E-02  .5637E-02  .2204E-02  .0000E+00

SLOPEO     .0000E+00  -.1430E-01  -.2856E-01  -.4316E-01  -.5850E-01

*** CONVERGENCE

 P = -.3388E+02
  YA    .0000E+00  .2623E+00  .5079E+00  .7207E+00  .8801E+00
**** STIFFNESS MATRIX IS SINGULAR
CALLED IN PSBUCK

---ELASTIC UNLOADING AND TENSIONING---
   X(I)     .0000E+00   .1558E+02   .3116E+02   .4674E+02   .6232E+02
     YA     .0000E+00   .1793E+01   .3540E+01   .5192E+01   .6425E+01

P = -.9188E+01
     YB     .0000E+00   .1784E+01   .3524E+01   .5171E+01   .6401E+01
   **************************************************************************************
     BM    .0000E+00   .1691E+02   .3333E+02   .4876E+02   .6018E+02
    PHI    .1348E-10  -.1815E-03  -.3610E+03  -.8212E+03  -.1200E+01
     EO   -.2439E-03  -.2451E-03  -.2540E-03  -.6540E-03  -.9420E-02
   ALPH   -.4727E-03  -.2826E-02  -.5990E-02  -.2671E-01  -.1580E+00
```

```
ALPHEO   -.3796E-02  -.3828E-02  -.4465E-02  -.2105E-01  -.1240E+00
 SLOPE        .1145E+00  .1117E+00  .1057E+00  .7898E-01  .0000E+00
SLOPEO     .0000E+00    -.1063E+00    -.2079E+00 -.3079E+00  -.4291E+00

----CONVERGENCE
P = -.7188E+01
     YB      .0000E+00    .1713E+01    .3391E+01    .4994E+01    .6208E+01
     YB      .0000E+00    .1703E+01    .3372E+01    .4970E+01    .6182E+01
     BM      .0000E+00    .1265E+02    .2500E+02    .3672E+02    .4552E+02
    PHI      .8134E-11  -.1364E-03  -.2728E-03  -.6936E-03  -.1185E-01
     EO     -.1925E-03  -.1937E-03  -.2026E-03  -.6026E-03  -.9368E-02
   ALPH     -.3544E+03  -.2126E-02  -.4619E-02  -.2474E-01  -.1556E+00
ALPHEO      -.2995E-02  -.3028E-02  -.3665E-02  -.2025E-01  -.1232E+00
 SLOPE          .1093E+00  1072E+00  .1025E+00  .7780E-01  .0000E+00
SLOPEO       .0000E+00  -.9634E-01  -.1894E+00  -.2835E+00  -.4024E+00

----CONVERGENCE
~~~~~~~~~~~~~~~~~~~~~~~~~~~~~~~~~~~~~~~~~~~~~~~~~~~~~~~~~~~~~~~~~~~~~~~~
PLASTIC TENSIONING

  P = .1081E+02
    YA    .0000E+00   .8819E+00   .1764E+00   .2646E+01   .3527E+01
    YB    .0000E+00   .1179E+01   .2384E+01   .3634E+01   .4690E+01
  P = .1447E+02
    YA    .0000E+00   .8825E+00   .1775E+01   .2687E+01   .3527E+01
    YB    .0000E+00   .1016E+01   .2067E+01   .3182E+01   .4159E+01
  P = .1547E+02
    YA    .0000E+00   .8710E+00   .1762E+01   .2691E+01   .3527E+01
    YB    .0000E+00   .6381E+00   .1312E+01   .2057E+01   .2745E+01
  P = .1497E+02
    YA    .0000E+00   .8523E+00   .1733E+01   .2667E+01   .3527E+01
    YB    .0000E+00   .9278E+00   .1890E+01   .2917E+01   .3829E+01
          ********************************************************************************
    BM    .0000E+00 -.1347E+02 -.2726E+02 -.4166E+02 -.5468E+02
   PHI    .1244E-10   .1404E-03   .2811E-03   .3058E-03 -.9075E-02
    EO    .3779E-03   .3765E-03   .3674E-03   .1706E-03 -.6908E-02
  ALPH    .3646E-03   .2188E-02   .4229E-02 -.7448E-02 -.1170E+00
ALPHEO    .5891E-02   .5857E-02   .5481E-02 -.6277E-02 -.8925E-01
 SLOPE        .5955E-01   .6174E-01   .6597E-01   .5852E-01    .0000E+00
    YB    .0000E+00   .9278E+00   .1890E+01   .2917E+01   .3829E+01
SLOPEO    .0000E+00 -.2177E-01 -.4583E-01 -.8016E-01 -.1546E+00

CONVERGENCE
~~~~~~~~~~~~~~~~~~~~~~~~~~~~~~~~~~~~~~~~~~~~~~~~~~~~~~~~~~~~~~~~~~~~~~~~
  P = .4560E+02
    YA    .0000E+00  .8462E-02  .2419E-01  .5333E-01  .1000E+00
****STIFFNESS MATRIX IS SINGULAR
                               CALLED IN TENSPL

  P = .4560E+02
   YA    .0000E+00  .8462E-02  .2419E-01  .5333E-01  .1000E+00
****STIFFNESS MATRIX IS SINGULAR
                               CALLED IN TENSPL

  P = .4560E+02
   YA    .0000E+00  .8462E-02  .2419E-01  .5333E-01  .1000E+00
****STIFFNESS MATRIX IS SINGULAR
                               CALLED IN TENSPL
****M IS GREATER THAN 25
**************************************************************************************************
```

4.3.4 User's Manual for the Program APCYCL

Input manual of the program APCYCL for pin-ended columns are given below. The program uses the approximated M-P-Φ and P-M-ϵ_0 relations. All inputs except Line 1 are written in free format and are to be given in sequence.

```
Line 1: Title and Control Data
   Column  1-60. Title of a job
```

```
   Column 61-62. IND=Controller for a type of column
         IND = 1 . . . Unsymmetric column, 1-cycle of loading
             = 2 . . . Symmetric column, 1-cycle of loading
             = 3 . . . Unsymmetric column, Buckling behavior
             = 4 . . . Symmetric column, Buckling behavior
             = 5 . . . Symmetric column, Lateral load
                       increasing, 1-cycle of loading
             = 6 . . . Symmetric column, Lateral load
                       increasing, Buckling behavior
  Column 63-64 . . . IWR = Controller for a type of
                       M-P-Φ curve
        IWR = 0 . . . No residual stress and no hydrostatic
                       pressure, Doubly symmetric curve
                       for unloading
             = 1 . . . With residual stress and no hydrostatic
                       pressure, Doubly symmetric curve
                       for unloading
             = 2 . . . With residual stress and hydrostatic
                       pressure, Doubly symmetric curve
                       for unloading
             = 3 . . . Curve fitting to Sherman's test result
                       (Sherman, 1979), Doubly
                       symmetric curve for unloading
             = 4 . . . No residual stress, symmetric curve
                       for unloading
             = 5 . . . With residual stress, symmetric curve
                       for unloading
             = 6 . . . No residual stress, Elastic-perfectly
                       plastic for unloading
             = 7 . . . With residual stress, Elastic-
                       perfectly plastic for unloading
    Column 65-68. QR = Normalized hydrostatic pressure
    Column 69-72. SPRG = End-restraint (kip-in./rad)

Line 2: Member Data
      AL = Length of a column (ft.)
       D = Diameter of a column (in.)
       T = Thickness of a column (in.)
      SY = Yield stress (ksi)
       E = Young's modulus (ksi)
     NEL = Number of segments
     Note: For a symmetric column, the total length is 2*AL

Line 3. Loading Data
      (When IND ≠ 5 or 6)
      PP = Starting value of axial load (kips)
```

PDELT = Increment of axial load (kips)
NP = Number of increment
EMOS = End-moment $(=M/M_y)$
QS = Lateral load at midspan $(=Q/Q_y)$
NQS = QS is loaded at NQS'th station
YLMT = Maximum limiting deflection (in.)
(When IND = 5 or 6)
QQ = Starting value of lateral load (kips)
QDELT = Increment of lateral load (kips)
NQ = Number of increments
EMOS = End-moment $(=M/M_y)$
PS = Axial force $(=P/P_y)$
YLMT = Maximum limiting deflection (in.)

Note: 1) $P = PP+NP*PDELT > P_{ultimate}$
2) Loading turns back at deflection = YLMT for cyclic behavior.

Line 4: Initial Imperfection
NCW = 1; Initial imperfection is a half sine.
= 2; Initial imperfections are to be input at each station.
WL(i) = Initial imperfections normalized by the member length (=W(i)/L)

Following the input manual described above, the input and the corresponding output data for a sample calculation are shown in the next section.

INPUT DATA OF THE PROGRAM "APCYCL" FOR A SAMPLE PROBLEM

```
BEHAVIOR OF BEAM-COLUMNS      2 1 0
5.1935 4.5 0.09375 36 30000 4
25 4 5 0 0 5 2
1
0.001 0.0477 0.0881 0.1152 0.1246
```

OUTPUT DATA OF THE PROGRAM "APCYCL" FOR A SAMPLE PROBLEM

```
BEHAVIOR OF BEAM-COLUMNS
  IND= 2 IWQ= 1  QR=  .00
  AL=  .5194E+01  D=  .4500E+01 T= .9375E-01 SY=  .3600E+02 E=  .3000E+05  NEL= 4
   P=  .2500E+02  PDELT=  .4000E+01  NP= 5 EMOS=  .0000E+00 QS=  .0000E+00  NQS= 5
       Y(LIMIT)=  .2000E+01

   INITIAL OUT-OF-STRAIGHTNESS   NCW= 1
   WL(I)=  .000E+00  .477E-01  .881E-01  .115E+00  .125E+00

   SI=  .3151E+01 SA=  .1298E+01 PHIV=  .5333E-03 PES=  .1285E+01 PY=  .4672E+02
        EMY=  .5041E+02
    WOQM=  .0000E+00  SRT=  .7999E+02  WYILD=  .8395E+00 DYILD=  .1496E+00
```

```
P=  -.2500E+02
  X(I)     .0000E+00    .1558E+062    .3116E+02    .4674E+02    .6232E+02

    YA     .0000E+00    .3402E-01    .6286E-01    .8214E-01     8890E-01
    BM     .0000E+00    .2043E+01    .3775E+01    .4932E+01    .5339E+01
    YB     .0000E+00    .3403E-01    .6287E-01    .8215E-01    .8891E-01
    YB     .0000E+00    .3403E-01    .6287E-01    .8215E-01    .8891E-01
      ************************************************************************************
    BM     .0000E+00    .2043E+01    .3775E+01    .4932E+01    .5339E+01
   PHI     .0000E+00   -.2161E-04   -.3994E-04   -.5218E-04   -.5648E-04
    EO    -.6421E-03   -.6421E-03   -.6421E-03   -.6421E-03   -.6421E-03
  ALPH    -.5826E-04   -.3325E-03   -.6143E-03   -.8027E-03   -.8688E-03
ALPHEO    -.1000E-01   -.1000E-01   -.1000E-01   -.1000E-01   -.1000E-01
 SLOPE          .2184E-02    .1851E-02    .1237E-02    .4344E-03    .0000E+00

SLOPEO     .0000E+00   -.1004E-01   -.2007E-01   -.3009E-01   -.4010E-01
CONVERGENCE

P =  -.2900E+02
  X(I)     .0000E+00    .1558E+02    .3116E+02    .4674E+02    .6232E+02
    YA     .0000E+00    .4455E-01    .8232E-01    .1076E+00    .1164E+00
    BM     .0000E+00    .2675E+01    .4943E+01    .6459E+01    .6991E+01
    YB     .0000E+00    .4455E-01    .8233E-01     1076E+00    .1164E+00
    YB     .0000E+00    .4455E-01    .8233E-01     1076E+00    .1164E+00
      ************************************************************************************
    BM     .0000E+00     .2675E+0    .4943E+01     6459E+01    .6991E+01
   PHI     .0000E+00   -.2830E-04   -.5229E-04   -.6832E-04   -.7395E-04
    EO    -.7449E-03   -.7449E-03   -.7449E-03   -.7449E-03   -.7449E-03
```

```
---POST-BUCKLING---

P = -.3700E+02
    YA     .0000E+00    .1285E+00    .2438E+00    .3315E+00     3652E+00
    BM     .0000E+00    .6519E+01    .1228E+02    .1652E+02    .1812E+02
    YB     .0000E+00     2303E+00    .4426E+00    .6049E+00    .6716E+00

P = -.2988E+02
    YA     .0000E+00     1285E+00    .2438E+00    .3315E+00    .3652E+00
    BM     .0000E+00    .5265E+01    .9918E+01    .1335E+02     1464E+02
    YB     .0000E+00    .9093E-01    .1685E+00    .2208E+00    .2393E+00

P = -.3344E+02
    YA     .0000E+00    .1285E+00    .2438E+00    .3315E+00    .3652E+00
    BM     .0000E+00    .5892E+01    .1110E+02    .1494E+02    .1638E+02
    YB     .0000E+00    .1157E+00    .2164E+00    .2883E+00    .3159E+00
```

```
---ELASTIC UNLOADING AND TENSIONING---
  X(I)     .0000E+00    .1558E+02    .3116E+02    .4674E+02    .6232E+02

    YA     .0000E+00    .6231E+00    .1207E+01    .1704E+01    .1990E+01

P = -.2188E+02
    BM     .0000E+00    .1468E+02    .2835E+02    .3981E+02    .4625E+02
    YB     .0000E+00    .6191E+00    .1200E+01    .1694E+01    .1978E+01
      ************************************************************************************
    BM     .0000E+00    .1468E+02    .2835E+02    .3981E+02    .4625E+02
   PHI     .0000E+00   -.1552E-03   -.3399E-03   -.7375E-03   -.2661E-02
    EO    -.5620E-03   -.5620E-03   -.5822E-03   -.9867E-03   -.3949E-02
  ALPH    -.3841E-03   -.2457E-02   -.5572E-02   -.1347E-01   -.3646E-01
ALPHEO    -.8729E-02   -.8782E-02   -.9570E-02   -.1869E-01   -.5383E-01
 SLOPE          .3973E-0     .3728E-01    .3170E-01    .1823E-0     .0000E+00

SLOPEO     .0000E+00   -.2106E-01   -.4106E-01   -.6303E-01   -.1019E+00

====CONVERGENCE

P =  -.1788E+02
    BM     .0000E+00    .1199E+02    .2316E+02    .3253E+02    .3780E+02
```

```
    YB    .0000E+00    .5699E+00    .1108E+01    .1573E+01    .1847E+01
    BM    .0000E+00    .1104E+02    .2139E+02    .3019E+02    .3524E+02
```

```
---PLASTIC TENSIONING---

P =  .4578E+02
    YA    .0000E+00    .8462E-02    .2419E-01    .5333E-01    .1000E+00
    BM    .0000E+00   -.2571E+01   -.5143E+01   -.7714E+0    -.1029E+02
***EM LESS THAN -EMPC

                         CALLED IN TENSPL

P=  .3595E+02
    YA    .0000E+00    .8462E-02    2419E-01     .5333E-01    .1000E+00
    BM    .0000E+00   -.2019E+01   -.4038E+01   -.6057E+01   -.8077E+01
    YB    .0000E+00    .3091E+00    .6225E+00    .9318E+00    .1148E+01

P=  .4087E+02
    YA    .0000E+00    .8462E-02    .2419E-01    .5333E-01    .1000E+00
    BM    .0000E+00   -.2295E+01   -.4590E+01   -.6886E+01   -.9181E+01
    YB    .0000E+00    .2776E+00    .5606E+00    .8449E+00    .1047E+01

P=  .4333E+02
    YA    .0000E+00    .1155E-01    .3020E-01    .6110E-01    .1000E+00
    BM    .0000E+00   -.2567E+01   -.5127E+01   -.7637E+01   -.9733E+01
***EM LESS THAN -EMPC

                         CALLED IN TENSPL

P=  .4574E+02
    YA    .0000E+00   -.2221E-01   -.4058E-01   -.5787E-01   -.8368E-01
    BM    .0000E+00   -.1166E+01   -.2175E+01   -.2620E+01   -.1874E+01
***EM LESS THAN -EMPC

                         CALLED IN TENSPL
```

```
P=  .4574E+02
    YA    .0000E+00   -.2221E-01   -.4058E-01   -.5787E-01   -.8368E-01
    BM    .0000E+00   -.1168E+01   -.2175E+01   -.2620E+01   -.1874E+01
***EM LESS THAN -EMPC

                         CALLED IN TENSPL

P=  .4572E+02
    YA    .0000E+00   -.2221E-01   -.4058E-01   -.5787E-01   -.8368E-01
    BM    .0000E+00   -.1165E+01   -.2174E+01   -.2619E+01   -.1873E+01
***EM LESS THAN -EMPC

                         CALLED IN TENSPL

***M IS GREATER THAN 25
```

4.3.5 Numerical Results

Comparison Between Exact and Approximate Solutions

Figure 4.11 compares the cyclic behavior of beam-column solutions based on the exact and approximate M-P-Φ and P-M-ϵ_0 relations. The solid line is the exact load-shortening solution based on the exact M-P-Φ and M-P-ϵ_0 relations. The dashed line is the exact, as well as the approximation of load-deflection solution using the doubly symmetric M-P-Φ relation for unloading branches. The dotted line and dashed-dotted line represent the solutions based on the symmetric M-P-Φ and the

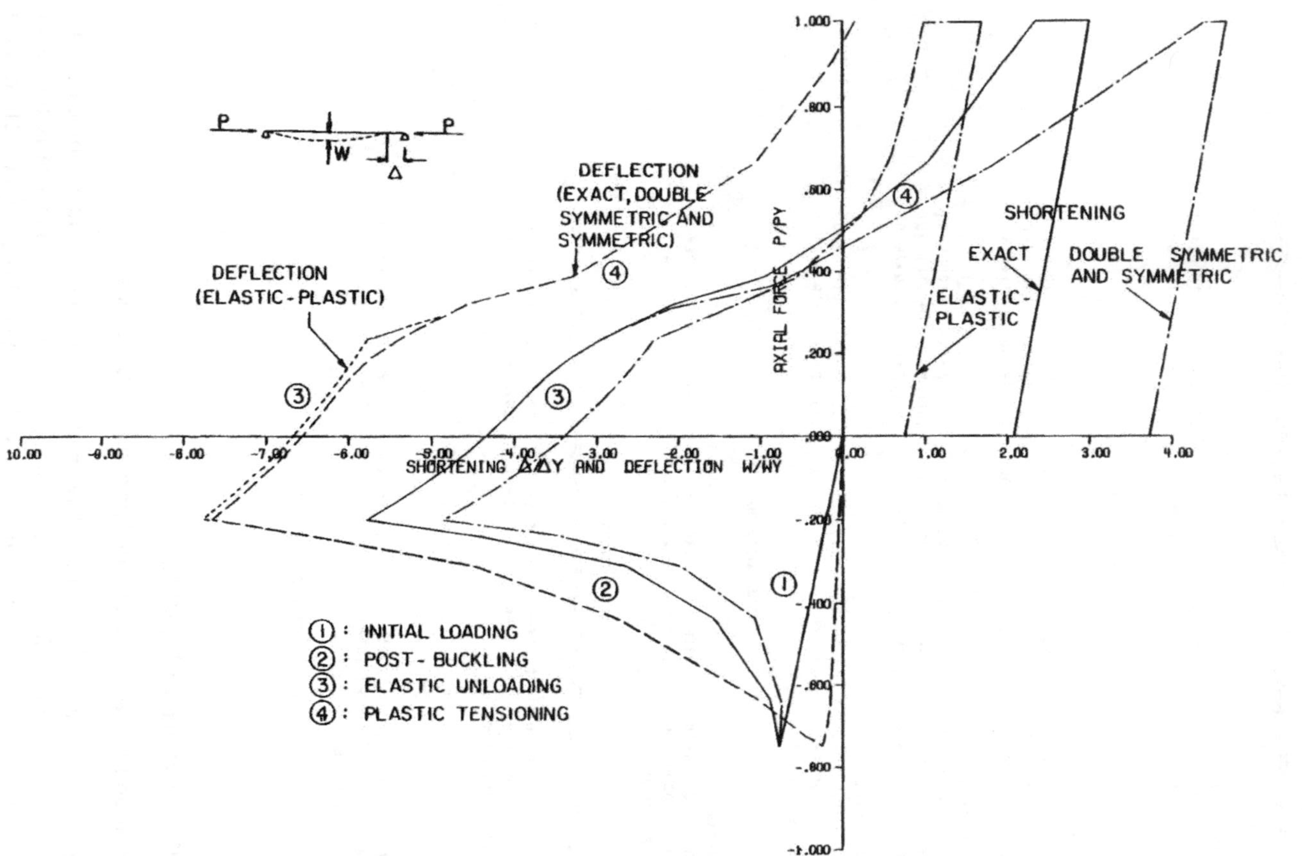

FIGURE 4.11 Comparison of Newmark's method between exact and approximated solutions.

elastic–perfectly plastic M-P-Φ relations for unloading. As for the P-M-ϵ_O relation, the symmetric curve approximation is used in all cases except when it is combined with the elastic–perfectly plastic M-P-Φ. In this case, the axial strain is assumed to be the elastic–perfectly plastic type. Note that in the initial loading including the post-buckling branch, the same expressions of M-P-Φ are used for all cases.

The comparison shows that there is no significant difference in the load-deflection behavior. Only the solution based on the elastic–perfectly plastic M-P-Φ approximation exhibits a small cusp because of the sharp change in slope of the M-P-Φ relation at the transition zone. The behavior agrees generally well with the exact solution. It is interesting to note that the doubly symmetric and symmetric M-P-Φ approximations result in a very similar load-deflection behavior. This fact implies that the exact shape of the transition zone from elastic to plastic in the unloading branch of the M-P-Φ relation has no significant effect on the load-deflection behavior.

The load-shortening behavior is seen somewhat differently in the plastic tension zone for the three solutions studied. The elastic assumption obviously gives the most conservative value. The symmetric P-M-ϵ_O approximation results in a large shortening comparing with the exact solution. Here, as in the load-deflection behavior, the exact shape of transition of the unloading P-M-ϵ_O curve does not affect much of the load-shortening behavior.

Typical Cyclic Behavior of Pin-Ended Columns

Figure 4.12a shows a typical cyclic behavior of a pin-ended column with $L/r = 80$. In this case, the theory uses the approximate generalized stress-strain relationships of the doubly symmetric type. The solid line and the dashed line represent the load-shortening and load-deflection relation, respectively. An initial out-of-straightness of 0.1% is used here. Four different reversed loading paths starting from the post-buckling branch are calculated. The reversed load-deflection curves (dashed lines) are almost parallel to each other in the elastic unloading-tensioning range, and after a plastic hinge is formed at the center cross section under tension, the slope of the curves becomes flatter. Eventually, all the reversed load-deflection curves merge to the same curve, as shown in Fig. 4.12a.

The load-shortening curves show a similar behavior only in the elastic unloading-tensioning range for different reversed loading paths. In the plastic tension range, the curves scatter widely but converge to the same yield strength. It is found that the larger the negative axial strain is attained, the larger the positive strain for the reversed loading can be applied. Note that the smaller the displacements of the curves, the closer the curves are to the elastic slopes.

Figure 4.12b shows the effect of lateral load Q on the cyclic behavior of pin-ended columns. The lateral load is seen to reduce the buckling strength drastically. After reversing the axial loading, the curves are seen essentially parallel to each other. Note that in the plastic tension range, the convergence in a numerical solution is not always possible, because of the almost zero slope of the M-P-Φ curve in this range. This makes the convergence of a solution highly sensitive.

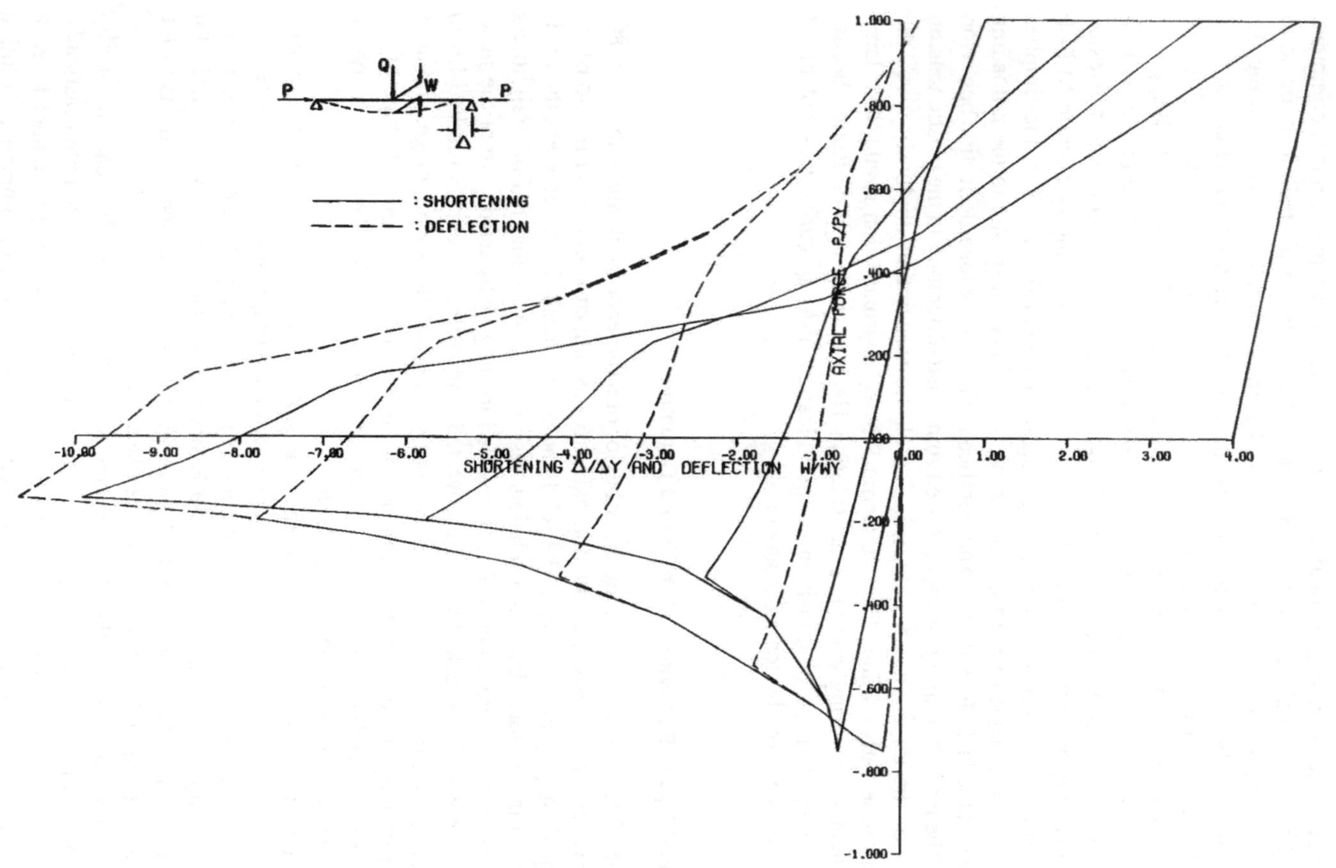

FIGURE 4.12a Cyclic behavior of pin-ended column, L/r = 80.

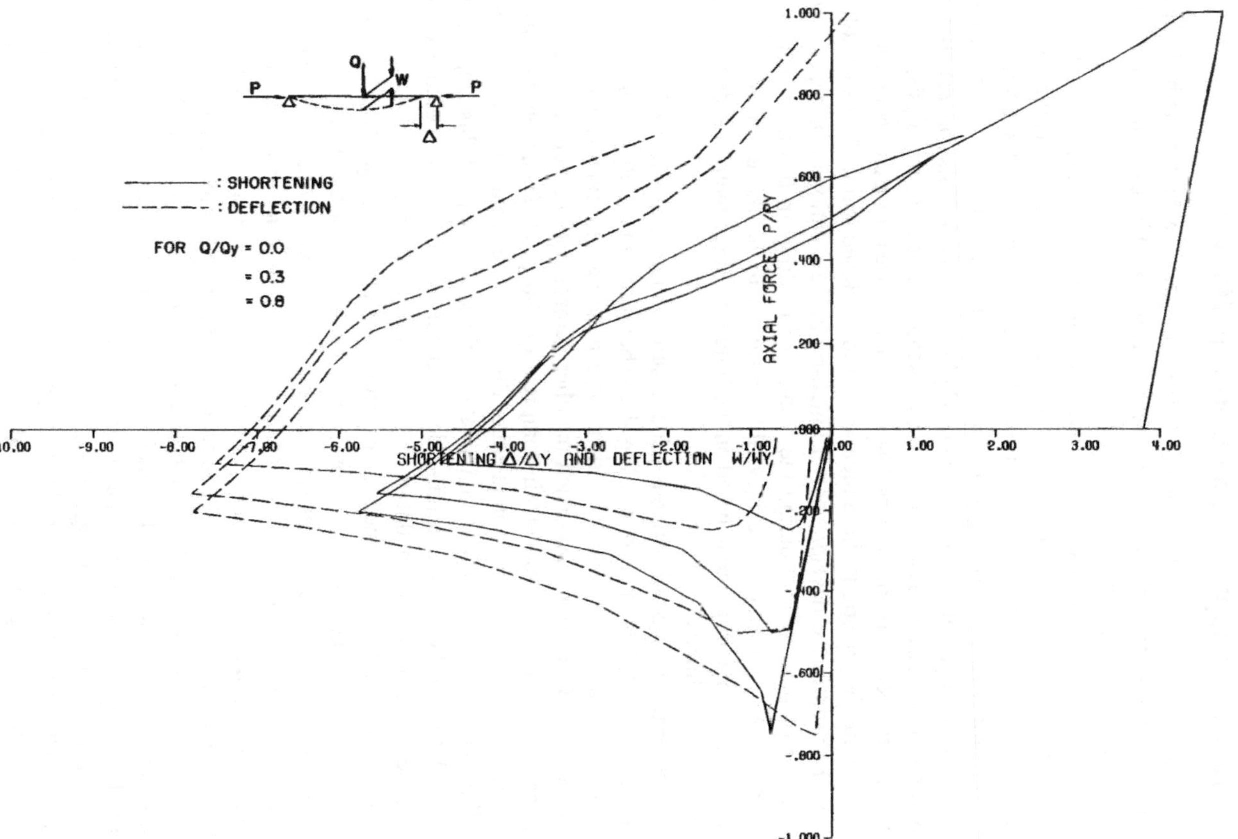

FIGURE 4.12b Effect of lateral load on cyclic behavior of pin-ended column, L/r = 80.

4.4 CYCLIC ANALYSIS OF FIXED-ENDED COLUMNS

4.4.1 Basic Concept — Hinge-by-Hinge Method

Since a fixed-ended column is statically indeterminate, Newmark's method used previously for pin-ended columns is not suitable. Here, the hinge-by-hinge method is applied to obtain solutions of fixed-ended columns.

The hinge-by-hinge method is often used in the study of elastic-plastic behavior of steel structures. When the load is increased gradually in a framed structure, plastic hinges at critical sections will develop in sequence, and the degree of indeterminacy of the structure reduces accordingly. In this approach, the plastic hinge is treated as a mechanical hinge with a constant value of plastic moment. This concept allows the solution of an elastic–plastic structure as a sequence of elastic analysis.

The elastic beam-column theory leads, in many cases, to exact solutions by directly solving the governing differential equation in closed form. The exact solutions can be used in the development of hinge-by-hinge solutions that depend on the locations of plastic hinge during the history of loading. In this approach, the M-P-ϕ and P-M-ϵ_O relations are inevitably assumed to be the elastic–perfectly plastic type. A computer program based on the hinge-by-hinge method is developed for fixed-ended columns.

The general behavior of a fixed-ended column under one cycle of reversed loading can be divided into nine stages. Fig. 4.13 shows schematically the nine stages of the load-deformation behavior, and their corresponding deformation patterns are sketched in Fig. 4.14. The fixed-ended member subjected to combined axial and lateral loads is first loaded elastically in compression (Stage 1). Then a plastic hinge with moment capacity m_{pc} will form at either the ends or the center of the member, depending on the length and load combination (Stage 2). Beyond this stage, the fixed-ended column behaves as a pin-ended column or a cantilever column with plastic bending moment, m_{pc}, maintained at the hinge locations.

When three hinges are formed at the ends and center (Stage 3), a two-bar mechanism is developed. The deflection is now controlled by the plastic moment at the hinges. This response corresponds to the post-buckling branch of the P-Δ curve shown in Fig. 4.13, where the axial load decreases as axial deformation increases. From Stage 2 to 3, relative permanent rotation occurs at the hinges.

When the axial load is reduced and reversed, the initial unloading behavior is elastic in compression (Stage 4) and then loaded in tension (Stage 5). During the elastic unloading and reloading, the relative rotations at the hinges do not change until plastic hinges are formed again at the ends or center of the column by bending combined with axial tension (Stage 6). At this stage of loading, the direction of plastic moments is reversed. When three hinges are formed, a two-bar mechanism under axial tension is again developed (Stage 7). Finally the column is stretched into a uniaxial state of yielding (Stage 8) and the axial tension reaches the value P_y. An unloading from this stage is obviously elastic until the axial load is reversed sufficiently in compression to permit the development of plastic hinges in compression again (Stage 9).

In outline form, the key features included in this part of work are as follows:

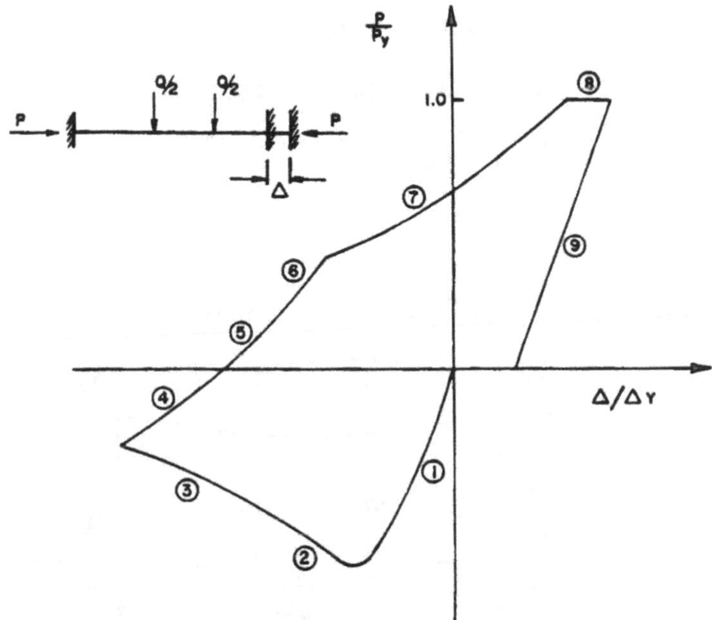

1. INITIAL ELASTIC LOADING IN COMPRESSION
2. ONE HINGE FORMED IN COMPRESSION
3. BOTH HINGES FORMED IN COMPRESSION
4. ELASTIC UNLOADING IN COMPRESSION
5. ELASTIC LOADING IN TENSION
6. ONE HINGE FORMED IN TENSION
7. BOTH HINGES FORMED IN TENSION
8. YIELDING IN TENSION
9. ELASTIC UNLOADING IN TENSION

FIGURE 4.13 Stages of fixed-ended column behavior under 1-cycle of loading.

1. Loads: cyclic axial compression-tension with constant lateral load
2. End Conditions: fully fixed
3. Method of Analysis: hinge-by-hinge method
4. M-P-ϕ Curves: elastic–perfectly plastic for initial loading and unloading
5. P-M-ϵ_o Curves: elastic–perfectly plastic
6. Results: P-w (load-deflection) curves and P-Δ (load-shortening) curves
7. Developed Computer Program: FIXCYCL

① Initial Elastic Loading in Compression

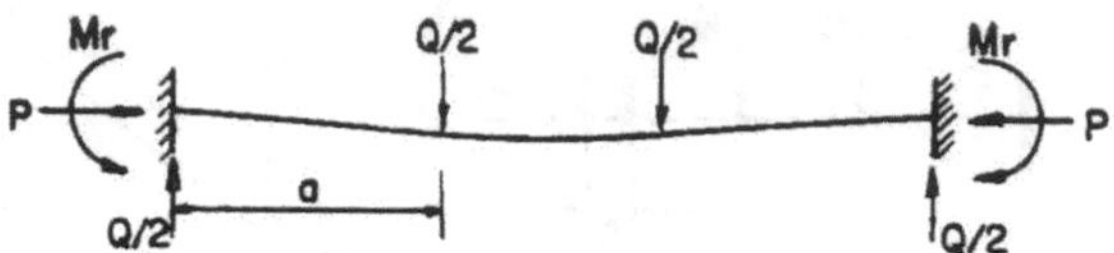

② One Hinge Formed in Compression

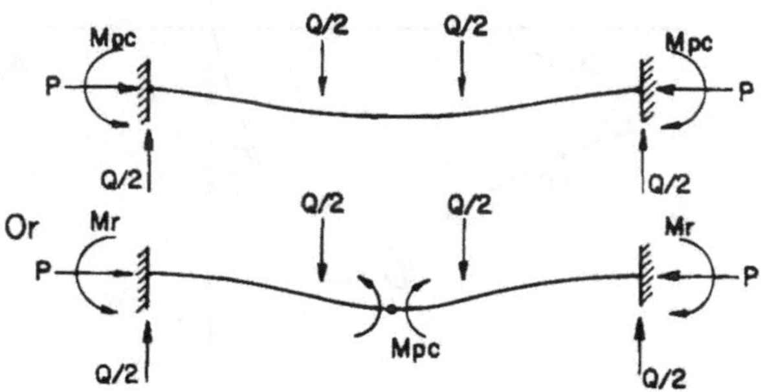

③ Both Hinges Formed in Compression

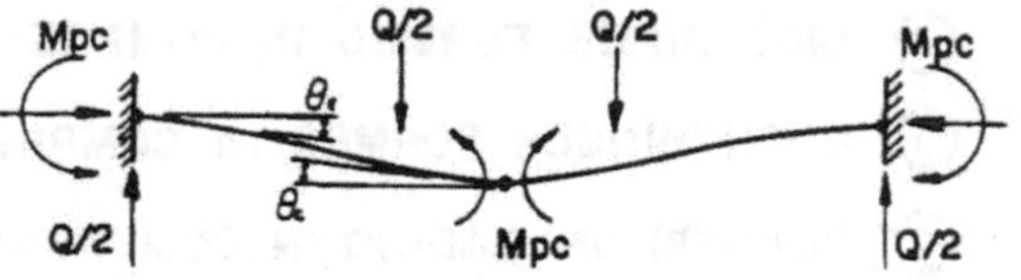

④ Elastic Unloading in Compression

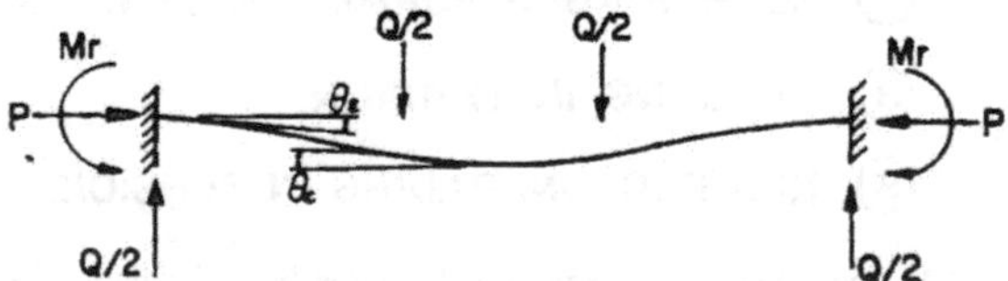

FIGURE 4.14 Behavior of fixed-ended column under 1-cycle of loading.

4.4.2 Compressive Axial Force (Stages 1 to 4)

The analytical concept used in this section for fixed beam-columns is called the "hinge-by-hinge method." In the following, general formulations and solutions will be given for each stage of loading described in Figs. 4.13 and 4.14.

⑤ Elastic Loading in Tension

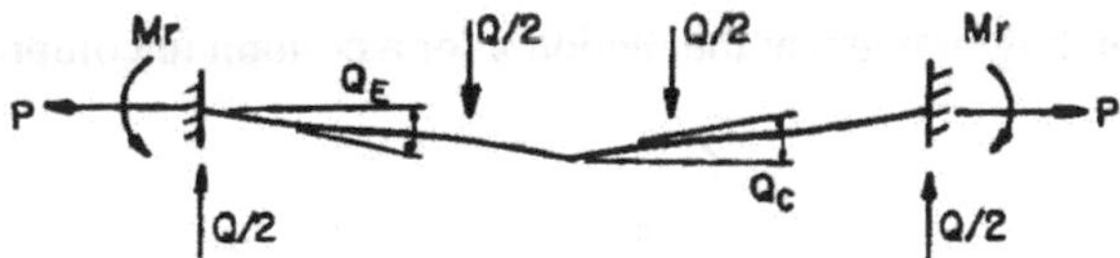

⑥ One Hinge Formed in Tension

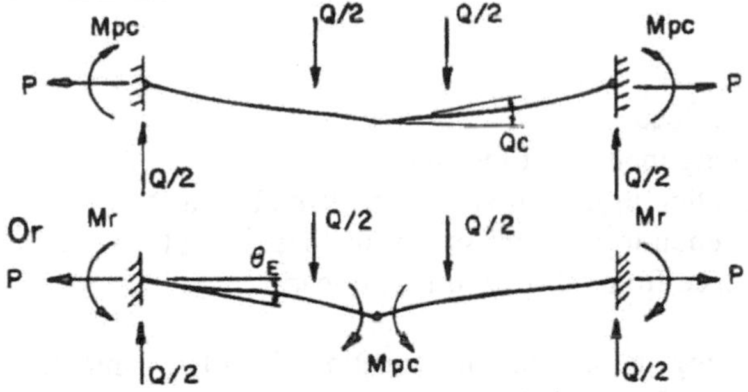

⑦ Both Hinges Formed in Tension

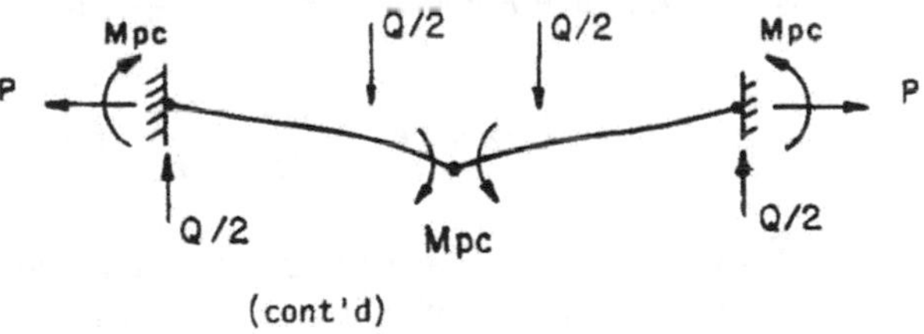

FIGURE 4.14 *Continued*

In elastic range, the curvature at a cross section is related to the bending moment by the relation:

$$\frac{d^2y}{dx^2} = -\frac{M}{EI} \tag{4.17}$$

in which

y = lateral deflection
x = distance from one end
M = bending moment at the section x
I = moment of inertia of the section

Solving this governing equation for each stage of loading, the corresponding deflection can be obtained.

The bending moment at the section x for a column in compression has the value

$$M = \begin{cases} -M_E + \dfrac{Q}{2}x + Py_1 & \text{for} \quad x \leq a \\ -M_E + \dfrac{Qa}{2} + Py_2 & \text{for} \quad x > a \end{cases} \tag{4.18}$$

in which

Q = lateral load
M_E = bending moment at the ends
y_1 = deflection to the left side of the lateral load
y_2 = deflection to the right side of the lateral load
a = distance from left end to lateral load

Substituting the equilibrium equation Eq. (4.18) into the moment-curvature relation Eq. (4.17), the deflections can be expressed in the general forms as

$$\begin{aligned} y_1 &= A \sin kx + B \cos kx - \frac{Q}{2P}x + \frac{M_E}{P} \quad \text{for} \quad x \leq a \\ y_2 &= C \sin kx + D \cos kx - \frac{Qa}{2P} + \frac{M_E}{P} \quad \text{for} \quad x > a \end{aligned} \tag{4.19}$$

in which $k = \sqrt{P/EI}$ and A, B, C, and D are the integration constants. These constants are to be determined in the following by appropriate boundary conditions.

Initial Elastic Loading in Compression (Stage 1)

There are five unknowns, including the fixed end moments, M_E. Therefore, five boundary conditions are required:

$$\begin{aligned} &1.\ y_1 = 0 && \text{at} \quad x = 0 \\ &2.\ dy_1/dx = 0 && \text{at} \quad x = 0 \\ &3.\ y_1 = y_2 && \text{at} \quad x = 0 \\ &4.\ dy_1/dx = dy_2/dx && \text{at} \quad x = 0 \\ &5.\ dy_2/dx = 0 && \text{at} \quad x = L/2 \end{aligned} \tag{4.20}$$

Using Eq. (4.19) and the first four boundary conditions of Eq. (4.20), the integration constants are determined.

$$A = \frac{Q}{2kP}$$

$$B = -\frac{M_E}{P} \tag{4.21}$$

$$C = \frac{Q}{2kP}(1 - \cos ka)$$

$$D = \frac{Q}{2kP}\sin ka - \frac{M_E}{P}$$

Using Eq. (4.21) and the last boundary condition of Eq. (4.20), we find the fixed-end moment.

$$M_E = \frac{Q}{2k}\left(\cos ka \cot\frac{kL}{2} + \sin ka - \cot\frac{kL}{2}\right) \tag{4.22}$$

Thus, the deflections for the initial elastic loading case in compression are determined by Eqs. (4.19), (4.21), and (4.22) for any given combination of P and Q.

One Hinge Formed at Ends in Compression (Stage 2)

When plastic hinges are formed at both ends, the bending moment along the column is given by Eq. (4.18) with the end moment being

$$M_E = M_{pc} \tag{4.23}$$

in which M_{pc} = plastic moment with axial force.

The four unknowns are determined from the four boundary conditions:

$$\begin{array}{lll} 1.\ y_1 = 0 & \text{at} & x = 0 \\ 2.\ dy_2/dx = 0 & \text{at} & x = L/2 \\ 3.\ y_1 = y_2 & \text{at} & x = a \\ 4.\ dy_1/dx = dy_2/dx & \text{at} & x = a \end{array} \tag{4.24}$$

from which we find

$$A = \left(\frac{Q}{2kP}\cos ka + \sin ka \cot\frac{kL}{2}\right) - \frac{M_{pc}}{P}\tan\frac{kL}{2}$$

$$B = -\frac{M_{pc}}{P} \tag{4.25}$$

$$C = \left(\frac{Q}{2kP}\sin ka - \frac{M_{pc}}{P}\right)\tan\frac{kL}{2}$$

$$D = \frac{Q}{2kP}\sin ka - \frac{M_{pc}}{P}$$

One Hinge Formed at Mid-Span in Compression (Stage 2)

When a plastic hinge is formed first at the center of a column, the bending moment along the column is expressed by Eq. (4.18) with

$$M_E = -M_{pc} + Py_c + \frac{Qa}{2} \tag{4.26}$$

in which y_c = deflection at the center of a column.

The necessary boundary conditions are

$$\begin{aligned} &1.\ y_1 = 0 && \text{at} \quad x = 0 \\ &2.\ dy_1/dx = 0 && \text{at} \quad x = 0 \\ &3.\ y_1 = y_2 && \text{at} \quad x = a \\ &4.\ dy_1/dx = dy_2/dx && \text{at} \quad x = a \end{aligned} \tag{4.27}$$

The integration constants are found to have the same forms as those given in Eq. (4.21). The unknown deflection at the center of the column is y_c. Once this deflection is determined, the fixed-end moment M_E can be calculated directly from Eq. (4.26). To obtain y_c, we substitute the integration constants, Eq. (4.21), into the second equation of Eq. (4.19) and solve for y_c.

$$\begin{aligned} y_c = \frac{Q}{2kP}\left(\tan\frac{kL}{2} - \cos ka \tan\frac{kL}{2} + \sin ka - ka\right) \\ - \frac{M_{pc}}{P}\left(\frac{1}{\cos\frac{kL}{2}} - 1\right) \end{aligned} \tag{4.28}$$

Both Hinges Formed in Compression (Stage 3)

Since plastic hinges are formed at both ends as well as at the center, the fixed-end moments have the value

$$M_E = M_{pc} \tag{4.23}$$

In this case, the boundary conditions are

$$\begin{aligned}
&1.\ y_1 = 0 && \text{at} \quad x = 0 \\
&2.\ d^2y_2/dx^2 = -M/(EI) && \text{at} \quad x = L/2 \\
&3.\ y_1 = y_2 && \text{at} \quad x = a \\
&4.\ dy_1/dx = dy_2/dx && \text{at} \quad x = a
\end{aligned} \tag{4.29}$$

Using Eqs. (4.19) and (4.29), the integration constants are determined:

$$\begin{aligned}
A &= \frac{Q}{2kP}\left(\cos ka - \sin ka \cot\frac{kL}{2}\right) + \frac{M_{pc}}{P}\left(\frac{1}{\sin\frac{kL}{2}} + \cot\frac{kL}{2}\right) \\
B &= -\frac{M_{pc}}{P} \\
C &= -\frac{Q}{2kP}\sin ka \cot\frac{kL}{2} + \frac{M_{pc}}{P}\left(\frac{1}{\sin\frac{kL}{2}} + \cot\frac{kL}{2}\right) \\
D &= \frac{Q}{2kP}\sin ka - \frac{M_{pc}}{P}
\end{aligned} \tag{4.30}$$

The slopes or relative rotations at the ends and at the center, which are needed in later calculations, are found by taking the derivative of the deflections.

$$\begin{aligned}
\theta_E &= \left(\frac{dy_1}{dx}\right)_{x=0} = Ak - \frac{Q}{2P} \\
\theta_c &= \left(\frac{dy_2}{dx}\right)_{x=L/2} = k\left(C\cos\frac{kL}{2} - D\sin\frac{kL}{2}\right)
\end{aligned} \tag{4.31}$$

It should be noted that these slopes or rotations do not change when the column behaves elastically.

Elastic Unloading in Compression (Stage 4)

The boundary conditions for the five unknowns in Eq. (4.19) are

$$\begin{aligned}
&1.\ y_1 = 0 && \text{at} \quad x = 0 \\
&2.\ dy_1/dx = \theta_E && \text{at} \quad x = 0
\end{aligned}$$

$$\begin{aligned} &3.\ y_1 = y_2 && \text{at} \quad x = a \\ &4.\ dy_1/dx = dy_2/dx && \text{at} \quad x = a \\ &5.\ dy_2/dx = \theta_c && \text{at} \quad x = L/2 \end{aligned} \tag{4.32}$$

From the first four boundary conditions and Eq. (4.19), we find the four constants:

$$\begin{aligned} A &= \frac{Q}{2kP} + \frac{\theta_E}{k} \\ B &= -\frac{M_E}{P} \\ C &= \frac{Q}{2kP}(1 - \cos ka) + \frac{\theta_E}{k} \\ D &= \frac{Q}{2kP}\sin ka - \frac{M_E}{P} \end{aligned} \tag{4.33}$$

Substituting these constants and Eq. (4.19) into the last condition in Eq. (4.32), the end moment is determined.

$$M_E = \frac{P}{k \sin \frac{kL}{2}} \left[\frac{Q}{2P} \left(\sin ka \sin \frac{kL}{2} + \cos ka \cos ka \cos \frac{kL}{2} - \cos \frac{kL}{2} \right) + \theta_c - \theta_E \cos \frac{kL}{2} \right] \tag{4.34}$$

The equations derived in this section are valid for the range where the axial load remains compressive. Table 4.2 summarizes the end moment and integration constants for each of the four stages of compressive loadings.

4.4.3 Tensile Axial Force (Stages 5 to 9)

The governing differential equation is the same as that of the compression case.

$$\frac{d^2y}{dx^2} = -\frac{M}{EI} \tag{4.17}$$

Referring to Fig. 4.14, the bending moment at section x has the same form as that of the compressive force case except that the sign for the axial load term is reversed.

TABLE 4.2 Deflections of Each Loading Stage in Compression

$$y_1 = A \sin kx + B \cos kx - \frac{Q}{2P}x + \frac{M_E}{P} \text{ for } x \leq a, \quad y_2 = C \sin kx + D \cos kx - \frac{Qa}{2P} + \frac{M_E}{P} \text{ for } x \geq a$$

Stages	M_E	A	B	C	D
(1) Elastic Loading 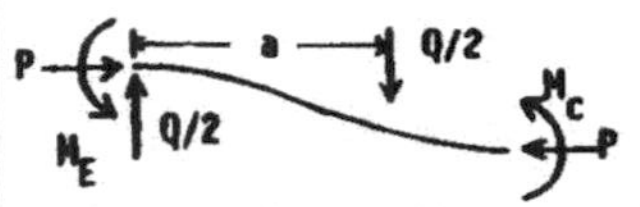	$\frac{Q}{2k}\left(\cos ka \cot \frac{kL}{2} + \sin ka - \cot \frac{kL}{2}\right)$	$\frac{Q}{2kP}$	$-\frac{M_E}{P}$	$\frac{Q}{2kP}(1 - \cos ka)$	$\frac{Q}{2kP} \sin ka - \frac{M_E}{P}$
(2) A. One Hinge at Ends 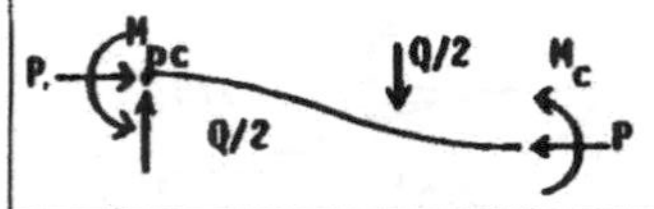	M_{pc}	$C + \frac{Q}{2kP} \cos ka$	$-\frac{M_{pc}}{P}$	$\left(\frac{Q}{2kP} \sin ka - \frac{M_{pc}}{P}\right) \times \tan \frac{kL}{2}$	$\frac{Q}{2kP} \sin ka - \frac{M_{pc}}{P}$
(2) B. One Hinge at Center 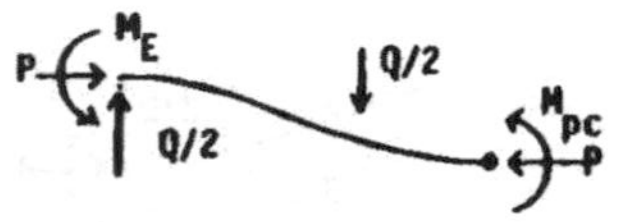	$-M_{pc} + \frac{Qa}{2} + Py_c$	$\frac{Q}{2kP}$	$-\frac{M_E}{P}$	$\frac{Q}{2kP}(1 - \cos ka)$	$\frac{Q}{2kP} \sin ka - \frac{M_E}{P}$
	$y_c = -\frac{M_{pc}}{P}\left(\frac{1}{\cos \frac{kL}{2}} - 1\right) + \frac{Q}{2kP}\left(\tan \frac{kL}{2} - \cos ka \tan \frac{kL}{2} + \sin ka - ka\right)$				

(3) Both Hinges	M_{pc}	$\frac{Q}{2kP}\left(\cos ka - \sin ka \cot \frac{kL}{2}\right) + \frac{M_{pc}}{P}\left(\frac{1}{\sin \frac{kL}{2}} + \cot \frac{kL}{2}\right)$	$-\frac{M_{pc}}{P}$	$A - \frac{Q}{2kP} \cos ka$	$\frac{Q}{2kP} \sin ka - \frac{M_{pc}}{P}$
(4) Unloading	$\frac{P}{k \sin \frac{kL}{2}}\left[\frac{Q}{2P}\left(\sin ka \sin \frac{kL}{2} + \cos ka \cos \frac{kL}{2} - \cos \frac{kL}{2}\right) + \theta_c - \theta_E \cos \frac{kL}{2}\right]$	$\frac{Q}{2kP} + \frac{\theta_E}{k}$	$-\frac{M_E}{P}$	$\frac{Q}{2kP}(1 - \cos ka) + \frac{\theta_E}{k}$	$\frac{Q}{2kP} \sin ka - \frac{M_E}{P}$

$$M = \begin{cases} -M_E + \dfrac{Q}{2}x - Py_1 \\ -M_E + \dfrac{Qa}{2} - Py_2 \end{cases} \tag{4.35}$$

The substitution of Eq. (4.35) into Eq. (4.17) results in the general solution for deflections.

$$\begin{aligned} y_1 &= Ee^{kx} + Fe^{-kx} + \frac{Q}{2P}x - \frac{M_E}{P} \quad \text{for} \quad x \leq a \\ y_2 &= Ge^{kx} + He^{-kx} + \frac{Qa}{2P} - \frac{M_E}{P} \quad \text{for} \quad x > a \end{aligned} \tag{4.36}$$

in which E, F, G, and H are integration constants.

Elastic Tensioning (Stage 5)

The boundary conditions are the same as those of Stage 4 and given by Eq. (4.32). Substitution of the deflection equations into the first four boundary conditions of Eq. (4.32) results in

$$\begin{aligned} E &= \frac{M_E}{2P} - \frac{Q}{4kP} + \frac{\theta_E}{2k} \\ F &= \frac{M_E}{2P} + \frac{Q}{4kP} - \frac{\theta_E}{2k} \\ G &= E + \frac{Q}{4kP}e^{-ka} \\ H &= F - \frac{Q}{4kP}e^{ka} \end{aligned} \tag{4.37}$$

Using Eqs. (4.37) and (4.36) and the last boundary condition of Eq. (4.32), the end moment is determined.

$$M_E = \frac{1}{e^{kL/2} - e^{-kL/2}} \left\{ \frac{2P}{k}\theta_c + \frac{Q}{2k}\left[(1 - e^{ka})e^{-kL/2} + (1 - e^{-ka})\, e^{kL/2} \right] - \frac{P}{k}\theta_E(e^{kL/2} + e^{-kL/2}) \right\} \tag{4.38}$$

Since the rotations of the hinges are given by Eq. (4.31), the deflections corresponding to various load combinations of P and Q in the elastic tensioning range are calculated from Eq. (4.36).

One Hinge Formed at Ends in Tension (Stage 6)

In Eq. (4.35), the end moment is simply replaced by the plastic moment for the present case.

$$M_E = -M_{pc} \tag{4.39}$$

The boundary conditions are

$$\begin{aligned} &1.\ y_1 = 0 && \text{at} \quad x = 0 \\ &2.\ y_1 = y_2 && \text{at} \quad x = a \\ &3.\ dy_1/dx = dy_2/dx && \text{at} \quad x = a \\ &4.\ dy_2/dx = \theta_c && \text{at} \quad x = L/2 \end{aligned} \tag{4.40}$$

in which the second condition of Eq. (4.32) is deleted because of the formation of plastic hinges at both ends.

The integration constants are:

$$\begin{aligned} E &= \frac{1}{e^{kL/2} + e^{-kL/2}} \left[\frac{\theta_c}{k} - \frac{M_{pc}}{P} e^{-kL/2} - \frac{Q}{4kP} (e^{-ka} e^{kL/2} + e^{ka} e^{-kL/2}) \right] \\ F &= -E - \frac{M_{pc}}{P} \\ G &= E + \frac{Q}{4kP} e^{-ka} \\ H &= F - \frac{Q}{4kP} e^{ka} \end{aligned} \tag{4.41}$$

Now, all the unknowns in Eq. (4.36) have been determined.

One Hinge Formed at Mid-Span in Tension (Stage 6)

In this case, the end moment is not known, and five boundary conditions are needed. Instead of using the slope condition at the center of the column in Eq. (4.32), we use the curvature condition.

$$\begin{aligned} &1.\ y = 0 && \text{at} \quad x = 0 \\ &2.\ dy_1/dx = \theta_E && \text{at} \quad x = 0 \end{aligned}$$

$$
\begin{aligned}
&3.\ y_1 = y_2 && \text{at} \quad x = a \\
&4.\ dy_1/dx = dy_2/dx && \text{at} \quad x = a \\
&5.\ d^2y_2/dx^2 = M_{pc}/(EI) && \text{at} \quad x = L/2
\end{aligned}
\tag{4.42}
$$

The end moment and the integration constants are

$$
M_E = \frac{1}{e^{kL/2} + e^{-kL/2}} \left\{ 2\,M_{pc} + \frac{Q}{2k}\left[(1 - e^{-ka})e^{kL/2} - (1 - e^{ka})\,e^{-kL/2} \right] - \frac{P}{k}\,\theta_E(e^{kL/2} - e^{-kL/2}) \right\} \tag{4.43}
$$

and

$$
\begin{aligned}
E &= \frac{M_E}{2P} - \frac{Q}{4kP} + \frac{\theta_E}{2k} \\
F &= \frac{M_E}{2P} + \frac{Q}{4kP} - \frac{\theta_E}{2k} \\
G &= \frac{M_E}{2P} - \frac{Q}{4kP}(1 - e^{-ka}) + \frac{\theta_E}{2k} \\
H &= \frac{M_E}{2P} + \frac{Q}{4kP}(1 - e^{ka}) - \frac{\theta_E}{2k}
\end{aligned}
\tag{4.44}
$$

Note that Eq. (4.44) is identical to Eq. (4.37).

Both Hinges Formed in Tension (Stage 7)

Since plastic hinges are formed at both ends as well as at the center of the column in Stage 7, the second condition in Eq. (4.42) is deleted. Thus, the boundary conditions are identical to the case of both hinges formed in compression (Eq. 4.29).

$$
\begin{aligned}
&1.\ y_1 = 0 && \text{at} \quad x = 0 \\
&2.\ y_1 = y_2 && \text{at} \quad x = a \\
&3.\ dy_1/dx = dy_2/dx && \text{at} \quad x = a \\
&4.\ d^2y_2/dx^2 = M_{pc}/(EI) && \text{at} \quad x = L/2
\end{aligned}
\tag{4.45}
$$

The constants are

$$E = \frac{1}{e^{kL/2} + e^{-kL/2}} \left[\frac{M_{pc}}{P} (1 + e^{-kL/2}) + \frac{Q}{4kP} (e^{ka} e^{-kL/2} - e^{-ka} e^{kL/2}) \right]$$

$$F = -E - \frac{M_{pc}}{P} \tag{4.46}$$

$$G = E + \frac{Q}{4kP} e^{-ka}$$

$$H = F - \frac{Q}{4kP} e^{ka}$$

The end moment and the integration constants for each loading stage from 5 to 7 in the axial tension range are summarized in Table 4.3. The solutions for Stages 8 and 9 can be solved in a similar manner. However, the solutions are neglected here since they do not play an important role in cyclic behavior of fixed-ended beam-columns.

4.4.4 Load-Shortening Relation

The axial shortening of a column is caused by the axial compressive strain and lateral deflection. In the elastic range, the lateral deflection is small so that the axial compressive strain dominates the shortening of the column. When the deflection becomes large, the shortening due to lateral deflection is more important. This implies that the inelastic contribution to the total axial strain does not contribute significantly to total shortening of a member. Further, most parts of an elastic–plastic column remain in the elastic range when plastic hinges are formed at critical sections. The difference between elastic and inelastic shortenings due to axial strain can be seen in Fig. 4.11 for the case of a pin-ended column.

The shortening due to lateral deflection can be calculated numerically from the exact shape of deflection as given by Eqs. (4.19) or (4.36). However, a two-straight-line simplification might be sufficient for the study of cyclic behavior of a member. In this simplification, we assume a two-straight-line deflected shape knuckled at the ends and center of the column. Thus, the total shortening is the sum of the two contributions due to axial strain and geometrical change, which can be expressed in the simple form as

$$\Delta = \frac{P}{2\pi Etr} + \left(1 - \cos\frac{\pi L}{2}\right)L \tag{4.47}$$

This is used in the following numerical studies.

TABLE 4.3 Deflections of Each Loading Step in Tension

$$y_1 = Ee^{kx} + Fe^{-kx} + \frac{Q}{2P}x - \frac{M_E}{P} \text{ for } x < a, \qquad y_2 = Ge^{kx} + He^{-kx} + \frac{Qa}{2P} - \frac{M_E}{P} \text{ for } x > a$$

Stages	M_E	E	F	G	H
(5) Elastic Tension 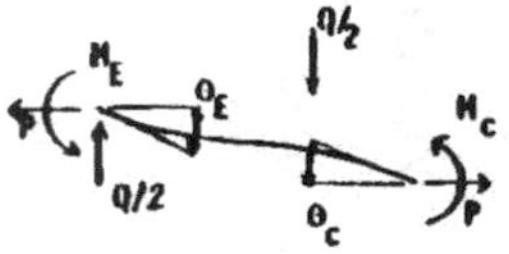	$\frac{1}{e^{kL/2} - e^{-kL/2}}\left\{\frac{2P}{k}\theta_c + \frac{Q}{2k}\left[(1 - e^{ka})e^{-kL/2} + (1 - e^{-ka})e^{kL/2}\right] - \frac{P}{k}\theta_E(e^{kL/2} + e^{-kL/2})\right\}$	$\frac{M_E}{2P} - \frac{Q}{4kP} + \frac{\theta_E}{2k}$	$\frac{M_E}{2P} + \frac{Q}{4kP} - \frac{\theta_E}{2k}$	$E + \frac{Q}{4kP}e^{-ka}$	$F - \frac{Q}{4kP}e^{ka}$
(6) A. One Hinge at Ends 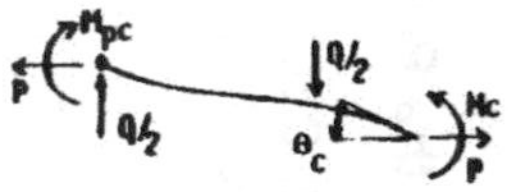	$-M_{pc}$	$\frac{1}{e^{kL/2} + e^{-kL/2}}\left[\frac{\theta_c}{k} - \frac{M_{pc}}{P}e^{-kL/2} - \frac{Q}{4kP}(e^{-ka}e^{kL/2} + e^{ka}e^{-kL/2})\right]$	$-E - \frac{M_{pc}}{P}$	$E + \frac{Q}{4kP}e^{-ka}$	$F - \frac{Q}{4kP}e^{ka}$
(6) B. One Hinge at Center 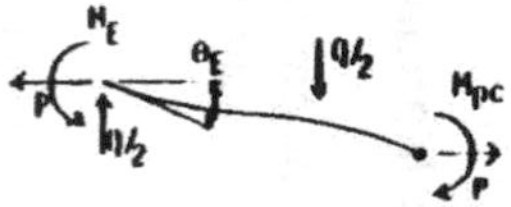	$\frac{1}{e^{kL/2} + e^{-kL/2}}\left\{2M_{pc} + \frac{Q}{2k}[(1 - e^{-ka})e^{kL/2} - (1 - e^{ka})e^{-kL/2}] - \frac{P}{k}\theta_E(e^{kL/2} - e^{-kL/2})\right\}$	$\frac{M_E}{2P} - \frac{Q}{4kP} + \frac{\theta_E}{2k}$	$\frac{M_E}{2P} + \frac{Q}{4kP} - \frac{\theta_E}{2k}$	$E + \frac{Q}{4kP}e^{-ka}$	$F - \frac{Q}{4kP}e^{ka}$
(7) Both Hinges 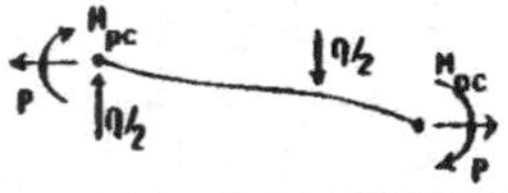	$-M_{pc}$	$\frac{1}{e^{kL/2} - e^{-kL/2}}\left[\frac{M_{pc}}{P}(1 + e^{-kL/2}) + \frac{Q}{4kP}(e^{ka}e^{-kL/2} - e^{-ka}e^{kL/2})\right]$	$-E - \frac{M_{pc}}{P}$	$E + \frac{Q}{4kP}e^{-ka}$	$F - \frac{Q}{4kP}e^{ka}$

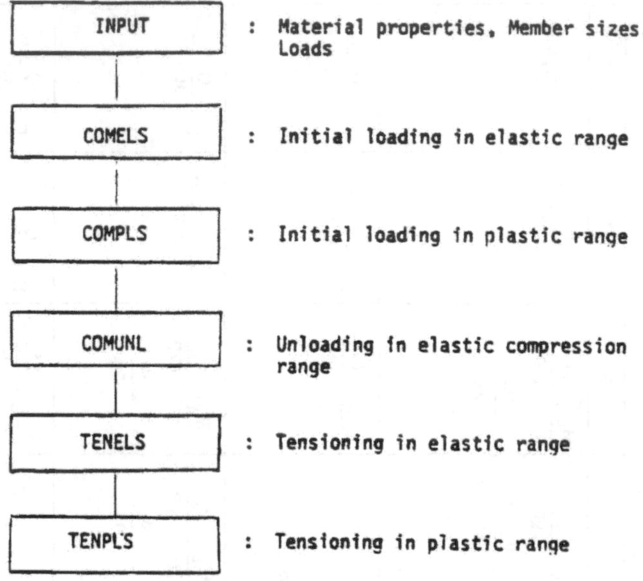

FIGURE 4.15 Flow chart of FIXCYCL.

4.4.5 Computer Implementation

Based on the previous formulations, a computer program has been developed to obtain numerical results. The program named "FIXCYCL" consists of three subroutines: one in compression and two in tension. A brief flow chart is given in Fig. 4.15.

Starting with an elastic loading in compression, the axial load P is increased with a constant lateral load Q. For a given combination of P and Q, the deflection and shortening are calculated, and the corresponding bending moments at the ends and center of the column are checked. If the moments exceed the fully plastic value M_{pc}, the calculations shift to the next stage. In this way, hinge-by-hinge solutions are traced for the member, and eventually, a plastic mechanism develops. The maximum deflection controls the turning point of a cycle. The computing flow for the axial tension state is similar to that of compression case.

4.4.6 User's Manual for the Program FIXCYCL

Input manual of the program FIXCYCL for fixed-ended beam-columns are given below. All data are written in free format unless otherwise noted and to be input in sequence.

```
Line 1: Title of the Job
      Column 1-68. Title of the Job
```

```
Line 2: MCT = Controller for plastic moment
        MCT = 0 . . . Full plastic moment
            = 1 . . . Average plastic moment (see Eq.
                      (3.48) in Sec. 3.8.2)

Line 3: Material and Member Information
      AL = One-half of a column length (ft.)
       D = Diameter of a column (in.)
```

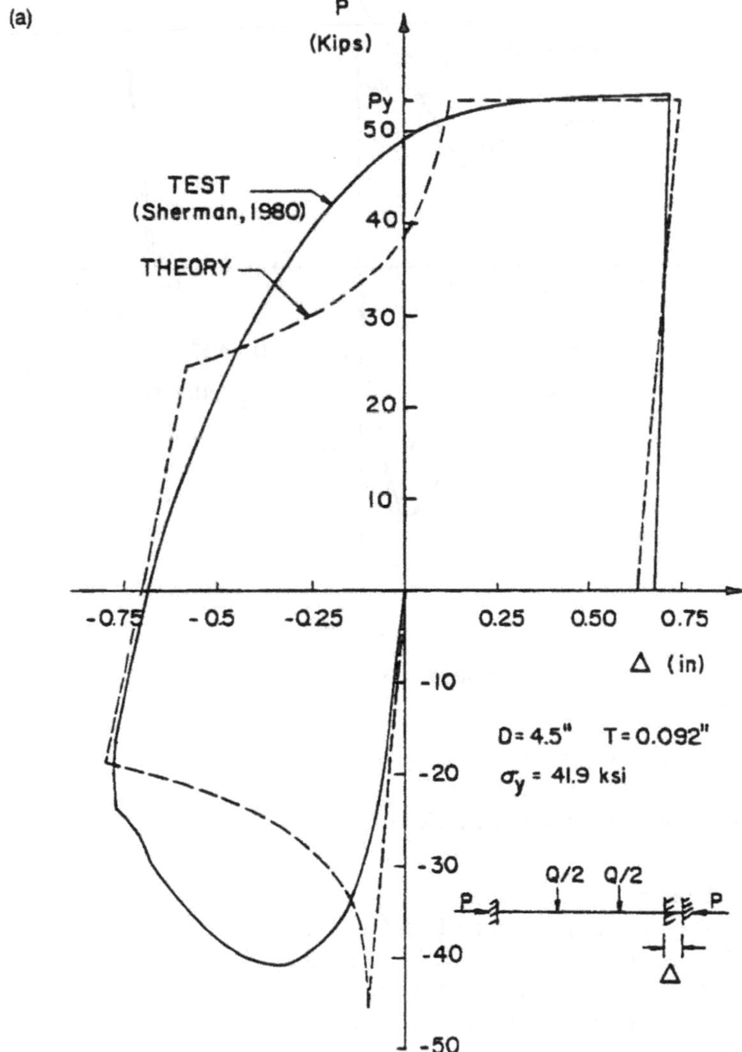

FIGURE 4.16 (a) Comparison of hinge-by-hinge method with test, KL/r = 29. (b) Comparison of hinge-by-hinge method with test, KL/r = 72.

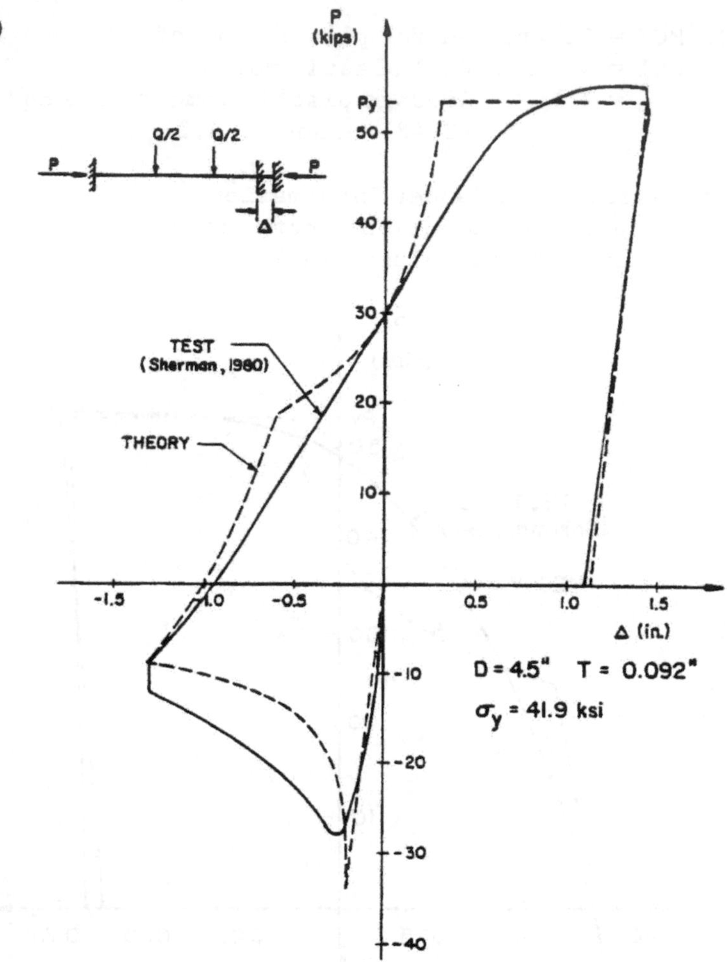

FIGURE 4.16 *Continued*

T = Thickness of a column (in.)
ST = Yield stress (ksi)
YONG = Young's modulus (ksi)

Line 4: Loads
PP = Starting value of axial load (kips)
PDELT = Increment of axial load (kips)
QQ = Starting value of lateral load (kips)
QDELT = Increment of lateral load (kips)
NQ = Number of loading cases for lateral load
AX = Position of lateral load from end (ft.)
YLMT = Limiting deflection for unloading (in.)

Remarks:

1) Repeat Lines 1 through 3 to continue another job.
2) All lines except Line 1 are in free format.

Following the input manual described above, the input and the corresponding output data for a sample problem are shown in the next section.

INPUT DATA OF THE PROGRAM "FIXCYCL" FOR A SAMPLE PROBLEM

```
CHECK WITH THE ASSUMED DEFLECTION METHOD
0
10.4 4.5 0.092 36 30000
1 2 0.47626 0.47626 2 10.4 7.5
```

OUTPUT DATA OF THE PROGRAM "FIXCYCL" FOR A SAMPLE PROBLEM

```
CHECK WITH THE ASSUMED DEFLECTION METHOD
 MCT = 0  AL=  .1040E+02 D=  .4500E+01 T=  .9200E-01 SY=  .3600E+02 YONG=  .3000E+05
 PP= .1000E+01 PDELT=  .2000E-01  QQ=  .4763E+00 QDELT=  .4763E+00 NQ= 2  AX=  .1040E+02
      YLMT=  .7500E+01
 SI= .3096E+01 SA= .1274E+01 PY= .4587E+02 EMY= .4953E+02 QY= 1588E+01  PBUCK= .5885E+02
      SL=  .8006E+02
```

(1) . . . INITIAL ELASTIC LOADING IN COMPRESSION

Q	P	Y	S	MR	MC	MPC	KP
.4763E+00	-.1000E+01	.4224E+00	-.7960E-02	.1507E+02	.1507E+02	.6431E+02	1
.4763E+00	-.3000E+01	.4372E+00	-.2112E-01	.1552E+02	.1552E+02	.6402E+02	2
.4763E+00	-.5000E+01	.4534E+00	-.3430E-01	.1599E+02	.1599E+02	.6344E+02	3
.4763E+00	-.7000E+01	.4707E+00	-.4749E-01	.1651E+02	.1651E+02	.6257E+02	4
.4763E+00	-.9000E+01	.4893E+00	-.6069E-01	.1706E+02	.1706E+02	.6142E+02	5
.4763E+00	-.1100E+02	.5095E+00	-.7392E-01	.1766E+02	.1766E+02	.5997E+02	6
.4763E+00	-.1300E+02	.5315E+00	-.8716E-01	.1831E+02	.1831E+02	.5824E+02	7
.4763E+00	-.1500E+02	.5555E+00	-.1004E+00	.1903E+02	.1903E+02	.5622E+02	8
.4763E+00	-.1700E+02	.5813E+00	-.11373+00	.1980E+02	.1980E+02	.5391E+02	9
.4763E+00	-.1900E+02	.6107E+00	-.1271E+00	.2066E+02	.2066E+02	.5131E+02	10
.4763E+00	-.2100E+02	.6426E+00	-.1404E+00	.2161E+02	.2161E+02	.4842E+02	11
.4763E+00	-.2300E+02	.6782E+00	-.1539E+00	.2266E+02	.2266E+02	.4525E+02	12
.4763E+00	-.2500E+02	.7179E+00	-.1674E+00	.2383E+02	.2383E+02	.4178E+02	13
.4763E+00	-.2700E+02	.7626E+00	-.1810E+00	.2515E+02	.2515E+02	.3803E+02	14
.4763E+00	-.2900E+02	.8133E+00	-.1947E+00	.2665E+02	.2665E+02	.3399E+02	15
.4763E+00	-.3100E+02	.8713E+00	-.2085E+00	.2836E+02	.2836E+02	.2986E+02	16
4763E+00	-.3300E+02	.9382E+00	-.2226E+00	.3034E+02	.3034E+02	.2584E+02	17

(3) . . . BOTH HINGES FORMED AT ENDS AND CENTER IN COMPRESSION

Q	P	Y	S	THE	THC	MPC	KP
.4763E+00	-.3300E+02	.6654E+00	-.2191E+00	-.1070E-02	-.1070E-02	.2584E+02	18
.4763E+00	-.3200E+02	.8122E+00	-.2143E+00	-.3718E-03	-.3718E-03	.2785E+02	18
.4763E+00	-.3100E+02	.9677E+00	-.2099E+00	.4050E-03	.4050E-03	.2986E+02	18
.4763E+00	-.3000E+02	.1134E+01	-.2062E+00	.1268E-02	.1268E-02	.3187E+02	19
.4763E+00	-.2900E+02	.1319E+01	-.2033E+00	.2260E-02	.2260E-02	.3399E+02	20
.4763E+00	-.2800E+02	.1512E+01	-.2012E+00	.3341E-02	.3341E-02	.3605E+02	21
.4763E+00	-.2700E+02	.1714E+01	-.1999E+00	.4515E-02	.4515E-02	.3803E+02	22
.4763E+00	-.2600E+02	.1932E+01	-.1996E+00	.5789E-02	.5789E-02	.3994E+02	23
.4763E+00	-.2500E+02	.2154E+01	-.2004E+00	.7174E-02	.7174E-02	.4178E+02	24
.4763E+00	-.2400E+02	.2391E+01	-.2025E+00	.8681E-02	.8681E-02	.4355E+02	25
.4763E+00	-.2300E+02	.2643E+01	-.2062E+00	.1032E-01	.1032E-01	.4525E+02	26
.4763E+00	-.2200E+02	.2912E+01	-.2115E+00	.1211E-01	.1211E-01	.4687E+02	27
.4763E+00	-.2100E+02	.3197E+01	-.2190E+00	.1407E-01	.1407E-01	.4842E-02	28
.4763E+00	-.2000E+02	.3515E+01	-.2290E+00	.1623E-01	.1623E-01	.4990E-02	29
.4763E+00	-.1900E+02	.3817E+01	-.2420E+00	.1859E-01	.1859E-01	.5131E+02	30

.4763E+00	-.1800E+02	.4199E+01	-.2588E+00	.2121E-01	.2121E-01	.5265E+02	31
.4763E+00	-.1700E+02	.4594E+01	-.2801E+00	.2412E-01	.2412E-01	.5391E+02	32
.4763E+00	-.1600E+02	.5030E+01	-.3072E+00	.2738E-01	.2738E-01	.5510E+02	33
.4763E+00	-.1500E+02	.5515E+01	-.3416E+00	.3104E-01	.3104E-01	.5622E+02	34
.4763E+00	-.1400E+02	.6058E+01	-.3855E+00	.3519E-01	.3519E-01	.5727E+02	35
.4763E+00	-.1300E+02	.6674E+01	-.4417E+00	.3994E-01	.3994E-01	.5824E+02	36
.4763E+00	-.1200E+02	.7381E+01	-.5147E+00	.4543E-01	.4543E-01	.5914E+02	37
.4763E+00	-.1203E+02	.7358E+01	-.5122E+00	.4525E-01	.4525E-01	.5912E+02	38

(4) . . . ELASTIC UNLOADING IN COMPRESSION

Q	P	Y	S	MR	MC	MPC	KP
.4763E+00	-.1203E+02	.7358E+01	-.5122E+00	.5912E+02	.5912E+02	.5912E+02	39
.4763E+00	-.9625E+01	.7048E+01	-.4608E+00	.4878E+02	.4878E+02	.6100E+02	40
.4763E+00	-.7218E+01	.6768E+01	-.4140E+00	.3929E+02	.3929E+02	.6246E+02	41
.4763E+00	-.4812E+01	.6512E+01	-.3711E+00	.3053E+02	.3053E+02	.6351E+02	42
.4763E+00	-.2406E+01	.6277E+01	-.3314E+00	.2241E+02	.2241E+02	.6413E+02	43

(5) . . . ELASTIC LOADING IN TENSION

Q	P	Y	S	MR	MC	MPC	KP
.4763E+00	.2000E+01	.5896E+01	-.2654E+00	.8963E+01	.8963E+01	.6420E+02	44
.4763E+00	.4000E+01	.5740E+01	-.2379E+00	.3379E+01	.3379E+01	.6376E+02	45
.4763E+00	.6000E+01	.5594E+01	-.2115E+00	-.1923E+01	-.1923E+01	.6304E+02	46
.4763E+00	.8000E+01	.5456E+01	-.1863E+00	-.6966E+01	-.6966E+01	.6203E+02	47
.4763E+00	.1000E+02	.5327E+01	-.1620E+00	-.1177E+02	-.1177E+02	.6073E+02	48
.4763E+00	.1200E+02	.5204E+01	-.1386E+00	-.1636E+02	-.1636E+02	.5914E+02	49
.4763E+00	.1400E+02	.5088E+01	-.1160E+00	-.2076E+02	-.2076E+02	.5727E+02	50
.4763E+00	.1600E+02	.4978E+01	-.9407E-01	-.2497E+02	-.2497E+02	.5510E+02	51
.4763E+00	.1800E+02	.4874E+01	-.7279E-01	-.2901E+02	-.2901E+02	.5265E+02	52
.4763E+00	.2000E+02	.4775E+01	-.5207E-01	-.3289E+02	-.3289E+02	.4990E+02	53
.4763E+00	.2200E+02	.4681E+01	-.3187E-01	-.3663E+02	-.3663E+02	.4687E+02	54
.4763E+00	.2400E+02	.4591E+01	-.1215E-01	-.4023E+02	-.4023E+02	.4355E+02	55
.4763E+00	.2600E+02	.4506E+01	.7153E-02	-.4371E+02	-.4371E+02	.3994E+02	56

(7) . . . BOTH HINGES FORMED AT ENDS AND CENTER IN TENSION

Q	P	Y	S	THE	THC	MPC	KP
.4763E+00	.2600E+02	.4216E+01	.2741E-01	.4214E-01	.4214E-01	.3994E+02	57
.4763E+00	.2997E+02	.3122E+01	.1177E+00	.3163E-01	.3163E-01	.3192E+02	58
.4763E+00	.3395E+02	.2286E+01	.1798E+00	.2323E-01	.2323E-01	.2394E+02	59
.4763E+00	.3792E+02	.1626E+01	.2265E+00	.1627E-01	.1627E-01	.1596E+02	60
.4763E+00	.4189E+02	.1090E+01	.2640E+00	.1035E-01	.1035E-01	.7981E+01	61
.4763E+00	.4587E-02	.6480E+00	.2962E+00	.5192E-02	.5192E-02	.0000E+00	62

****

(1) . . . INITIAL ELASTIC LOADING IN COMPRESSION

Q	P	Y	S	MR	MC	MPC	KP
.9525E+00	-.1000E+01	.8448E+00	-.1225E-01	.3014E+02	.3014E+02	.6431E+02	1
.9525E+00	-.3000E+01	.8747E+00	-.2572E-01	.3103E+02	.3103E+02	.6402E+02	2
.9525E+00	-.5000E+01	.9068E+00	-.3924E-01	.3199E+02	.3199E+02	.6344E+02	3
.9525E+00	-.7000E+01	.9413E+00	-.5281E-01	.3301E+02	.3301E+02	.6257E+02	4
.9525E+00	-.9000E-01	.9786E+00	-.6645E-01	.3412E+02	.3412E+02	.6142E+02	5
.9525E+00	-.1100E+02	.1019E+01	-.8016E-01	.3532E+02	.3532E+02	.5997E+02	6
.9525E+00	-.1300E+02	.1063E+01	-.9395E-01	.3663E+02	.3663E+02	.5824E+02	7
.9525E+00	-.1500E+02	.1111E+01	-.1078E+00	.3805E+02	.3805E+02	.5622E+02	8
.9525E+00	-.1700E+02	.1164E+01	-.1219E+00	.3961E+02	.3961E+02	.5391E+02	9
.9525E+00	-.1900E+02	.1221E+01	-.1360E+00	.4132E+02	.4132E+02	.5131E+02	10
.9525E+00	-.2100E+02	.1285E+01	-.1504E+00	.4321E+02	.4321E+02	.4842E+02	11
.9525E+00	-.2300E+02	.1356E+01	-.1649E+00	.4532E+02	.4532E+02	.4525E+02	12

(3) . . . BOTH HINGES FORMED AT ENDS AND CENTER IN COMPRESSION

Q	P	Y	S	THE	THC	MPC	KP
.9525E+00	-.2300E-02	.1350E+01	-.1648E+00	-.3093E-04	-.3093E-04	.4525E+02	13
.9525E+00	-.2200E-02	.1559E+01	-.1632E+00	.1290E-02	.1290E-02	.4687E+02	13
.9525E+00	-.2100E-02	.1782E+01	-.1626E+00	.2735E-02	.2735E-02	.4842E+02	14

---

(7) . . . BOTH HINGES FORMED AT ENDS AND CENTER IN TENSION

Q	P	Y	S	THE	THC	MPC	KP
.9525E+00	.2800E-02	.4697E+01	.6064E-02	.4514E-01	.4514E-01	.3605E+02	47

.9525E+00	.3157E+02	.3701E+01	.9644E-01	.3558E-01	.3558E-01	.2871E+02	48
.9525E+00	.3515E+02	.2916E+01	.1614E+00	.2778E-01	.2778E-01	.2153E+02	49
.9525E+00	.3872E+02	.2277E+01	.2123E+00	.2115E-01	.2115E-01	.1435E+02	50
.9525E+00	.4229E+02	.1745E+01	.2518E+00	.1543E-01	.1543E-01	.7177E+01	51
.9525E+00	.4587E+02	.1296E+01	.2861E+00	.1038E-01	.1038E-01	.0000E+00	52

4.4.7 Numerical Results

The theoretical predictions based on the hinge-by-hinge method are compared with tests reported by Sherman (1980). Fig. 4.16 shows typical comparisons for KL/r = 29 and 72 (Toma and Chen, 1982). Since an elastic–perfectly plastic M-P-Φ relation is used in the present study, there is a sharp peak near the ultimate compressive strength. After reaching the critical compressive load, the axial load drops sharply in the present theory, while the change observed in the test is much more gradual. In the tension branch, again, the theory has discontinuities at the points where plastic hinges are just formed or where the entire cross section is fully yielded.

Although some discrepancy is observed locally between the theory and the test, theoretical predictions based on the hinge-by-hinge method can predict the cyclic behavior of beam-columns with good accuracy.

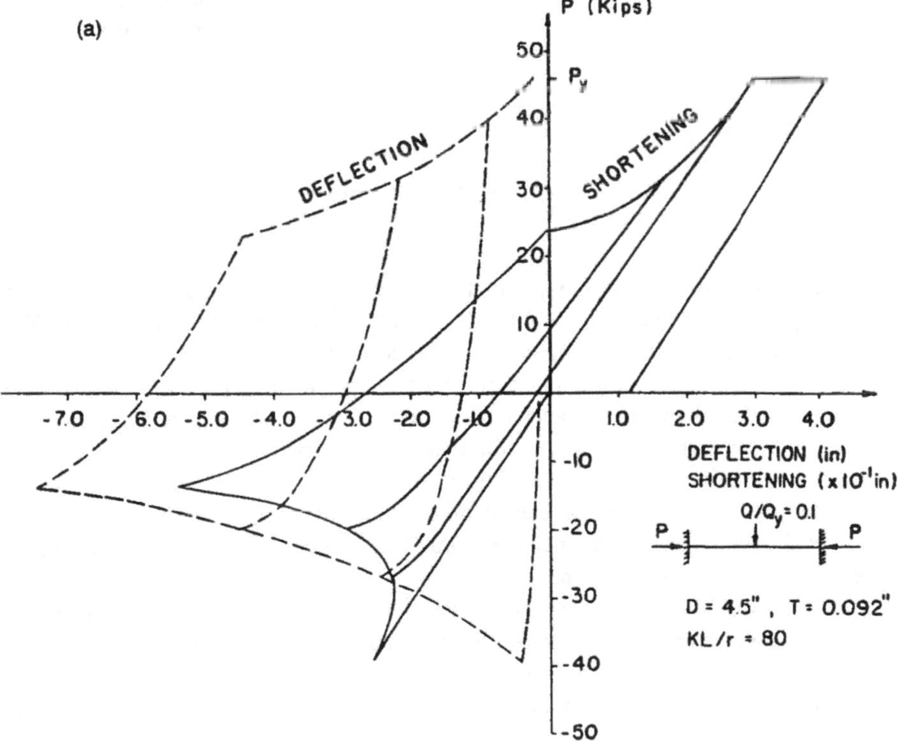

FIGURE 4.17 (a) Cyclic behavior of fixed-ended column, KL/r = 80. (b) Effect of lateral load on cyclic behavior of fixed-ended column.

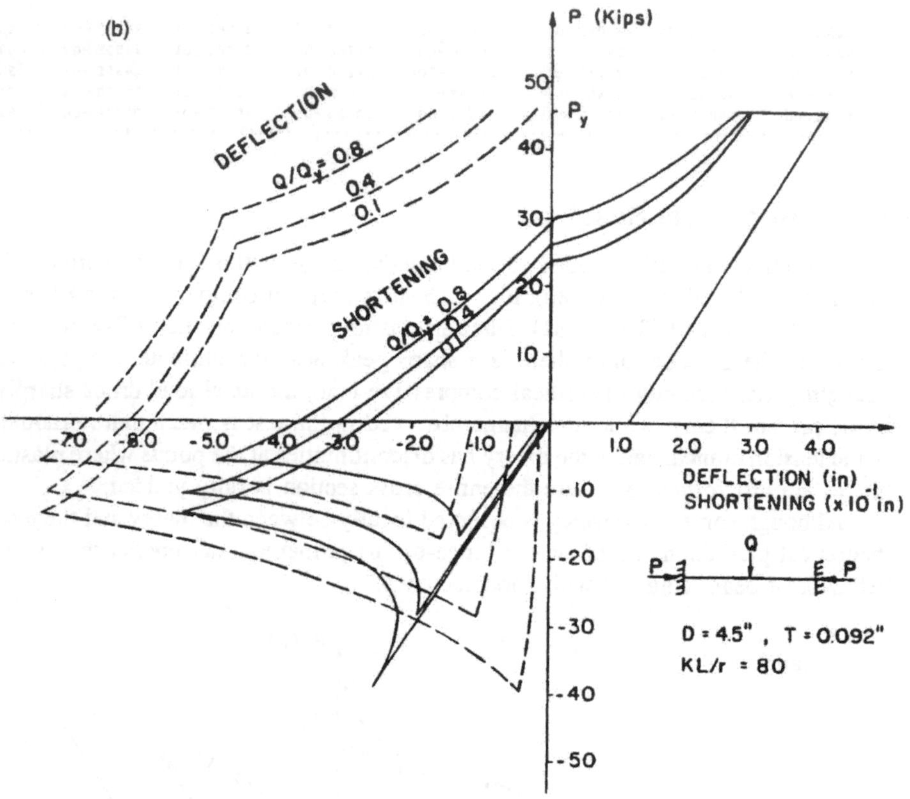

FIGURE 4.17 *Continued*

Some typical cyclic behaviors of fixed-ended columns are studied here by the hinge-by-hinge method. Figure 4.17 shows the behavior of the column with KL/r = 80 subjected to different unloading paths corresponding to the same lateral load $Q/Q_y = 0.1$. Since an elastic–perfectly plastic type of M-P-Φ and P-M-ϵ_o is used, the load-displacement curves behave in an idealized manner compared to the solution based on Newmark's method in which the nonlinearity of M-P-Φ and P-M-ϵ_o is included. It can be seen from the figure that the post-buckling and plastic tension branches are fixed, while the elastic unloading-tensioning branch connects these two fixed envelopes. The sooner the axial load is reversed, the closer the slope of the curve is to the elastic slope.

Figure 4.17b shows the effect of lateral load on the cyclic load-displacement behavior of a column. In the reversed load branches the original curve is seen simply shifted upward by the increase of lateral load Q.

Comparing to Newmark's method used for the pin-ended columns, the hinge-by-hinge method behaves closely to the schematical sketch of Fig. 4.13. This is because the hinge-by-hinge method assumes that the generalized stress-strain relationships are elastic–perfectly plastic.

References

Chen, W. F. (1971) Further Studies of Inelastic Beam-Column Problem, *Journal of the Structural Division, ASCE*, 97 (ST2), 529–44.

Chen, W. F. and Ross, D. A. (1977) Tests of Fabricated Tubular Columns, *Journal of the Structural Division, ASCE*, 103 (ST3), 619–34.

Higginbotham, A. B. and Hanson, R. D. (1976) Axial Hysteretic Behavior of Steel Members, *Journal of the Structural Division, ASCE*, 102 (ST7), Proc. Paper 12245, 1365–81.

Jain, A. K., Goel, S. C. and Hanson, R. D. (1978) Inelastic Response of Restrained Steel Tubes, *Journal of the Structural Division, ASCE*, (ST6), Proc. Paper 13832, 946–60.

Sherman, D. R. (1979) Experimental Study of Post Local Buckling Behavior in Tubular Portal Type Beam-Columns, Report to Shell Oil Company, University of Wisconsin-Milwaukee.

Sherman, D. R. (1980) Post Local Buckling Behavior of Tubular Strut Type Beam-Columns: An Experimental Study, Report to Shell Oil Company, University of Wisconsin-Milwaukee.

Toma, S. and Chen, W. F. (1979) Analysis of Fabricated Tubular Columns, *Journal of the Structural Division, ASCE*, 105 (ST11), 2343–66.

Toma, S. and Chen, W. F. (1982) Cyclic Analysis of Fixed-Ended Steel Beam-Columns, *Journal of the Structural Division, ASCE*, 108 (ST6), 1385–99.

5: Analysis Considering Local Buckling Effects

I. S. Sohal,
Department of Civil and Environmental Engineering, Rutgers University, Piscataway, New Jersey

W. F. Chen
School of Civil Engineering, Purdue University, West Lafayette, Indiana

5.1 INTRODUCTION

In the usual steel cylindrical beam-column analysis, the cross section of the tubes is assumed to remain the same during the entire loading history. However, at large deformations, significant local buckling or distortion may occur in thin-walled cylindrical cross sections (Fig. 5.1). This local buckling or distortion of the cross section will have a significant effect on the maximum load-carrying capacity and the post-buckling behavior of steel cylindrical beam-columns.

The distortion of cylindrical sections may be accounted for by using shell theory or empirical relations based on experiments. The finite element analysis by the elastic–plastic large-deformation shell element is overly complicated and may not be appropriate for an engineering design. Further, very few experiments have been performed on fabricated cylindrical members for investigating this local buckling behavior under the combined action of bending and thrust. The data available to date are not sufficient to develop a direct statistical, empirical relationship that includes the effect of the local buckling. Therefore, the problem is handled by the strength-of-material approach, which combines the basic principles of stress analysis with available experimental data.

In this chapter, a simple kinematic model, similar to that of "plane section remains plane" in beam theory is proposed to describe the changing shape of the locally buckled or distorted cross section during loading. Based on this kinematic model, the moment-curvature (M-Φ) relationships for locally buckled thin-walled sections are developed for monotonic and reversed loading. The developed M-Φ relationships for the section are used to derive load-deflection relationships for the beam-column. These load-deflection relationships are implemented in a computer program, "BRACE," which can be used to trace one-cycle behavior of beam-columns. The user's manual and solutions of example problems are presented in later parts of the chapter.

0-8493-8282-3/96/$0.00+$.50

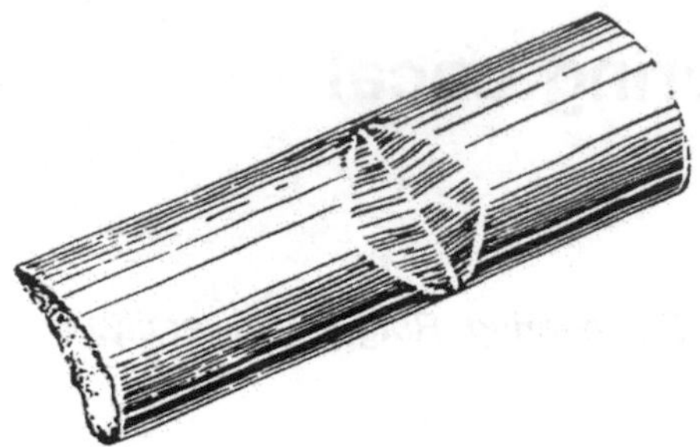

FIGURE 5.1 Local buckling of a tubular member.

5.2 KINEMATIC MODEL FOR CROSS SECTIONAL DISTORTION

In this model, the locally buckled section is assumed to consist of an ideal shape with a flat, straight, distorted portion and a circular, distorted portion (Fig. 5.2). This sudden change in cross-sectional geometry occurs when the compressive strain at the extreme fiber reaches the critical strain $\epsilon_{o,cr}$. This critical strain is determined from experimental results. The distorted shape at a given curvature, Φ, can be determined on the basis of the following kinematic assumptions.

1. The height of the distorted section on the compressive side of the neutral axis, B_c, Fig. 5.3, is determined from the condition that the compressive strain, ϵ_c, at the extreme fiber, is equal to critical strain, $\epsilon_{o,cr}$.
2. The height of the section on the tension side of the neutral axis, B_t, Fig. 5.3, is determined from the equilibrium condition in axial direction.

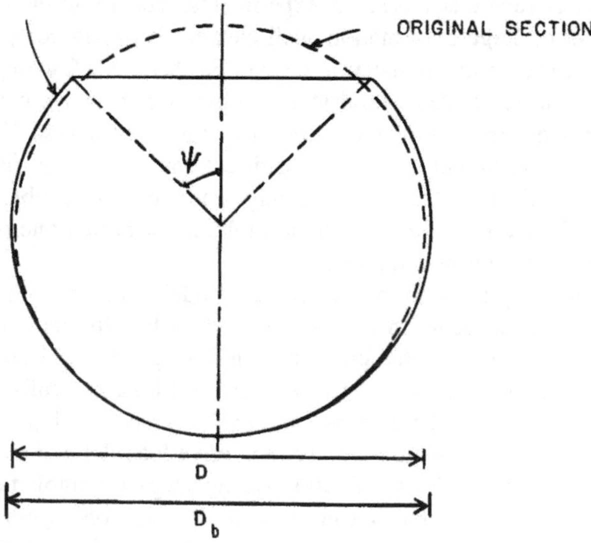

FIGURE 5.2 Modeled geometry of locally buckled section.

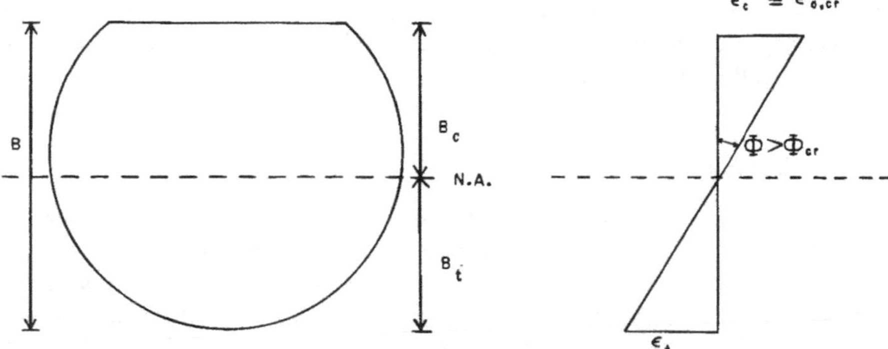

FIGURE 5.3 Height of locally buckled section at given curvature.

3. The change in the length of the circumference of the distorted cross section is negligible as compared to the circumference of the original cross section.

During a reversed loading, there may be some rebound of the distorted section (reduction of the flat portion) on the tension side of the neutral axis and some local distortion (another flat portion) may occur on the compression side of the neutral axis. Since the reversed moment occurs during the tensile axial load, it is reasonable to assume that further distortion of the cross section will be small under the combined action of tension and bending.

5.3 M-P-Φ ANALYSIS OF SECTION

The first step in beam-column analysis is to obtain section behavior or M-P-Φ relation of a section. Here we present a formulation for numerical analysis and closed-form expressions for M-P-Φ relationships for various regimes of one complete cycle of loading including reversed loading with local buckling effects.

5.3.1 Pre-Local-Buckling Analysis

The procedure for computing pre-local-buckling moment-thrust-curvature relationships has been presented in Chapter 2. Closed-form expressions have been fitted to the numerical M-P-Φ results obtained by this procedure (Sohal and Chen, 1984; Chen and Sohal, 1988). For the pre-local-buckling regime, these closed form expressions can be employed to study the local buckling effects.

5.3.2 Post-Local-Buckling Analysis

The post-local-buckling part of M-P-Φ is obtained in two steps. The first step is to determine the critical curvature at which the local buckling starts, i.e., to determine the point up to which the expressions developed previously are still applicable. The second step is to determine the moment capacity of a section subjected to a curvature greater than the critical curvature.

Critical Curvature

Experimental results (Korol, 1978; Sherman, 1983) indicate that the normalized critical strain ϵ_{cr}, required to determine the distorted shape and critical curvature, is mainly a function of the diameter-to-thickness ratio of the tube. This normalized critical strain is determined from the following equation curve fitted conservatively with the available experimental data (Korol, 1978; Sherman, 1983):

$$\epsilon_{cr} = 4.1 - 500\left(\frac{t}{D}\right) + 22{,}500\left(\frac{t}{D}\right)^2 \tag{5.1}$$

Once the normalized critical strain ϵ_{cr} is known, the critical curvature can be determined with the knowledge of simple strength of material as

$$\phi_{cr} = \epsilon_{cr} - \epsilon \tag{5.2}$$

in which $\epsilon = \epsilon_o/\epsilon_y$ = the axial strain normalized with respect to the yield strain due to axial thrust (Chen and Han, 1985).

Moment Capacity of Locally Buckled Section

The reduction in the moment carrying capacity of a locally buckled section is due mainly to the change in its cross section shape. At a given curvature, $\phi > \phi_{cr}$, the geometry of the distorted cross section can be determined by using the described kinematic model. Once the locally buckled shape is determined, the moment carrying capacity of this section can be calculated in the usual manner by summing up the moments about the centroid of the section.

5.3.3 Reversed Loading Analysis

For the given shape of the cross section, the load carrying capacity of a section is usually determined by the following two equilibrium equations.

$$P = \int_A \sigma \, dA \tag{5.3}$$

$$M = \int_A \sigma \, y_c \, dA \tag{5.4}$$

in which σ is the stress in the area dA, and y_c is the distance of the centroid of this area from the centroid of the section. The stress σ at a point in the cross section can be determined by using the assumption of plane section remains plane and the known stress-strain relation. For a monotonically increasing curvature, stress distribution in the section can be expressed as follows:

For $-\sigma_y \leq \sigma \leq \sigma_y$

$$\sigma = E(\epsilon_o + \Phi y_c) = E(\Phi y_n) \tag{5.5}$$

For $\sigma \geq \sigma_y$

$$\sigma = \sigma_y \tag{5.6}$$

For $\sigma \leq -\sigma_y$

$$\sigma = -\sigma_y \tag{5.7}$$

in which ϵ_o is the axial strain due to axial load, Φ is the curvature, y_n is the distance of a given point from the neutral axis, which is located by satisfying Eq. (5.3).

After strains are reversed, the loading history must be included in Eqs. (5.3 to 5.7). For the reversed loading case, the current stress σ in an element of the cross section, can be expressed as (Sohal and Chen, 1988b; Chen and Sohal, 1988)

For $-\sigma_y \leq \sigma \leq \sigma_y$

$$\sigma = \sigma_r + E[\epsilon_o - \epsilon_r + (\Phi - \Phi_r)y_c] \tag{5.8}$$

or

$$\sigma = \sigma_r + E[(\Phi - \Phi_r)y_{nc}] \tag{5.9}$$

For $\sigma \geq \sigma_y$

$$\sigma = \sigma_y \tag{5.10}$$

For $\sigma \leq -\sigma_y$

$$\sigma = -\sigma_y \tag{5.11}$$

in which ϵ_r is the axial strain due to an axial load at the strain reversal, Φ_r is the curvature at the strain reversal, y_{nc} is the distance of the centroid of an element from the current neutral axis, and σ_r is the stress in the element at the strain reversal and can be expressed as

$$\sigma_r = E(\epsilon_r + \Phi_r y_c) = E\Phi_r y_r \tag{5.12}$$

where y_r is the distance of the centroid of an element from the neutral axis at the strain reversal. It should be noted that after local buckling, the section becomes asymmetric. For an asymmetric section loaded in the elastic–plastic regime, the neutral axis varies from the elastic neutral axis position (centroid axis) to the fully plastic neutral axis position even for the pure bending case.

5.3.4 Closed-Form Expression for a Complete Cycle of Loading

To trace the load-deflection curve of a beam-column, it is more efficient to use closed-form M-P-Φ expressions than those generated numerically. To this end, closed-form M-P-Φ expressions are curve-fitted with the numerical data. In the following, first, the expressions for the monotonic loading case are presented. Next, three types of expressions for reversed loading are described. The reversed loading expressions for local buckling cases are simple extensions of those without local buckling, presented in Chapter 2, originally proposed by Toma and Chen (1982).

Monotonic Loading

For the monotonic loading case the M-P-Φ curve is divided into four parts: elastic, primary plastic, secondary plastic, and post-local-buckling regimes (Fig. 5.4). The expressions for these four regimes are (Sohal and Chen, 1987b; Chen and Sohal, 1988):

For $\phi \leq \phi_1$

$$m = a\phi \tag{5.13}$$

For $\phi_1 \leq \phi \leq \phi_2$

$$m = b - \frac{c}{\sqrt{\phi}} \tag{5.14}$$

For $\phi_2 \leq \phi \leq \phi_{lb}$

$$m = m_{pc} - \frac{f}{(\phi + d)^2} \tag{5.15}$$

For $\phi \geq \phi_{lb}$

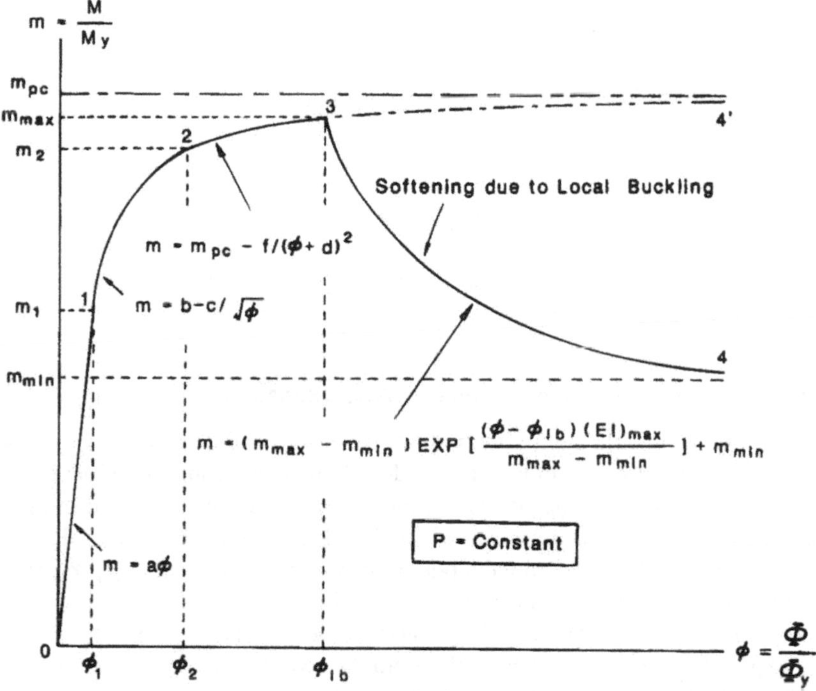

FIGURE 5.4 Closed-form expressions for M-P-Φ curve for monotonic loading.

$$m = (m_{max} - m_{min})\exp\left[\frac{(\phi - \phi_{1b})EI_{max}}{m_{max} - m_{min}}\right] + m_{min} \tag{5.16}$$

In the above expressions, m is the moment normalized with respect to the yield moment M_y; m_{pc} is the plastic moment capacity of the section reduced for the axial thrust P and normalized with respect to M_y; ϕ is the curvature normalized with respect to the yield curvature $\Phi_y = 2\sigma_y/ED$; D is the diameter of the tube; m_{max} is the maximum moment capacity of the section at the start of local buckling, normalized with respect to M_y; m_{min} is the minimum moment capacity after local buckling, normalized with respect to M_y; EI_{max} is the initial slope of the softening curve at the start of the local buckling; and ϕ_{1b} is the initial curvature at the start of local buckling normalized with respect to Φ_y. The constants a, b, c, d, and f are determined by using the following expressions derived from the continuity conditions between the regimes and the maximum plastic moment capacity of cross section m_{pc} (Sohal and Chen, 1984; Chen and Sohal, 1988):

$$a = \frac{m_1}{\phi_1} \tag{5.17}$$

$$b = \frac{m_2\sqrt{\phi_2} - m_1\sqrt{\phi_1}}{\sqrt{\phi_2} - \sqrt{\phi_1}} \tag{5.18}$$

$$c = \frac{m_2 - m_1}{1/\sqrt{\phi_1} - 1/\sqrt{\phi_2}} \tag{5.19}$$

$$d = \frac{4(m_{pc} - m_2)\phi_2^{1.5}}{c} - \phi_2 \tag{5.20}$$

$$f = (m_{pc} - m_2)\,(\phi_2 + d)^2 \tag{5.21}$$

in which m_1, m_2, ϕ_1, and ϕ_2 are the boundaries of the regimes and are expressed in terms of the axial thrust p.

Reversed Loading — Doubly Symmetric Approximation

For a doubly symmetric approximation, the reversed loading branch has the same shape as the initial loading branch but is double in size. From the unloading point onward, the reversed loading branch is doubly symmetric to the initial branch. To obtain the expressions for the reversed loading branch for which unloading starts in the pre-local-buckling regime, the boundary values of moment (e.g., m_1, m_2, and m_{pc}) and curvature (e.g., ϕ_1 and ϕ_2) were doubled.

For the cases in which unloading starts in the post-local-buckling branch, the same expressions are used but with the boundary values of moment multiplied by the factor of ($A_m m_r/m_{max}$) and the boundary values of curvature multiplied by the factor ($A_m m_{max}/m_r$). A_m is evaluated from the following equation

$$A_m = \frac{m_{pcr} + m_{pc}}{m_{pcr}} \tag{5.22}$$

in which m_r is the normalized moment-carrying capacity of section at the point of curvature reversal (Fig. 5.5) and m_{pcr} is the normalized plastic moment capacity of the section reduced for the axial load at the point of curvature reversal. The factor A_m becomes equal to 2 for the cases in which axial load remains constant. The resulting expressions for the reversed moment-curvature relationship are (Sohal and Chen, 1988b; Chen and Sohal, 1988):

For $\phi'_1 \leq \phi \leq \phi_r$

$$m = m_r + a\left(\frac{m_r}{m_{max}}\right)^2 (\phi - \phi_r) \tag{5.23}$$

For $\phi'_2 \leq \phi \leq \phi'_1$

$$m = m_r - A_m b_r\left(\frac{m_r}{m_{max}}\right) + \frac{A_m c_r \sqrt{A_m (m_r/m_{max})}}{\sqrt{\phi_r - \phi}} \tag{5.24}$$

For $\phi \leq \phi'_2$

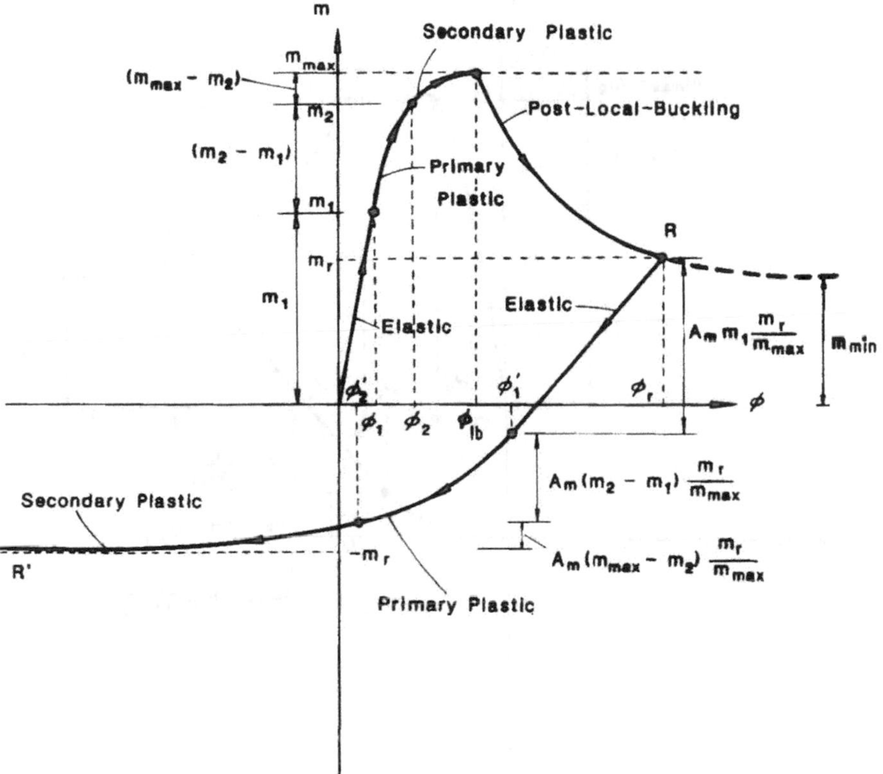

FIGURE 5.5 Doubly symmetric approximation of cyclic M-P-Φ curve.

$$m = -m_{pcr}A_m\left(\frac{m_r}{m_{max}}\right) + m_r + \frac{A_m^3 f_r(m_{max}/m_r)}{[\phi_r - \phi + A_m d_r(m_{max}/m_r)]^2} \tag{5.25}$$

in which

$$\phi_1' = \phi_r - A_m\left(\frac{m_{max}}{m_r}\right)\phi_{1r} \tag{5.26}$$

$$\phi_2' = \phi_r - A_m\left(\frac{m_{max}}{m_r}\right)\phi_{2r} \tag{5.27}$$

where ϕ_r is the normalized curvature at the reversed loading point. The constants b_r, c_r, d_r, and f_r in Eqs. (5.24) to (5.27) are determined by using the axial load at the point of curvature reversal.

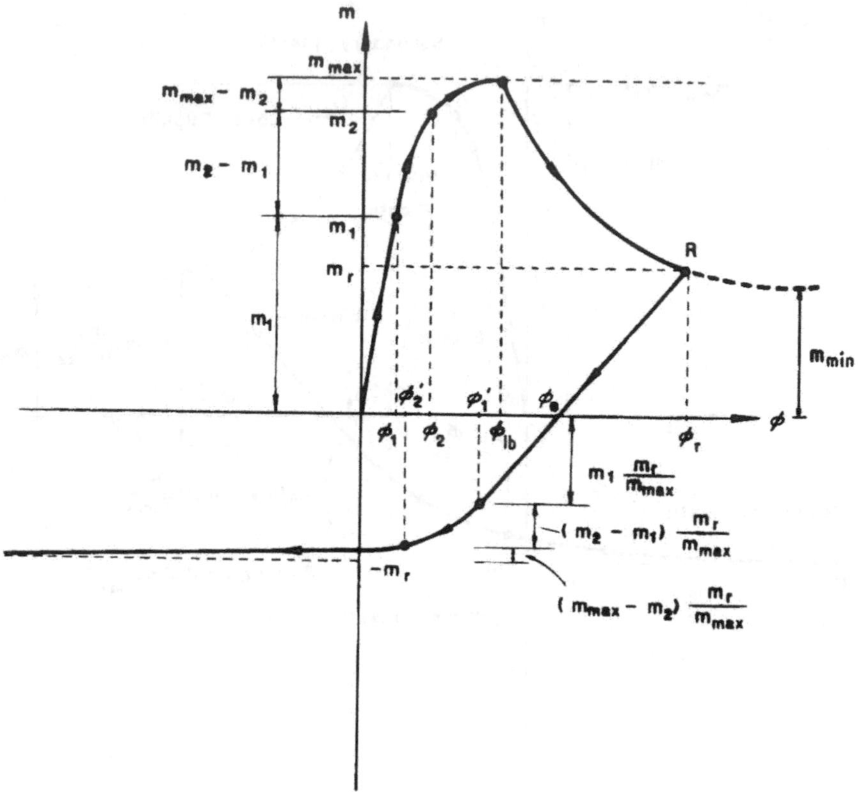

FIGURE 5.6 Symmetric approximation of cyclic M-P-Φ curve.

Reversed Loading — Symmetric curve

For the symmetric approximation, the unloading branch up to the zero moment is always elastic. From the zero moment point onward, the reversed curve is identical to the initial loading curve. To obtain the expressions for the case in which unloading starts in the post-local-buckling branch, the boundary values of moment are multiplied by the factor (m_r/m_{max}) and the boundary values of curvature are multiplied by the factor (m_{max}/m_r) (Fig. 5.6). The resulting reversed moment-curvature expressions are (Sohal and Chen, 1988b; Chen and Sohal, 1988):

For $\phi'_1 \leq \phi \leq \phi_r$

$$m = a\left(\frac{m_r}{m_{\max}}\right)^2(\phi - \phi_0) \tag{5.28}$$

For $\phi'_2 \leq \phi \leq \phi'_1$

$$m = -b\left(\frac{m_r}{m_{max}}\right) + \frac{c\sqrt{(m_r/m_{max})}}{\sqrt{\phi_0 - \phi}} \tag{5.29}$$

For $\phi \leq \phi'_2$

$$m = -m_{pc}\left(\frac{m_r}{m_{max}}\right) + \frac{f(m_{max}/m_r)}{[\phi_0 - \phi + d(m_{max}/m_r)]^2} \tag{5.30}$$

in which

$$\phi_0 = \phi_r - \frac{m_r}{a(m_r/m_{max})^2} \tag{5.31}$$

$$\phi'_1 = \phi_0 - \left(\frac{m_{max}}{m_r}\right)\phi_1 \tag{5.32}$$

$$\phi'_2 = \phi_0 - \left(\frac{m_{max}}{m_r}\right)\phi_2 \tag{5.33}$$

The constants a, b, c, d, and f are the same as given by Eqs. (5.17) to (5.21).

Reversed Loading—Elastic-Perfectly Plastic Approximation

This is the most simplified approximation for the reversed loading branch of the M-P-Φ curve. In this approximation, the unloading branch is assumed either linearly elastic or perfectly plastic. To obtain the expressions for the case in which the unloading starts in the post-local-buckling branch, the plastic moment capacity m_{pc} for the reversed loading is reduced by the factor (m_r/m_{max}) and the slope of the elastic branch is reduced by the factor $(m_r/m_{max})^2$ (Fig. 5.7). The resulting reversed moment-curvature expressions are (Sohal and Chen, 1988b; Chen and Sohal, 1988):

For $\phi'_1 \leq \phi \leq \phi_r$

$$m = a\left(\frac{m_r}{m_{max}}\right)^2(\phi - \phi_0) \tag{5.34}$$

For $\phi \leq \phi'_1$

$$m = -m_{pc}\left(\frac{m_r}{m_{max}}\right) \tag{5.35}$$

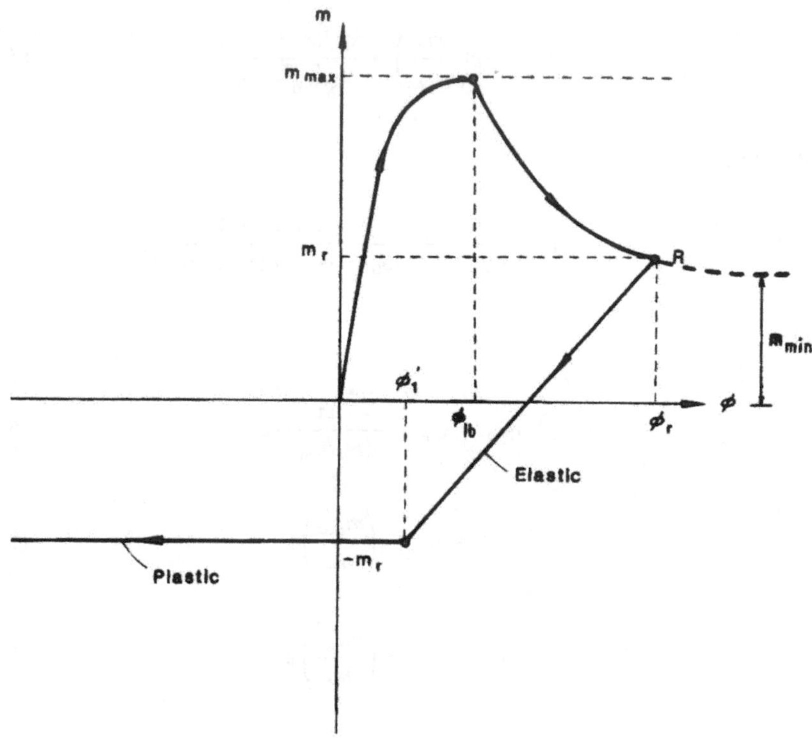

FIGURE 5.7 Elastic–perfectly plastic approximation of cyclic M-P-Φ curve.

in which

$$\phi_0 = \phi_r - \frac{m_r}{a(m_r/m_{\max})^2} \tag{5.36}$$

$$\phi_1' = \phi_0 - \frac{m_{pc}}{a(m_r/m_{\max})^2} \tag{5.37}$$

5.4 LOAD-DEFLECTION ANALYSIS OF MEMBERS

Axial Load–Lateral Defection expressions for a beam-column in various regimes of previously described M-P-Φ relationships are obtained by a modified assumed deflection method. In this section, the modified assumed deflection method is first briefly described. Then, expressions for elastic, primary yielded, secondary yielded, and post-local-buckling regimes are presented. At the end, expressions for three types of approximations for reversed loading branch are reported.

5.4.1 Modified Assumed Deflection Method

In this method, we need to consider equilibrium of external loads and internal resistance at only one critical section, which leads to analytical expressions of the load-deflection relation. These analytical expressions drastically reduce the computational time required for tracing the load-deflection and load-shortening relationships. The method has special significance because it does not require hinge length to treat the softening branch of the M-P-Φ curve.

The deflected shape for the beam columns is assumed to be the elastic deflected shape in the pre-buckling region. In the post-buckling region, the deflected shape is assumed to vary smoothly from the elastic deflected shape to the failure mechanism shape. The deflection W, at a given distance x, from the end of a beam-column can be written as (Sohal and Chen, 1988a; Chen and Sohal, 1988):

$$W = W_m[f_w F_e + (1 - f_w)F_m] \tag{5.38}$$

in which W_m is the maximum deflection at mid-span of the beam-column; F_e is a function representing the elastic deflected shape; F_m is a function representing the failure mechanism shape; and f_w is a weighting factor. The elastic deflection shape function F_e, for a pin-ended beam-column under axial load, can be written as

$$F_e = \sin \frac{\pi x}{L} \tag{5.39}$$

and for a fixed-ended beam-column it can be written as

$$F_e = \frac{1}{2}\left(1 - \cos \frac{2\pi x}{L}\right) \tag{5.40}$$

in which L is the length of the beam-column.

The failure mechanism shape for a beam-column is close to a bilinear shape (Fig. 5.8), and can be simulated by assuming the curvature distribution as follows:

$$\phi = \frac{\phi_m}{1 + \left(\dfrac{x/L - 1/2}{\beta_1}\right)^2} \tag{5.41}$$

in which ϕ_m is the maximum curvature at mid-span and β_1 is a parameter representing the spread of curvature at failure (Fig. 5.8). The failure mechanism shape function corresponding to the curvature distribution given by Eq. (5.41) can be integrated and written as

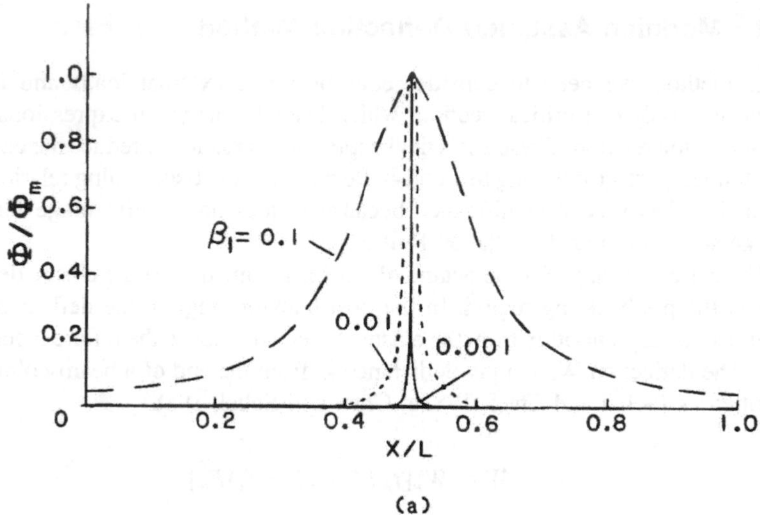

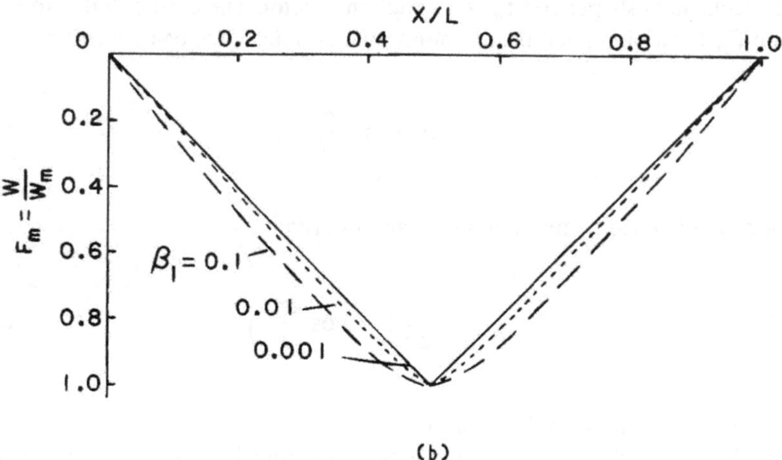

FIGURE 5.8 Distribution of curvature and displacement along beam-columns at failure: (a) curvature; (b) displacement.

$$F_m = 1 - \frac{\left(\dfrac{x - L/2}{\beta_1 L}\right)\tan^{-1}\left(\dfrac{x - L/2}{\beta_1 L}\right) - \dfrac{1}{2}\ln\left[1 + \left(\dfrac{x - L/2}{\beta_1 L}\right)^2\right]}{\left(\dfrac{1}{2\beta_1}\right)\tan^{-1}\left(\dfrac{1}{2\beta_1}\right) - \dfrac{1}{2}\ln\left(1 + \dfrac{1}{4\beta_1^2}\right)} \tag{5.42}$$

The curvature distribution and the failure mechanism function for $\beta_1 = 0.1$, 0.01, and 0.001, are shown in parts (a) and (b) of Fig. 5.8. The weighting factor f_w is equal to one in the pre-buckling regime. In the post-buckling regime, f_w decreases exponentially with the decrease in the load-carrying capacity of the beam-column and can be written as follows:

$$f_w = \exp\left[\left(1 - \frac{P_{max}}{P}\right)\beta_2\right] \tag{5.43}$$

where β_2 is a parameter representing the rate of change of the deflected shape. The P_{max} in Eq. (5.43) is the maximum axial capacity of the beam-column and is calculated by the average flow moment method as proposed previously by Chen and Atsuta (1972).

In order to determine the parameters β_1 and β_2, the load-deflection and load-shortening curves of several beam-columns obtained from the proposed deflected shape were compared with those obtained from Newmark's numerical integration method (Chen and Atsuta, 1976), a more rigorous method. For beam-columns with an effective slenderness ratio (KL/r) between 40 and 120, the most suitable values of β_1 and β_2 were found to be 0.04 and 0.3, respectively, where K = the effective length factor; L = the length of the beam-column; and r = the radius of gyration of the section. These values of β_1 and β_2 will be used from here on.

In the modified assumed deflection method, the load-deflection relationship is obtained by equating external moment to internal moment. In order to obtain the internal moment in terms of a given deflection, a relationship between the curvature and the displacement is needed. This relationship can be obtained by a double differentiation of the assumed deflected shape given by Eq. (5.38). For pin-ended beam-columns, the resulting relationship at mid-span for $\beta_1 = 0.04$ reduces to

$$\phi_P = W_m \frac{P_{cr}}{M_y} [f_w + 3.9(1 - f_w)] \tag{5.44}$$

in which P_{cr} is Euler's buckling load. For fixed-ended beam-columns, the resulting relationship is

$$\phi_P = \frac{W_m}{2} \frac{P_{cr}}{M_y} [f_w + 1.95(1 - f_w)] \tag{5.45}$$

where the parameter f_w takes the value 1 in the pre-buckling regime, and is given by Eq. (5.43) in the post-buckling regime.

5.4.2 Relation for Elastic Regime

The mid-span deflection W_O due to end moments and/or lateral load is calculated from the conventional beam theory. The additional deflection W_m due to axial load is determined by considering the moment equilibrium at the critical section.

For pin-ended beam-columns, the bending moment at the mid-span induced by external loads is

$$M_{ext} = M_{MQ} + P(W_i + W_o + W_m) \tag{5.46}$$

where $M_{MQ} = M_O + QL/4$ = bending moment due to the end moment M_O and/or the lateral load, Q; W_i = initial imperfection at mid-span; and P = axial load. In the nondimensionalized form, we have

$$m_{ext} = m_{MQ} + m_i + m_{oP} + \frac{P}{M_y} W_m \tag{5.47}$$

where $m_{ext} = M_{ext}/M_y$; $m_{MQ} = M_{MQ}/M_y$, $m_i = PW_i/M_y$; and $m_{op} = PW_O/M_y$. For fixed-ended beam-columns, nondimensionalized external moment can be expressed as

$$m_{ext} = m_{MQ} + m_i + m_{oP} + \frac{P}{M_y}\frac{W_m}{2} \tag{5.48}$$

In the elastic range, the internal resisting moment at mid-span is

$$m_{int} = a\phi_m = a(\phi_{MQ} + \phi_P) \tag{5.49}$$

where $m_{int} = M_{int}/M_y$; a = stiffness constant; $\phi_m = \Phi_m/\Phi_y$; and $\phi_{MQ} = m_{MQ}$. For pin-ended beam-columns, by equating Eqs. (5.48) and (5.49) and by substituting ϕ_P from Eq. (5.44), for $W_m \leq W_{1P}$, the mid-span deflection W_m can be expressed (Sohal and Chen, 1988a; Chen and Sohal, 1988) as

$$W_m = \frac{m_{MQ}(1 - a) + m_i + m_{oP}}{\left(a\, S_P - \dfrac{P}{P_{cr}}\right)\dfrac{P_{cr}}{M_y}} \tag{5.50}$$

in which

$$S_P = f_w + 3.9\,(1 - f_w) \tag{5.51}$$

and W_{1P} is the elastic limit deflections for a pin-ended beam-column. The following expression for this elastic limit deflection is obtained by substituting ϕ_P from Eq. (5.44) and $\phi_m = \phi_1$ in Eq. (5.49).

$$W_{1P} = (\phi_1 - m_{MQ})\left(\frac{M_y}{P_{cr}}\right)\frac{1}{S_P} \tag{5.52}$$

5.4.3 Relation for Primary Yield Regime

Similarly, by equating the external bending moment to the internal resisting moment in the primary yield range, the expressions for the mid-span displacement are obtained. For a pin-ended beam-column, for $W_{1P} \leq W_m \leq W_{2P}$ and $W_m \leq W_{1bP}$, the resulting expression is

$$\left[m_{MQ} + m_i + m_{oP} - b + \frac{P}{M_y}W_m\right]^2\left[m_{MQ} + \frac{P_{cr}}{M_y}W_mS_P\right] = c^2 \tag{5.53}$$

in which W_{2P} is the primary yield limit deflection for a pin-ended beam-column and W_{1bP} is the deflection at the start of the local buckling for a pin-ended beam-column. W_{2P} and W_{1bP} can be expressed as

$$W_{2P} = (\phi_2 - m_{MQ})\left(\frac{M_y}{P_{cr}}\right)\frac{1}{S_P} \tag{5.54}$$

$$W_{1bP} = (\phi_{1b} - m_{MQ})\left(\frac{M_y}{P_{cr}}\right)\frac{1}{S_P} \tag{5.55}$$

5.4.4 Relation for Secondary Yield Regime

For pin-ended beam-columns, for $W_{2P} \leq W_m \leq W_{1bP}$, we have

$$\left[m_{pc} - m_{MQ} - m_i - m_{0P} - \frac{P}{M_y}W_m\right]\left[m_{MQ} + d + \frac{P_{cr}}{M_y}W_mS_p\right]^2 = f \tag{5.56}$$

5.4.5 Relation for Post-Local-Buckling Regime

For pin-ended beam-columns, for $W_m \geq W_{1bP}$, we have

$$(m_{max} - m_{\min})\exp\left[\frac{\left(\frac{P_{cr}}{M_y}W_MS_P - \phi_{1b}\right)EI_{\max}}{m_{\max} - m_{\min}}\right] + m_{\min} = m_{MQ} + m_i + m_{oP} + \frac{P}{M_y}W_m \tag{5.57}$$

Equations (5.50), (5.53), (5.56), and (5.57) are used to determine both the pre- and post-maximum deflections of a beam-column with a given value of axial thrust. Equations (5.50), (5.53), and (5.56) can be solved directly without any iteration. However, a few iterations are required to solve Eq. (5.57). Equations (5.50) to (5.57) can also be used for fixed-ended beam-columns by simply replacing the mid-span displacements by half their values and by replacing S_P by S_F. The constant S_F defines the effect of changing deflected shape for fixed-ended beam-columns and can be written as

$$S_F = f_w + 1.95(1 - f_w) \tag{5.58}$$

5.4.6 Relation for Reversed Loading Regime

It is shown that the decrease in deflection due to reversal in axial loading can be assumed to have a sinusoidal shape (Sohal and Chen, 1987a). The load-deflection relationship for the reversed loading regime can therefore be obtained by simple modification of those for the monotonic loading as follows.

To determine the internal resisting moment during the reversed loading for a given deflection, we need a relationship between the deflection and the curvature. For the pin-ended beam-column, this relationship is obtained by a simple modification of Eq. (5.44) as

$$\phi - \phi_r = (W_m - W_r)\frac{P_e}{M_y}S_{Pr} \tag{5.59}$$

in which P_e is the Euler's critical load for the beam-column; ϕ is the current curvature at the mid-span of the beam-column, normalized with respect to the yield curvature Φ_y; ϕ_r is the curvature at the mid-span at the reversed loading point, normalized with respect to Φ_y; and S_{Pr} is a constant for the pin-ended beam-column, defining the effect of the changing deflected shape and can be written as

$$S_{Pr} = f_{wu} + 3.9(1 - f_{wu}) \tag{5.60}$$

in which the weighting factor f_{wu} can be assumed to vary linearly from one at the start of unloading regime to f_{wr} at the tensile yield load, where f_{wr} is the value of f_w at the reversed loading point. The internal resisting moment for a given curvature can be determined from one of the three types of the proposed closed-form approximations for the cyclic M-P-Φ relations. The load-deflection corresponding to these three approximations is derived as follows.

Doubly Symmetric Approximation

In the elastic unloading regime, the internal resisting moment can be written as

$$m_{int} = m_r + aR_L^2(\phi - \phi_r) \tag{5.61}$$

where m_{int} is the current internal resisting moment of the section, normalized with respect to M_y; m_r is the internal resisting moment of the section at the reversal point, normalized with respect to M_y; $R_L = m_r/m_{max}$ is a factor representing the effect of local buckling of the cross section; and m_{max} is the normalized maximum moment that could be applied to the section with the axial force equal to that at the reversal point.

For the pin-ended beam-column, substituting S_{Pr} from Eq. (5.60) in Eq. (5.59), using this equation to substitute the value of $(\phi - \phi_r)$ in Eq. (5.61), then equating resulting m_{int} to m_{ext} from Eq. (5.47) and simplifying for the mid-span deflection W_m, the following expression, for $W_m \geq W_{1P}$, is obtained (Sohal and Chen, 1987a; Chen and Sohal, 1988)

$$W_m = \frac{m_{MQ} + m_i + m_o P - m_r + W_r S_{Pr}(P_e/M_y)aR_L^2}{S_{Pr}(P_e/M_y)aR_L^2 - P/M_y} \tag{5.62}$$

where W_{1P} is the elastic limit deflection for the pin-ended beam-column. The following expression for this elastic limit is obtained by substituting ϕ in Eq. (5.59) by the elastic limit curvature for the reversed loading case.

$$W_{1P} = W_r - A_m \frac{\phi_{1r}}{R_L h} \frac{M_y}{P_e} \frac{1}{S_{Pr}} \tag{5.63}$$

Similarly, in the primary yield regime, by equating the external moments to the internal resistance, the following equation for the mid-span deflection for $W_{1P} \geq W_m \geq W_{2P}$ can be derived.

$$\left[m_{MQ} + m_i + m_{oP} + A_m b_r R_L - m_r + W_m \frac{P}{M_y}\right]^2 \times \left[(W_r - W_m)\frac{P_e}{M_y} S_{Pr}\right] = A_m^3 C_r^2 R_L \tag{5.64}$$

in which W_{2P} is the primary yield limit deflection in

$$W_{2P} = W_r - A_m \frac{\phi_{2r}}{R_L} \frac{M_y}{P_e} \frac{1}{S_{Pr}} \tag{5.65}$$

In the secondary yield regime, for $W_m \leq W_{2P}$, the equation for W_m is

$$\left[A_m m_{pcr} R_L + m_{MQ} + m_i + m_{oP} - m_r + W_m \frac{P}{M_y}\right] \times \left[(W_r - W_m)\frac{P_e}{M_y} S_{Pr} + \frac{A_m d_r}{R_L}\right]^2 = A_m^3 \frac{f_r}{R_L} \quad (5.66)$$

A_m in Eqs. (5.63) to (5.66) is determined from the following equation

$$A_m = \frac{m_{pcr} + m_{pc}}{m_{pcr}} \quad (5.67)$$

in which m_{pcr} is the normalized plastic moment capacity of the section reduced for the axial load at the point of curvature reversal.

Symmetric Approximation

In the elastic unloading regime, the use of the symmetric approximation for the internal moment resistance results in the same load-deflection equation as that for the doubly symmetric approximation, Eq. (5.62). However, for this case, the elastic limit deflection W_{1P} will be greater than that for the doubly symmetric approximation and can be written as

$$W_{1P} = W_r - \left(\frac{m_r}{aR_L^2} + \frac{\phi_1}{R_L}\right)\frac{M_y}{P_e}\frac{1}{S_{Pr}} \quad (5.68)$$

In the unloading primary yield regime, for $W_{1P} \geq W_m \geq W_{2P}$, the following equation for the mid-span displacement W_m can be derived as

$$\left[m_{MQ} + m_i + m_{OP} + bR_L + W_m \frac{P}{M_y}\right]^2 \times \left[(W_r - W_m)\frac{P_e}{M_y} S_{Pr} - \frac{m_r}{aR_L^2}\right] = c^2 R_L \quad (5.59)$$

in which the primary yield limit is given by

$$W_{2P} = W_r - \left(\frac{m_r}{aR_L^2} + \frac{\phi_2}{R_L}\right)\frac{M_y}{P_e}\frac{1}{S_{Pr}} \quad (5.70)$$

In the secondary yield regime, for $W_m \geq W_{2P}$, the equation for W_m is

$$\left[m_{pc}R_L + m_{MQ} + m_i + m_{oP} + W_m \frac{P}{M_y}\right] \times\left[(W_r - W_m)\frac{P_e}{M_y}S_{Pr} - \frac{m_r}{aR_L^2} + \frac{d}{R_L}\right]^2 = \frac{f}{R_L} \quad (5.71)$$

Elastic–Perfectly Plastic Approximation

In the elastic unloading regime, the use of the elastic–perfectly plastic approximation for internal moment resistance also results in the same load-deflection equation as that for the doubly symmetric approximation. However, elastic limit deflection W_{1P}, for this case, will be greater than that for the doubly symmetric and the symmetric approximation and can be written as

$$W_{1P} = W_r - \frac{(m_r + m_{pc}R_L)}{aR_L^2}\frac{M_y}{P_e}\frac{1}{S_{Pr}} \quad (5.72)$$

In the plastic regime, for $W_m \geq W_{1P}$, the load-deflection relationship can be expressed by the following linear relationship:

$$W_m = -\frac{(m_{pc}R_L + m_{MQ} + m_i + m_{oP})M_y}{P} \quad (5.73)$$

The above three types of load-deflection equations derived for the pin-ended beam-column can be used for the fixed-ended beam-column by replacing mid-span displacements by half their values and by replacing S_{Pr} by S_{Fr}. The constant S_{Fr} defines the effect of the changing deflected shape for the fixed-ended beam-column and can be written as

$$S_{Fr} = f_{wu} + 1.95(1 - f_{wu}) \quad (5.74)$$

5.5 LOAD-SHORTENING ANALYSIS OF MEMBER

The total axial shortening of a beam-column consists of two parts: (1) the axial shortening due to the axial strain and (2) the axial shortening due to the lateral deflection. Both of the shortenings are calculated by dividing the beam-column into a number of segments (N).

The axial shortening due to the axial strain is calculated by (Chen and Han, 1985)

$$\Delta_s = \sum_{i=1}^{N} \left(\frac{L}{N}\right)\epsilon_{oi} \tag{5.75}$$

in which ϵ_{oi} is the axial strain in segment i. The axial strain ϵ_{oi} is determined further in two additional steps. First, curvature is determined from the known mid-span deflection and weighting factor f_w. Then this curvature and the given axial thrust are used to determine the axial strain from the thrust-curvature-strain relationship.

The axial shortening due to the lateral deflection is calculated by (Chen and Han, 1985)

$$\Delta_g = \sum_{i=1}^{N} \left[\frac{L}{N} - \sqrt{\left(\frac{L}{N}\right)^2 - (W_{1i} - W_{2i})^2}\right] \tag{5.76}$$

in which W_{1_i} and W_{2_i} are the total deflections at the two ends of the segment i. The deflections W_{1_i} and W_{2_i} in Eq. (5.76) are determined by using Eq. (5.38). Note that in the determination of the axial shortening due to the axial strain, we have assumed that the thrust-axial strain (p − ϵ) relationship corresponding to a given curvature would be unaffected by the local buckling of the cross section.

5.6 COMPUTER IMPLEMENTATION

The theory described in the five preceding sections is implemented to develop a computer program, "BRACE." It can be used to obtain a one-cycle load-deflection and a one-cycle load-shortening relationship of a member including the effects of local buckling of a tubular cross section. The program consists of a main program and 25 subroutines. The arrangement of the program is shown in Fig. 5.9. The function of the main program and its subroutines are explained in the following.

main: to control all the subroutines

abcf: to determine constants a, b, c, d, and f for given boundaries of the three regimes of the moment-curvature curve

admcol: to compute deflection and shortening corresponding to each value of axial load

bklphi: to determine the local buckling curvature for a given value of axial load

closed: to determine the moment resistance of a section in the post-local-buckling branch by using the closed form equation for softening branch of moment-curvature curve

cubic: to solve cubic equation (used in "rotcub")

datage: to read input data, generate preliminary data, and echo the read and generated data

dsymm: to determine lateral deflection for a given value of axial load in the reversed

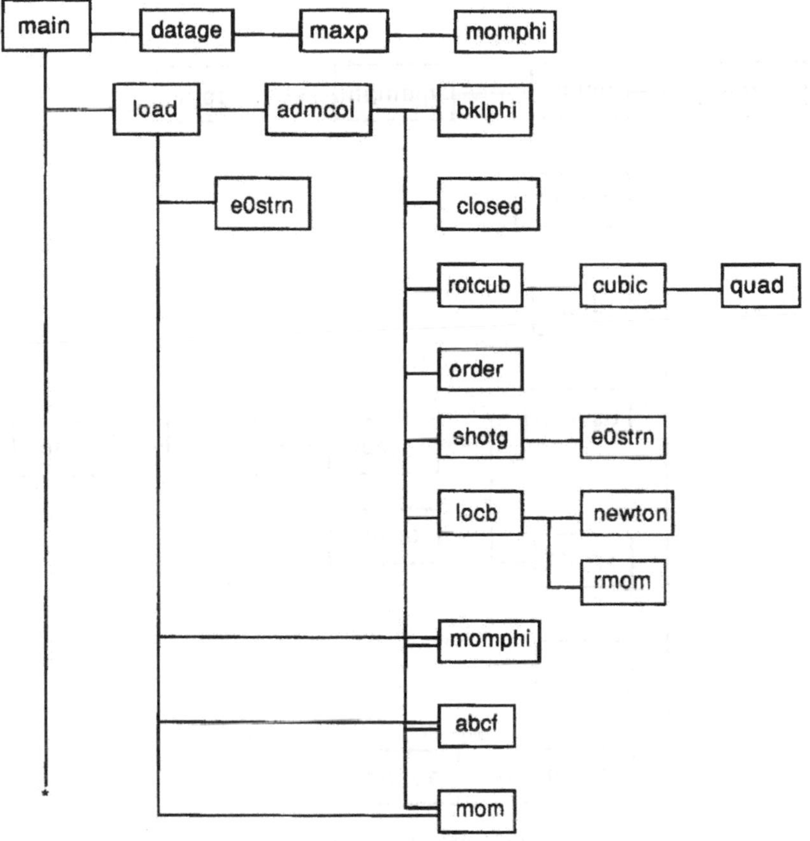

FIGURE 5.9 Structure of the program "Brace."

loading regime by using a doubly symmetric approximation for reversed loading branch of moment-curvature curve

e0strn: to determine axial strain for given values of axial load and curvature

load: to compute load-deflection and load-shortening curve in the loading regime

locb: to compute the shape of a locally buckled cross section

maxp: to estimate the ultimate axial load of the brace

mom: to determine moment for given curvature and constants a, b, c, d, f, ϕ_1, ϕ_2, and m_{pc}

momphi: to determine boundaries of the three regimes of moment-curvature curve for a given value of axial load

newton: Newton-Raphson method to solve a non-linear equation (used in 'locb')

order: to arrange 3 variables in ascending order

out: to print and prepare plot files of load-deflection or load-shortening relationships

pelas: to determine axial load up to which the brace will stay in the elastic range in the reversed loading regime

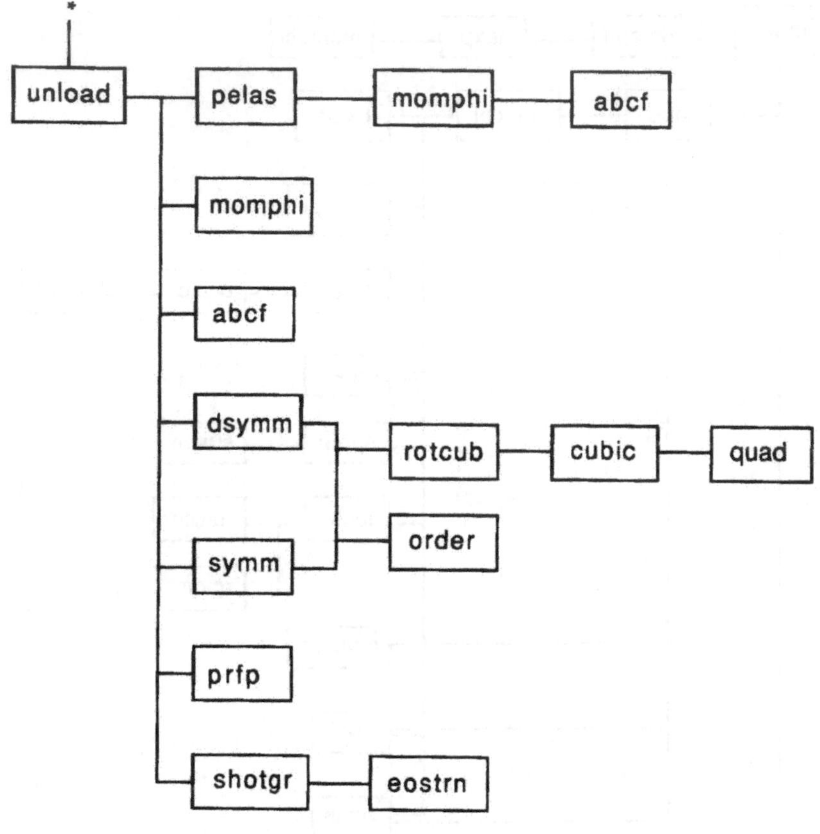

FIGURE 5.9 *Continued*

prfp: to determine lateral deflection for a given value of axial load in the reversed loading regime by using the elastic perfectly plastic approximation for reversed loading branch of moment-curvature curve

quad: to determine roots of a quadratic equation (used in "cubic")

rmom: to compute moment resistance of a locally buckled section

rotcub: to determine roots of a load-deflection equation (cubic-equation)

shotg: to compute shortening of the brace in the loading regime for a given value of axial load and lateral deflection

shotgr: to compute shortening of the brace in the reversed loading regime for a given value of axial load and lateral deflection

symm: to determine lateral deflection for a given value of axial load in the reversed loading regime by using symmetric approximation for reversed loading branch of moment-curvature curve

unload: to compute load-deflection and load shortening curve in the unloading/reversed loading regime

5.7 USER'S MANUAL FOR THE PROGRAM BRACE

The input data consists of seven lines, all of which have a free format. The description of the input data is as follows:

```
Line 1 (free format)
     ncase: number of cases to be analyzed
     nout: indicator for selection of output method
            1 = print the coordinates of points on curves
            2 = write the output in files 'lodef' and 'losh'
                which can be directly used with the
                graphics softwares to plot the desired
                load-deflection and load-shortening
                relationships, respectively.
            3 = both print and write in plot files

Line 2 (free format)
     irs: indicator for selection of boundary conditions
          and type of moment-curvature curve
          + = pin-ended
          - = fixed-ended
          abs(irs) = 100 = moment-curvature without
                           residual stresses
          abs(irs) = 101= moment-curvature with
                           residual stresses
          abs(irs) = 111 = moment-curvature with residual
                                 stresses and external
                                 pressure
          abs(irs) > 200 = local buckling included
          abs(irs) > 300 = closed form expressions for
                                 the softening branch of moment-
                                 curvature curve
    nond: indicator to specify whether output data should
          be non-dimensionalized or not
          1 = non-dimensionalize the output data with
              respect to corresponding yield values
          0 = print and/or plot without non-
              dimensionalization

Line 3 (free format)
      d: diameter of tube
      t: thickness of tube
      e: Young's modulus of material of tube
   sigy: yield stress of material of tube
```

Line 4 (free format)

- al: length of brace in feet
- wi: initial imperfection as % of length of brace
- nel: number of elements to be used for calculating the shortening

Line 5 (free format)

- pp: initial value of axial load
- pdelt: increment in axial load
- np: expected number of increments in axial load to reach maximum capacity of brace
- emos: end moments normalized by yield moments
- qs: lateral load at the mid-span of brace normalized with the yield lateral load

Line 6 (free format)

- wmr: lateral deflection at reversal normalized by yield lateral deflection under pure moment, controls load reversal only when it is positive
- smr: shortening at reversal normalized by yield axial shortening under pure axial load, controls load reversal only when it is positive
- smt: elongation at reversal in tensile region normalized by yield axial elongation under pure axial load

Line 7 (free format)

- nin: number of increments to be used for unloading regime (from compressive axial load at reversal in post-buckling/post-local-buckling regime to tensile yield axial load)
- irev: indicator to select the approximation to be used for reversed loading branch of moment-curvature relationship
- irev: 1 = doubly symmetric (recommended for use)
- irev: 2 = symmetric
- irev: 3 = elastic-perfectly plastic
- ishp: indicator to select the deflected shape to be used during the unloading regime
- ishp: 1 = unloading deflected shape linearly varying from sinusoidal at reversal point to mechanism at tensile yield point (recommended for use)
- ishp: 2 = unloading deflected shape sinusoidal
- ishp: 3 = unloading deflected shape mechanism shape

5.8 SOLUTION OF SAMPLE EXAMPLES

To demonstrate the use of the program, the input data and the resulting plots are presented in what follows for five examples. The first example compares the load-shortening relationships of a pin-ended column obtained with and without local buckling effects. The second and third examples, respectively, study the effects of D/t and KL/r on load-shortening behavior of fixed-ended columns. The fourth and fifth examples, respectively, study the effects of end moments and lateral loads on the load-shortening behavior of beam-columns. For the first example, input data, output data, load-shortening plots and description of plots are all presented. For the other examples, the large printout of load-shortening results are skipped and only plots of load-shortening curves along with their description and input data are provided.

5.8.1 Effects of Local Buckling on a Pin-Ended Column

The one-cycle load-shortening relationships of a pin-ended column, with the slenderness ratio of 80 and diameter-to-thickness ratio of 48, including local buckling effects, is compared with that without local buckling effects. The doubly symmetric approximation is used for the moment-curvature relationship for both cases. The input data and output data for this example are listed in the following

Input Data

```
2         1
311       1
15.       0.3125    30000.     36.
34.627    0.001     4
10.       10.       200        0.        0.
-1.0      3.0       2
50        1         1
111       1
15.       0.3125    30000.     36.
34.627    0.001     4
10.0      10.       200        0.        0.
-1.0      3.0       2.0
50        1         1
```

Output Data

```
total number of cases to be analyzed = 2
nout = 1
 input and preliminary data for case 1
    irs =  311          nond =   1
    d =   15.000    t =   .313    e =   30000.00
```

```
sigy =    36.00
length =  415.52   wi =  .4155   nel = 4
moi =   389.00   area = 14.42    rad. of gyr =   5.194
kl/r = 80.00
phiy =      .1600E-03
el.buck.load =   667.08    yld.axial load =   519.10
my =     1867.20     mp =    2531.62      sh fac = 1.356
pp =     10.00     pdelt = 10.00     np =  200
em0s =  .00      qs =  .00
w0mq =   .000      wyild = 2.799      syild = .499
max. axial capacity =   400.98
wmr = -2.80     smr =     1.50     smt =   1.00
nin = 50      irev =  1        ishp =  1
axial load at reversal =      48.98
m/my at reversal = .417
```

Input and Preliminary Data for Case 2

```
irs = 111          nond =  1
d =  15.000    t =    .313      e =  30000.00
sigy =    36.00
length =  415.52   wi =  .4155     nel = 4
moi =   389.00   area =  14.42   rad. of gyr =  5.194
kl/r = 80.00
phiy =      .1600E-03
el.buck.load =      667.08   yld.axial load =   519.10
my =     1867.20      mp =   2531.62      sh fac = 1.356
pp =     10.00      pdelt = 10.00      np =  200
em0s =  .00        qs =  .00
w0mq =   .000      wyild =  2.799     syild =   .499
max. axial capacity =   400.98
wmr =  -2.80     smr =    1.50     smt =  1.00
nin =   50     irev = 1        ishp =  1
axial load at reversal =     162.77
m/my at reversal =      1.101
```

	ax. load	defl	short
case = 1			
1	-.02	-.00	-.02
2	-.04	-.00	-.04
3	-.06	-.01	-.06
4	-.08	-.01	-.08
5	-.10	-.01	-.10
6	-.12	-.01	-.12
7	-.13	-.02	-.14
8	-.15	-.02	-.15

9	−.17	−.02	−.17
10	−.19	−.03	−.19
11	−.21	−.03	−.21
12	−.23	−.03	−.23
13	−.25	−.04	−.25
14	−.27	−.04	−.27
15	−.29	−.04	−.29
16	−.31	−.05	−.31
17	−.33	−.05	−.33
18	−.35	−.05	−.35
19	−.37	−.06	−.37
20	−.39	−.06	−.39
21	−.40	−.07	−.41
22	−.42	−.07	−.43
23	−.44	−.08	−.45
24	−.46	−.08	−.47
25	−.48	−.09	−.48
26	−.50	−.09	−.50
27	−.52	−.10	−.52
28	−.54	−.11	−.54
29	−.56	−.11	−.56
30	−.58	−.12	−.58
31	−.60	−.13	−.60
32	−.62	−.14	−.62
33	−.64	−.15	−.64
34	−.65	−.15	−.66
35	−.67	−.16	−.68
36	−.69	−.17	−.70
37	−.71	−.18	−.72
38	−.73	−.20	−.74
39	−.73	−.41	−.77
40	−.71	−.51	−.77
41	−.69	−.60	−.78
42	−.67	−.70	−.79
43	−.65	−.81	−.81
44	−.64	−.93	−.84
45	−.62	−1.05	−.87
46	−.60	−1.22	−.93
47	−.58	−1.37	−.99
48	−.56	−1.52	−1.06
49	−.54	−1.58	−1.07
50	−.52	−1.65	−1.08
51	−.50	−1.72	−1.10
52	−.48	−1.80	−1.12
53	−.46	−1.87	−1.14

54	−.44	−1.95	−1.17
55	−.42	−2.03	−1.19
56	−.40	−2.10	−1.21
57	−.39	−2.21	−1.25
58	−.37	−2.32	−1.30
59	−.35	−2.44	−1.34
60	−.33	−2.55	−1.39
61	−.31	−2.68	−1.44
62	−.29	−2.81	−1.49
63	−.27	−2.95	−1.55
64	−.25	−3.10	−1.61
65	−.23	−3.27	−1.68
66	−.21	−3.45	−1.75
67	−.19	−3.65	−1.85
68	−.17	−3.85	−1.94
69	−.15	−4.11	−2.07
70	−.13	−4.43	−2.25
71	−.12	−4.83	−2.51
72	−.10	−5.39	−2.94
73	−.09	−5.45	−3.00
74	−.09	−5.32	−2.73
75	−.07	−3.90	−1.64
76	−.05	−3.22	−1.24
77	−.03	−2.82	−1.03
78	−.01	−2.56	−.91
79	.02	−2.38	−.82
80	.04	−2.24	−.75
81	.06	−2.13	−.69
82	.08	−2.05	−.64
83	.10	−1.98	−.60
84	.12	−1.85	−.54
85	.15	−1.73	−.49
86	.17	−1.62	−.44
87	.19	−1.53	−.40
88	.21	−1.44	−.36
89	.23	−1.35	−.32
90	.26	−1.27	−.28
91	.28	−1.20	−.25
92	.30	−1.13	−.21
93	.32	−1.07	−.18
94	.34	−1.01	−.15
95	.37	−.94	−.12
96	.39	−.88	−.08
97	.41	−.82	−.05
98	.43	−.75	−.03

```
     99          .45          -.70          .00
    100          .47          -.64          .03
    101          .50          -.59          .06
    102          .52          -.54          .09
    103          .54          -.49          .11
    104          .56          -.44          .14
    105          .58          -.39          .16
    106          .61          -.36          .19
    107          .63          -.31          .21
    108          .65          -.27          .24
    109          .67          -.23          .26
    110          .69          -.20          .29
    111          .72          -.16          .31
    112          .74          -.13          .34
    113          .76          -.10          .36
    114          .78          -.07          .38
    115          .80          -.05          .40
    116          .82          -.02          .43
    117          .85           .01          .45
    118          .87           .03          .47
    119          .89           .05          .49
    120          .91           .07          .52
    121          .93           .09          .54
    122          .96           .11          .56
    123          .98           .13          .58
    124         1.00           .15          .60
    125         1.00           .15         2.00
    126          .00           .15         1.00
case = 2
      1         -.02          -.00         -.02
      2         -.04          -.00         -.04
      3         -.06          -.01         -.06
      4         -.08          -.01         -.08
      5         -.10          -.01         -.10
      6         -.12          -.01         -.12
      7         -.13          -.02         -.14
      8         -.15          -.02         -.15
      9         -.17          -.02         -.17
     10         -.19          -.03         -.19
     11         -.21          -.03         -.21
     12         -.23          -.03         -.23
     13         -.25          -.04         -.25
     14         -.27          -.04         -.27
     15         -.29          -.04         -.29
     16         -.31          -.05         -.31
```

17	−.33	−.05	−.33
18	−.35	−.05	−.35
19	−.37	−.06	−.37
20	−.39	−.06	−.39
21	−.40	−.07	−.41
22	−.42	−.07	−.43
23	−.44	−.08	−.45
24	−.46	−.08	−.47
25	−.48	−.09	−.48
26	−.50	−.09	−.50
27	−.52	−.10	−.52
28	−.54	−.11	−.54
29	−.56	−.11	−.56
30	−.58	−.12	−.58
31	−.60	−.13	−.60
32	−.62	−.14	−.62
33	−.64	−.15	−.64
34	−.65	−.15	−.66
35	−.67	−.16	−.68
36	−.69	−.17	−.70
37	−.71	−.18	−.72
38	−.73	−.20	−.74
39	−.73	−.41	−.77
40	−.71	−.51	−.77
41	−.69	−.60	−.78
42	−.67	−.70	−.79
43	−.65	−.81	−.81
44	−.64	−.93	−.84
45	−.62	−1.05	−.87
46	−.60	−1.22	−.93
47	−.58	−1.37	−.99
48	−.56	−1.52	−1.06
49	−.54	−1.69	−1.14
50	−.52	−1.86	−1.22
51	−.50	−2.03	−1.32
52	−.48	−2.21	−1.43
53	−.46	−2.40	−1.54
54	−.44	−2.59	−1.67
55	−.42	−2.80	−1.81
56	−.40	−3.02	−1.97
57	−.39	−3.26	−2.14
58	−.37	−3.52	−2.33
59	−.35	−3.81	−2.55
60	−.33	−4.12	−2.80
61	−.31	−4.36	−3.00

62	−.31	−4.36	−2.88
63	−.29	−4.24	−2.76
64	−.26	−4.14	−2.66
65	−.23	−4.04	−2.56
66	−.21	−3.96	−2.48
67	−.18	−3.88	−2.40
68	−.16	−3.81	−2.32
69	−.13	−3.74	−2.25
70	−.10	−3.68	−2.19
71	−.08	−3.62	−2.12
72	−.05	−3.57	−2.06
73	−.02	−3.52	−2.01
74	.00	−3.47	−1.95
75	.03	−3.43	−1.90
76	.05	−3.39	−1.85
77	.08	−3.35	−1.80
78	.11	−3.31	−1.75
79	.13	−3.25	−1.69
80	.16	−3.18	−1.63
81	.19	−3.10	−1.56
82	.21	−3.02	−1.49
83	.24	−2.94	−1.42
84	.26	−2.85	−1.35
85	.29	−2.75	−1.28
86	.32	−2.65	−1.21
87	.34	−2.55	−1.13
88	.37	−2.44	−1.06
89	.40	−2.33	−.99
90	.42	−2.22	−.92
91	.45	−2.11	−.85
92	.47	−1.98	−.79
93	.50	−1.85	−.72
94	.53	−1.71	−.65
95	.55	−1.57	−.58
96	.58	−1.43	−.52
97	.61	−1.30	−.47
98	.63	−1.16	−.41
99	.66	−1.03	−.36
100	.68	−.90	−.31
101	.71	−.78	−.27
102	.74	−.67	−.23
103	.76	−.57	−.19
104	.79	−.47	−.16
105	.82	−.37	−.13
106	.84	−.28	−.10

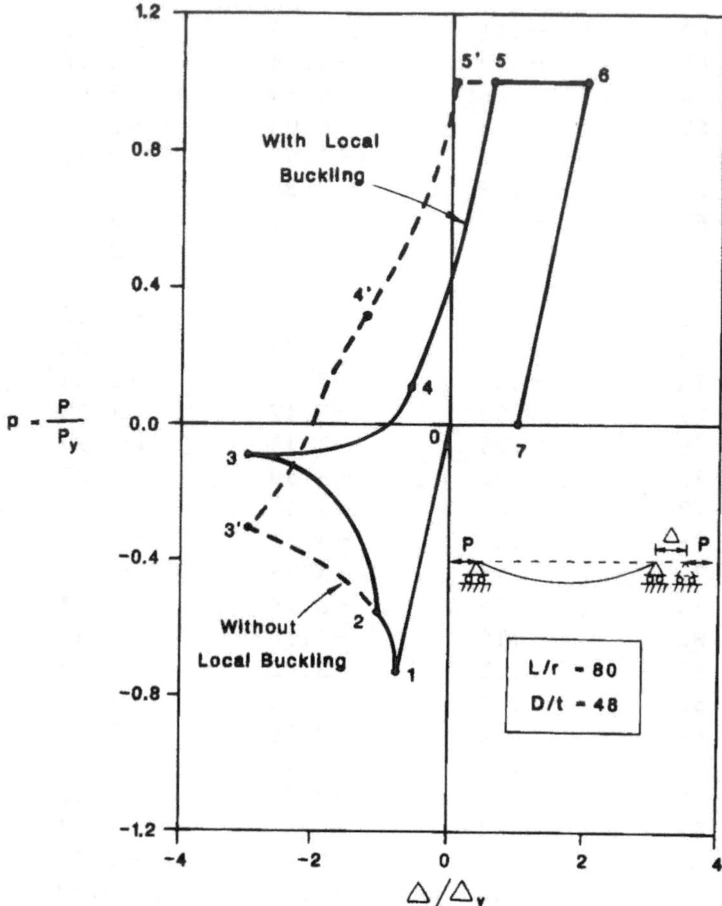

FIGURE 5.10 Effects of local buckling on the cyclic axial load-shortening relationship of a pin-ended column.

```
107        .87        -.20        -.07
108        .89        -.12        -.04
109        .92        -.05        -.01
110        .95         .02         .02
111        .97         .09         .04
112       1.00         .15         .07
113       1.00         .15        2.00
114        .00         .15        1.00
Stop—Program terminated.
```

The output data is plotted in Fig. 5.10. The solid line indicates the relationship in which the influence of local buckling of the cross section is considered and the dashed line shows the relationship in which the local buckling is not considered.

Branch 0–1 in Fig. 5.10 indicates the pre-buckling load-shortening relationship. The inelastic buckling takes place at Point 1 and there is a reduction in the load-carrying capacity of the column. At Point 2, the distortion of the cross section starts and the reduction in the load-carrying capacity of the column is accelerated. At Point 3, the axial deflection is reversed. The column is elastic in Branch 3–4, and the yielding starts at Point 4. At Point 5, the entire column is yielded and the load-carrying capacity of the column is the yield axial load for its section. There is a plastic elongation along Branch 5–6. The axial deflections are again reversed at Point 6. The Branch 6–7 indicates the elastic unloading.

For the load-shortening relationship obtained without considering the local buckling, the reversal point, the initial yield point in reversed loading regime and the fully yield point in reversed loading regime, are denoted by Points 3′, 4′, and 5′, respectively. Note that the local buckling of the cross section has significantly reduced the energy absorption capacity of the column.

5.8.2 Effects of Diameter-to-Thickness Ratio on Fixed-Ended Column

The axial load-axial shortening relationships of a fixed-ended column with KL/r = 80 are computed for D/t of 36 and 60. The input data is:

Input Data

```
2         1
-301      1
15.       0.4167    30000.    36.
69.254    0.001     4
10.       10.       200       0.      0.
-1.0      3.5       2.
50        1         1
-301      1
15.       0.250     30000.    36.
69.254    0.001     4
10.       10.       200       0.0     0.
-1.0      3.5       2.
50        1         1
```

The resulting load-shortening curves are plotted in Fig. 5.11. The increase in diameter-to-thickness ratio decreases the energy-absorption capacity of the column.

5.8.3 Effects of Slenderness Ratio on Fixed-Ended Column

The axial load-axial shortening relationship of a fixed-ended column with D/t = 48 is obtained for KL/r of 60, 80, and 100. The input data is:

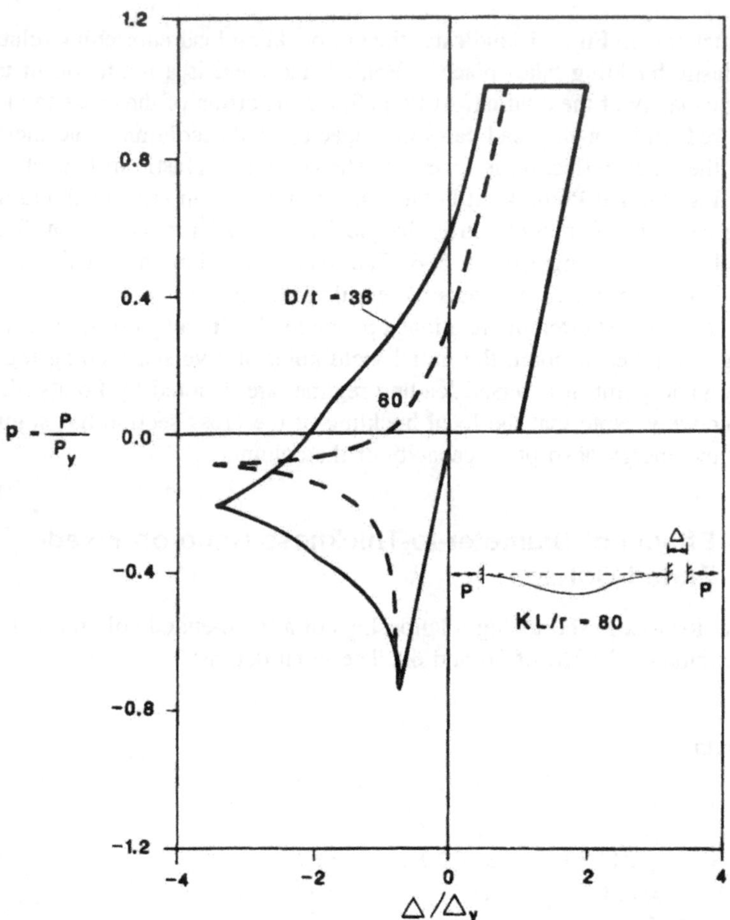

FIGURE 5.11 Effects of diameter-to-thickness ratio on the cyclic axial load-shortening relationship of a fixed-ended column.

Input Data

```
3          1
-301       1
15.0       .3125      30000.     36.
51.940     0.001      4
10.        10.        200        0.         0.
-1.0       3.0        2.
50         1          1
-301       1
15.        0.3125     30000.     36.
69.254     0.001      4
10.        10.        200        0.0        0
-1.0       3.0        2.
```

```
50        1         1
-301      1
15.       0.3125    30000.    36.
69.254    0.001     4
10.       10.       200       0.0       0.
-1.0      3.0       2.
50        1         1
-301      1
15.       0.3125    30000.    36.
86.568    0.001     4
10.       10.       200       0.0       0.
-1.0      3.0       2.
50        1         1
```

The resulting load-shortening curves are shown in Fig. 5.12. The increase in the slenderness ratio decreases the maximum compressive load-carrying capacity as well as the energy absorption capacity of the column.

5.8.4 Effects of End Moments on a Pin-Ended Beam-Column

The load-shortening behavior of a pin-ended beam-column (L/r = 80, D/t = 48) in the presence of end moments (M/M_y = 0, 0.2, and 0.4) is computed and compared. The input data for the example is:

Input Data

```
3         1
311       1
15.0      .3125     30000.    36.
34.627    0.001     4
10.       20.       200       0.        0.
-1.0      2.5       2.
50        1         1
311       1
15.       0.3125    30000.    36.
34.627    0.001     4
10.       20        200       0.2       0.
-1.0      2.5       2.
50        1         1
311       1
15.       0.3125    30000.    36.
34.627    0.001     4
10.       20.       200       0.4       0.
-1.0      2.5       2.
50        1         1
```

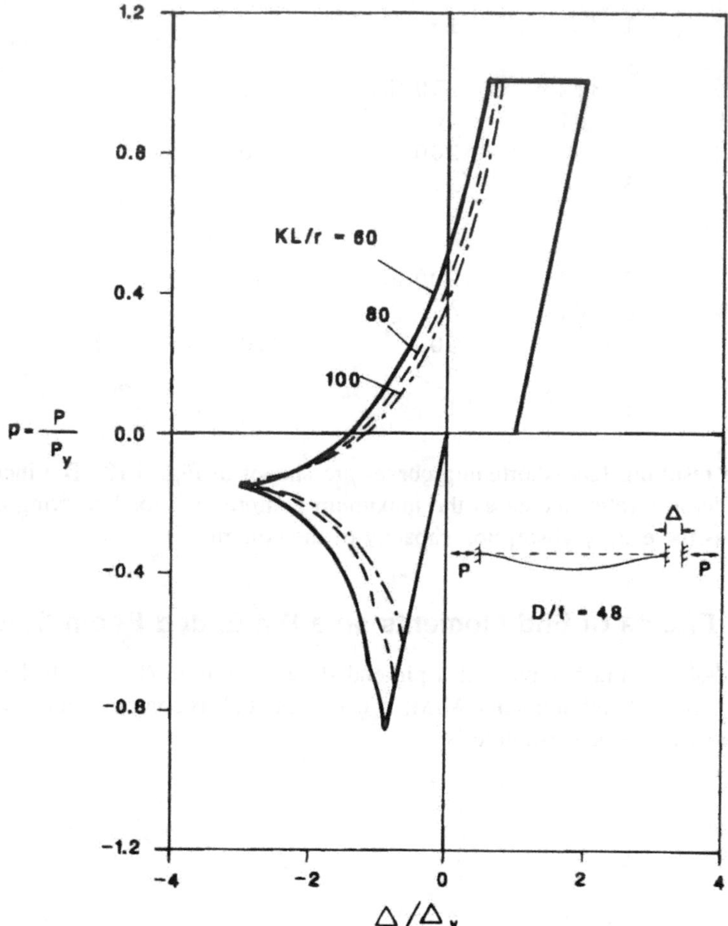

FIGURE 5.12 Effects of slenderness ratio on the cyclic axial load shortening relationship of a fixed-ended column.

The resulting load-shortening relationships are shown in Fig. 5.13. The end moments decrease the maximum compressive load-carrying capacity of the beam-column. However, energy absorption capacity is not significantly reduced by the presence of the end moments.

5.8.5 Effects of Lateral Loads on a Pin-Ended Beam-Column

The load-shortening relationship of a pin-ended beam-column (L/r = 80, D/t = 48) in the presence of mid-span lateral loads (Q/Q_y = 0, 0.2, 0.4) is computed and compared. The required input is:

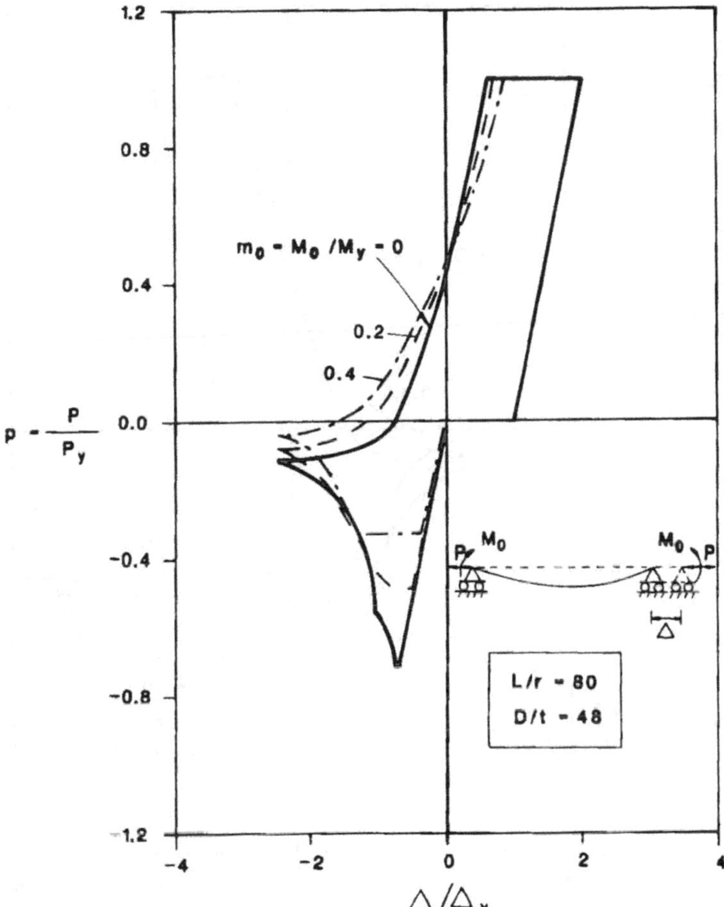

FIGURE 5.13 Effects of end moments on the cyclic axial load-shortening relationship of a pin-ended beam-column.

Input Data

```
3          2
311        1
15.        0.3125     30000.     36.
34.628     0.001      4
10.        20.        200        0.         0.
-1.0       2.5        2.
50         1          1
311        1
15.        0.3125     30000.     36.
34.628     0.001      4
10.        10.        200        0.2        0.
```

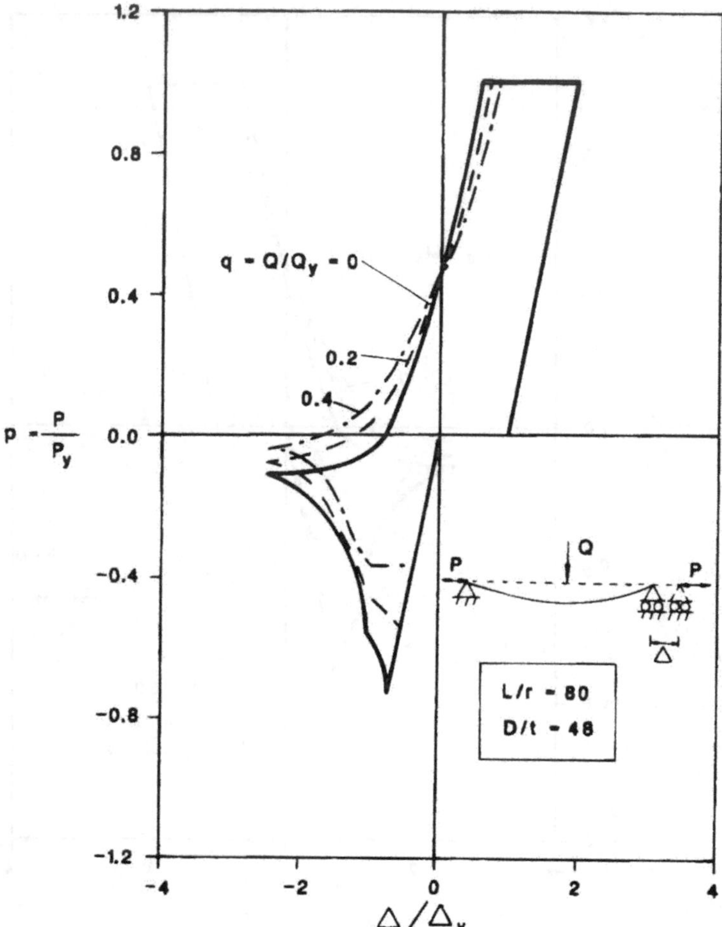

FIGURE 5.14 Effects of lateral load on the cyclic axial load-shortening relationship of a pin-ended beam-column.

```
-1.0      2.5       2.
50        1         1
311       1
15.       0.3125    30000.    36.
34.628    0.001     4
10.       10.       200       0.0       0.4
-1.0      2.5       2.
50        1         1
```

The computed load-shortening relationships are plotted in Fig. 5.14. The effects of lateral load are similar to those of the end moments.

References

Chen, W. F. and Sohal, I. S. (1988) Cylindrical members in offshore structures, *Thin-Walled Structures*, 6, 153–285.

Chen, W. F. and Han, D. J. (1985) *Tubular Members in Offshore Structures*, Pitman, London, U.K.

Chen, W. F. and Atsuta, T. (1976) *Theory of Beam-Columns, Vol. 1: In-Plane Behavior and Design*, McGraw-Hill, New York.

Chen, W. F. and Atsuta, T. (1972) Simple interaction equations for beam-columns, *Journal of Structural Engineering*, 98(7), 1413–26.

Korol, R. M. (1978) *Inelastic buckling of circular tubes under bending, Part II–Experimental Program*, Department of Civil Engineering, McMaster University, Hamilton, Ontario.

Sherman, D. R. (1983) *Report on bending capacity of fabricated pipes*, Department of Civil Engineering, University of Wisconsin, Milwaukee.

Sohal, I. S. and Chen. W. F. (1988a) Local and post-buckling behavior of tubular beam-columns, *Journal of Structural Engineering, ASCE*, 114 (5), 1073–90.

Sohal, I. S. and Chen. W. F. (1988b) Local buckling and inelastic cyclic behavior of tubular sections, *Thin-Walled Structures*, 6 (1), 63–80.

Sohal, I. S. and Chen. W. F. (1987a) Local buckling and inelastic cyclic behavior of tubular members, *Thin-Walled Structures*, 5 (6), 455–75.

Sohal, I. S. and Chen. W. F. (1987b) Local buckling and sectional behavior of fabricated tubes, *Journal of Structural Engineering, ASCE*, 113 (3), 519–33.

Sohal, I. S. and Chen. W. F. (1984) Moment-curvature expressions for fabricated tubes, *Journal of Structural Engineering, ASCE*, 110 (11), 2738–57.

Toma, S. and Chen, W. F. (1982) Inelastic cyclic analysis of pin-ended tubes, *Journal of Structural Engineering, ASCE*, 108 (10), 2279–94.

References

6: Analysis Considering Dent Damage Effect

L. Duan
Division of Structures, California Department of Transportation, Sacramento, California
W. F. Chen
School of Civil Engineering, Purdue University, West Lafayette, Indiana

NOTATIONS

The following symbols are used in this chapter:

- a, b, c, f, and k constants defining the shapes of moment-thrust-curvature curves and moment-thrust-axial strain curves
- D diameter of the tube
- dd dent depth
- F_y yield stress
- $f_1(R)$, $f_2(R)$ residual stress factors accounting for the effects on moment-curvature curves
- L length of member
- M bending moment
- M_p plastic moment
- M_u mean maximum moment capacity of an undented beam section
- M_y initial yield moment under pure bending of an undented beam section
- M_{ud} bending moment capacity of a dented section
- M_{pud} the moment capacity considering the effect of axial compression of a dented section
- m bending moment normalized by the yield moment
- $\overline{m}$ bending moment capacity normalized by the plastic moment
- m_{pc} moment capacity normalized by the yield moment considering the effect of axial compression of a undented section
- M_{pud} the moment capacity considering the effect of axial compression of a dented section
- m bending moment normalized by the yield moment
- $\overline{m}$ bending moment capacity normalized by the plastic moment
- m_{pc} moment capacity normalized by the yield moment considering the effect of axial compression of a undented section

0-8493-8282-3/96/$0.00+$.50

m_{pud} moment capacity normalized by the yield moment considering the effect of axial compression of a dented section
P axial compression
p axial compression normalized by axial yield strength
P_y axial yield strength
P_u mean compression yield strength including the effect of local buckling of an undented tubular section
r radius of gyration of the tube
S_d slope function of the descending branch of M-P-Φ curves of a dented section
t thickness of the tube
w deflection of member
w_{1i}, w_{2i} total resultant deflection at two ends of segment i
β dent directional angle
ϵ axial strain at the centroid of cross section normalized by the yield axial strain
ϵ_O axial strain at the centroid of cross section
ϵ_y initial yield axial strain
Φ curvature
Φ initial yield curvature under pure bending
Φ_{pud} peak curvature considering the effect of axial compression of a dented section
ϕ curvature normalized by the initial yield curvature
Δ axial-shortening of member
Δ_s axial-shortening due to axial strain
Δ_g axial-shortening due to geometric change

6.1 INTRODUCTION

Owing to their low drag coefficient in comparison to their structural shapes, cylindrical members are used extensively in offshore structures. These members are generally subjected to gravity, wind, wave, and current loads. Some members in the wave zone often experienced localized damage caused mainly by supply workboat collisions or dropped-object impacts (Fig. 6.1). In the last two decades, experimental and analytical research on structural tubes has made significant progress in establishing refined criteria for design of undamaged cylindrical members in offshore platforms (Marshall, 1970; Sherman, 1976; Chen and Ross, 1977; Toma and Chen, 1979; Sherman, 1982; Chen and Han, 1985; and Loh, 1990). However, available design specifications (API-RP-2A, 1989; API-RP-2A-LRFD, 1989; AISC-LRFD, 1993; and AISC-ASD, 1989) give no specific information on how such localized damage affects the behavior and strength of dented cylindrical members under field service conditions. To assess the fitness of these offshore structures in service, sound technical information is needed for these dented members in terms of both their behavior and ultimate strength. The prime object of this chapter is to address this need in the analytical study of dented cylindrical members.

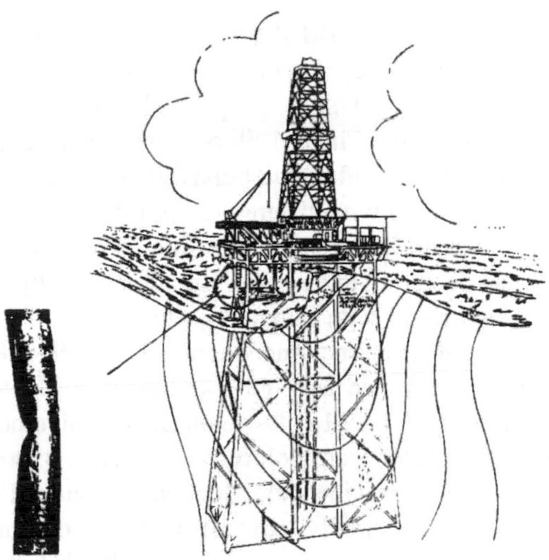

FIGURE 6.1 Damaged cylindrical members in offshore structures.

Damaged cylindrical members were first studied experimentally by Smith, Kirkwood, and Swan (1979). During the 1980s, a considerable amount of experimental and theoretical research about the effects of damage on the strength and behavior of cylindrical members has been conducted (Smith and Dow, 1981; Smith, Somerville, and Swan, 1981; Ellinas, 1984; Ueda and Rashed, 1985; Richards and Andronicou, 1985, Taby and Moan, 1985 and 1987; Yao, Taby, and Moan, 1986; Padula and Ostapenko, 1989; and Gu and Li, 1992). MacIntyre and Birkemoe (1989) used a nonlinear finite element shell analysis (ABAQUS) to investigate the denting and subsequent loading under axial compression of a dented cylindrical member. This is the most rigorous procedure among all existing analyses, but it requires a considerable computing effort.

Research reported in the open literature as in the past focused particular attention on damaged members subjected to axial compression combined with negative bending (compression at the dented side). In actual offshore structures, however, local damages may occur in any orientation and location along the member under axial compression combined with lateral loading. Little attention has been paid to dented members subjected to loads with different directions with respect to the dents. It is the purpose of this chapter to present a computer model for the analysis of dented cylindrical beam-columns subjected to biaxial bending with respect to the dents.

Several M-P-Φ-based numerical procedures, such as Newmark's Method and Assumed Deflection Method (Toma, 1980; Chen and Han, 1985), are available for analyzing the pre- and post-buckling behavior of undented cylindrical members without considering the effect of local bucking, i.e., the M-P-Φ curves have only the ascending branch. Sohal and Chen (1987, 1988) have developed an analysis method to include the effect of local buckling on the behavior of undented cylindrical members. This was described in the preceding chapter. However, their method is strictly applicable only for symmetrical members under symmetrical loadings.

The moment-thrust-curvature (M-P-Φ) relationships for dented cylindrical members have been developed by Duan, Loh, and Chen (1990a) from limited available experimental tests. Based on the proposed M-P-Φ relationships, an analytical procedure (Duan, Loh, and Chen, 1990b) and enhanced computer program, BCDENT, were developed to simulate the behavior of a general dented cylindrical beam-column subjected to loads with different directions with respect to the multiple dents. The computer program BCDENT so developed provides a practical tool for the analysis of multiple dented cylindrical beam-columns subjected to independent axial compression or combined with biaxial bending.

This chapter consists of seven sections. Section 6.2 briefly presents the modeling of M-P-Φ relationships for dented and undented cylindrical sections. Section 6.3 describes the development of an M-P-Φ-based analytical model to simulate the behavior of dented cylindrical beam-columns. Section 6.4 deals with computer implementation. Section 6.5 provides the documentation and User's Manual of computer program BCDENT. Finally, several solutions of undented and dented cylindrical member behavior predicted by program BCDENT are given in Sections 6.6 and 6.7.

6.2 M-P-Φ RELATIONSHIPS FOR DENTED CYLINDRICAL SECTIONS

In this section, a simplification of the M-P-Φ expressions for undented cylindrical sections developed by a Purdue research team (Chen, 1971; Saleeb, 1979; Toma and Chen, 1980; Sohal and Chen, 1984; Chen and Han, 1985) is first presented. The development of approximate and complete M-P-Φ relationships for dented cylindrical sections subjected to bending with different directions (D/t = 30 and 48, 10% and 20% dented depth) is then described. The further simplification of the M-P-ϵ_o expression is also discussed in this section. These relationships are valid for the range D/t ≤ 80, and dd/D ≤ 0.2. These recommended expressions will be used in Section 6.3 as the basic relationships for the analysis of dented cylindrical members.

6.2.1 Undented Cylindrical Sections

Using the following nondimensional quantities:

$$p = \frac{P}{P_y}, \qquad m = \frac{M}{M_y}, \qquad \phi = \frac{\Phi}{\Phi_y} \tag{6.1}$$

where P is axial compression; M is bending moment; Φ is curvature; P_y is axial yield strength = $F_y \pi t(D - t)$; M_y is initial yield moment under pure bending = $F_y \pi t(D - t)^2/4$; Φ_y is initial yield curvature under pure bending = $2F_y/(E(D - t))$ = $2\epsilon_y/(D - t)$; and F_y is yield stress.

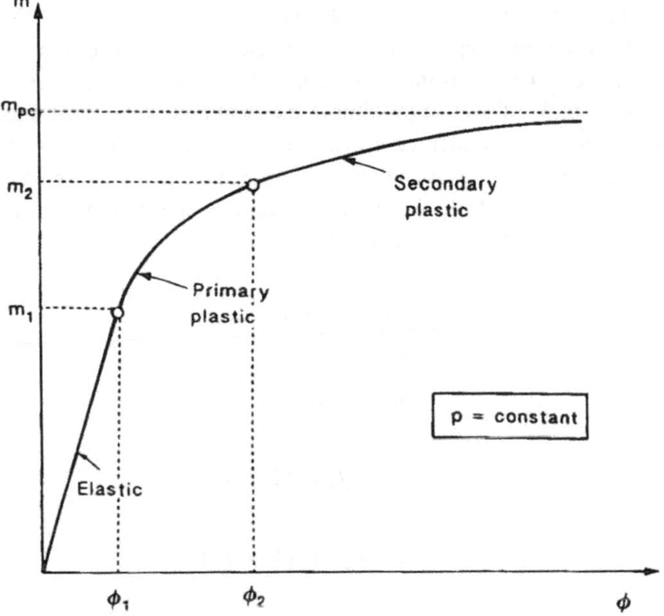

FIGURE 6.2 Moment-thrust-curvature curve for undented cylindrical sections without local buckling.

For an undented cylindrical section without local buckling, the nonlinear M-P-Φ relation shown in Fig. 6.2 can be represented by the following three-regime moment-curvature expressions, m = f(ϕ, p), developed previously by Chen (1971).

$$f(\phi, p) = \begin{cases} a\phi & \text{for} \quad \phi_1 \geq \phi \\ b - c/\sqrt{\phi} & \text{for} \quad \phi_1 < \phi \leq \phi_2 \\ m_{pc} - f/\phi^2 & \text{for} \quad \phi > \phi_2 \end{cases} \tag{6.2}$$

in which

$$a = \frac{m_1}{\phi_1} \tag{6.3}$$

$$b = \frac{m_2\sqrt{\phi_2} - m_1\sqrt{\phi_1}}{\sqrt{\phi_2} - \sqrt{\phi_1}} \tag{6.4}$$

$$c = \frac{m_2 - m_1}{1/\sqrt{\phi_1} - 1/\sqrt{\phi_2}} \tag{6.5}$$

$$f = (m_{pc} - m_2)\phi_2^2 \tag{6.6}$$

For a thin-walled cylindrical section, the boundaries of the three regimes, m_1, ϕ_1,

m_2, ϕ_2, and m_{pc} have been theoretically derived (Chen and Han, 1985). To develop simple M-P-Φ expressions, approximate expressions for the m-p-ϕ parameters were proposed by Saleeb (1979), Toma and Chen (1980), Sohal and Chen (1984), and Chen and Han (1985). These approximate expressions were used successfully in the numerical analysis of undented fabricated tubes. However, these formulas are still complicated. Some improvements for determining the boundaries of three-regime m_1, ϕ_1, m_2, ϕ_2, and m_{pc} in order to calculate the parameters, a, b, c, and f of m-p-ϕ curves are summarized as follows:

$$m_1 = f_1(R)(1 - p) \tag{6.7}$$

$$\phi_1 = f_1(R)(1 - p) \tag{6.8}$$

$$m_2 = f_1(R)(1 - p^2) \tag{6.9}$$

$$\phi_2 = f_2(R)(1 + p^2) \tag{6.10}$$

$$m_{pc} = 1.273\ (1 - p^{1.75}) \tag{6.11}$$

where $f_1(R)$ and $f_2(R)$ are residual stress factors that account for the effects of residual stresses on the m-p-ϕ curves

$$f_1(R) = \begin{cases} 1.0 & \text{without residual stresses} \\ 0.9 & \text{with residual stresses} \end{cases} \tag{6.12}$$

$$f_2(R) = \begin{cases} 1.0 & \text{without residual stresses} \\ 0.9 + 0.2p & \text{with residual stresses} \end{cases} \tag{6.13}$$

Using Eqs. (6.2) to (6.6) together with Eqs. (6.7) to (6.13), the approximate m-p-ϕ curves for the undented cylindrical cross-section with 1% out-of-roundness for cases of $p = 0$, 0.4, 0.6, and 0.8 are shown in Fig. 6.3 (without considering the effects of residual stresses) and in Fig. 6.4. (including the effects of residual stresses). In Figs. 6.3 and 6.4, M_p is the plastic moment capacity of an undented cylindrical beam section. The exact numerical m-p-ϕ curves obtained by Toma and Chen (1979) are also plotted in Figs. 6.3 and 6.4 for comparisons. It is seen that the agreement is very good.

A typical M-P-Φ curve of cylindrical sections considering the local buckling of cross sectional distortion as shown in Fig. 6.5 has an ascending branch and a sharp descending branch. Sohal and Chen (1987) studied the behavior of locally buckled tubes and proposed a closed form expression for the descending branch of the moment-curvature curve considering the effect of local buckling.

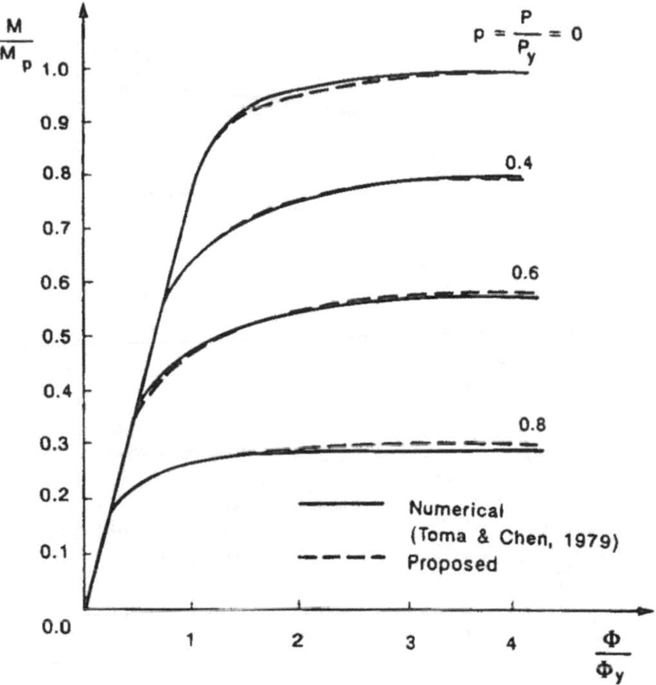

FIGURE 6.3 Moment-thrust-curvature curve for undented cylindrical sections with 1% out-of-roundness and without the effects of residual stresses.

For $\phi \geq \phi_{lb}$

$$m = (m_{max} - m_{min})\exp\left[\frac{(\phi - \phi_{lb})(EI)_{max}}{m_{max} - m_{min}}\right] + m_{min} \tag{6.14}$$

in which ϕ_{lb} is the curvature at which local buckling starts, the parameter m_{max} can be determined by using $\phi = \phi_{lb}$ in the expression Eq. (6.2). The parameters m_{min}, ϕ_{lb}, and $(EI)_{max}$ are functions of p and D/t. They can be found in Chapter 5 and in Sohal and Chen, 1987.

6.2.2 Dented Cylindrical Sections

The behavior of dented cylindrical members is much more complicated than that of undented tubes. It is difficult to use only one M-P-Φ expression to describe the behavior of both undented and dented cylindrical sections (Fig. 6.6). From practical viewpoints, the dent-thickness ratio dd/t = 1.0 is defined here as the limit of the dent effect, i.e., if dd/t is less than 1.0, the dent effect is neglected; otherwise, the dent effect must be considered. For a dented cylindrical section (dd/t ≥ 1.0), the following three-regime expression, $m = f^*(\phi, p)$, is used to approximate its M-P-Φ curve shown in Fig. 6.7:

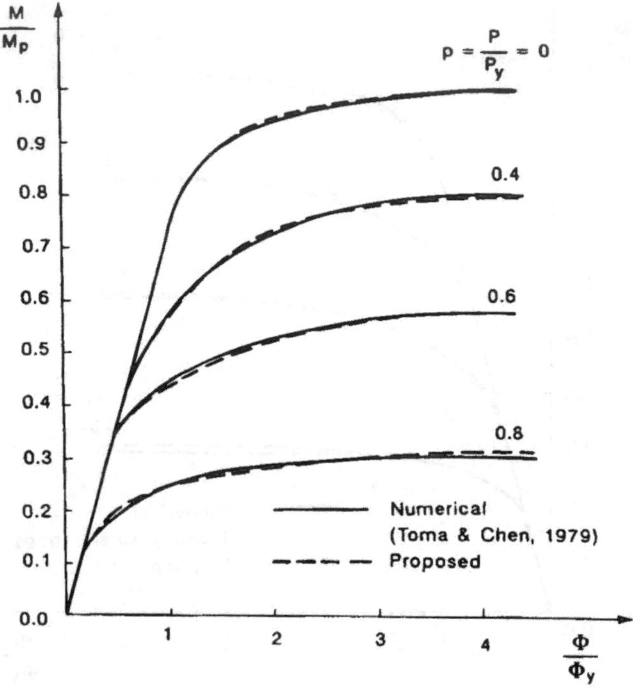

FIGURE 6.4 Moment-thrust-curvature curve for undented cylindrical sections with 1% out-of-roundess and with the effects of residual stresses.

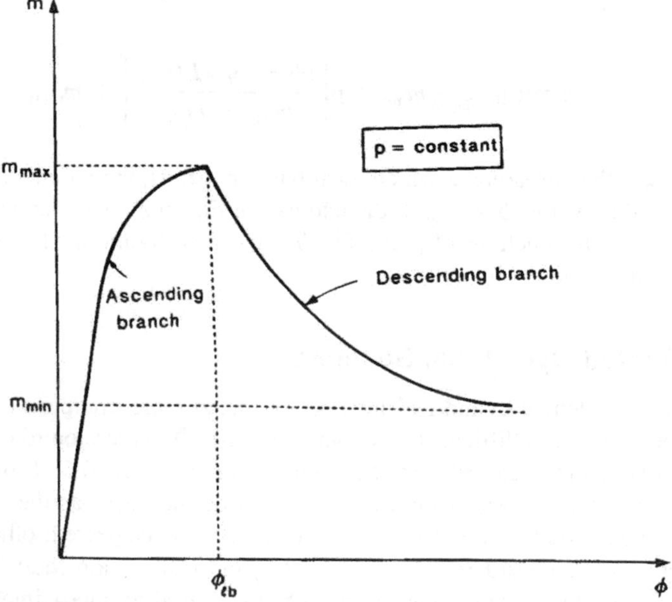

FIGURE 6.5 Moment-thrust-curvature curve for undented cylindrical sections with local buckling.

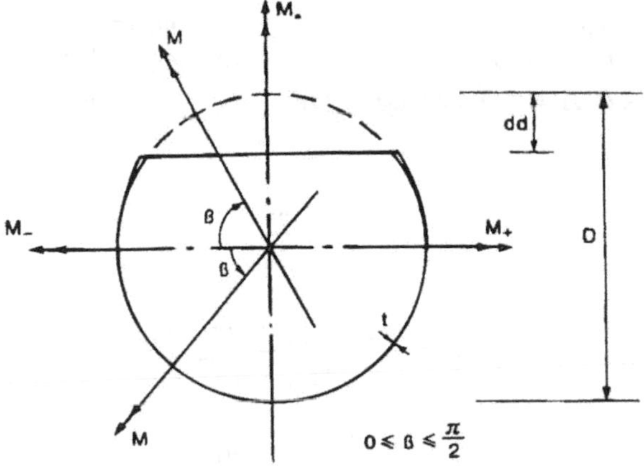

FIGURE 6.6 Dented cylindrical section with different bending moments.

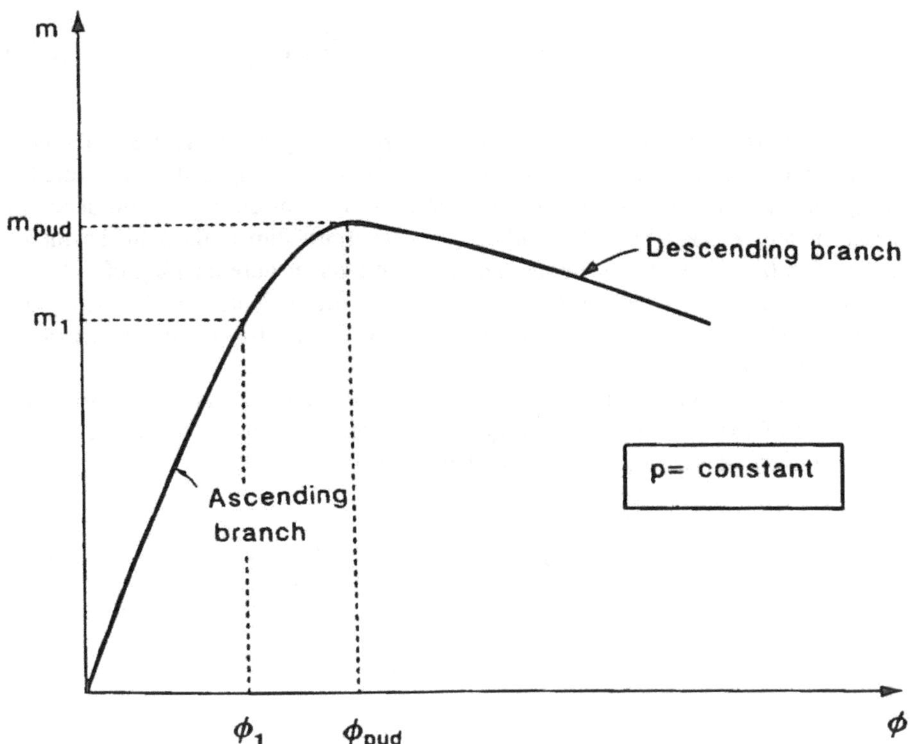

FIGURE 6.7 Moment-thrust-curvature curve for a dented cylindrical section.

$$f^*(\phi, p) = \begin{cases} a\phi & \text{for} \quad \phi_1 \geq \phi \\ b - c/\sqrt{\phi} & \text{for} \quad \phi_1 < \phi \leq \phi_{pud} \\ m_{pud}[1 - S_d(\phi/\phi_{pud} - 1)] & \text{for} \quad \phi_1 > \phi_{pud} \end{cases} \tag{6.15}$$

in which

$$a = \frac{m_1}{\phi_1} \tag{6.16}$$

$$b = \frac{m_{pud}\sqrt{\phi_{pud}} - m_1\sqrt{\phi_1}}{\sqrt{\phi_{pud}} - \sqrt{\phi_1}} \tag{6.17}$$

$$c = \frac{m_{pud} - m_1}{1/\sqrt{\phi_1} - 1/\sqrt{\phi_{pud}}} \tag{6.18}$$

For the ascending branch of an M-P-Φ curve of a dented cylindrical section, the boundary between the two regimes as defined by m_1 and ϕ_1 is proposed as follows:

$$m_1 = 0.8 f_1(R) m_{pud} \tag{6.19}$$

$$\phi_1 = f_1(R)(1 - p)\left(1 - \frac{dd}{D}\sin\beta\right) \tag{6.20}$$

where $f_1(R)$ is the residual stress function determined by Eq. (6.12); β is the dent directional angle defined in Fig. 6.6 ($\beta \leq \pi/2$ should be imposed on Eq. (6.20)); m_{pud} and ϕ_{pud} are the normalized moment capacity considering the influence of axial compression, and its corresponding normalized curvature of the dented cylindrical section (note that for the initial yield curvature under pure bending of a dented section = $(2\epsilon_y)/(D - dd|\cos\beta|)$, for simplicity, the undented Φ_y is used); and S_d is the slope function of the descending branch of M-P-Φ curves, which is discussed below.

For dented cylindrical beam-column sections, the moment capacity considering the effect of axial compression, M_{pud}, can be predicted by the following design interaction equation (Duan, Loh, and Chen, 1990a):

$$M_{pud} = M_{ud}\left[1 - \left(\frac{P}{P_{ud}}\right)^{\alpha}\right] \tag{6.21}$$

$$\alpha = 1.75 - 0.1\frac{dd}{t}\left(1 - \frac{2\beta}{\pi}\right) \geq 1.0 \tag{6.22}$$

where P_{ud} is the axial compressive strength of a dented section and M_{ud} is the bending moment capacity of a dented section. Landet's tests (Lander and Johnsen

1987) and Smith's tests (Smith, Kirkwood, and Swan, 1979; Richards and Andronicou, 1985) have shown that the dent-to-thickness ratio dd/t is the key factor in the determining of axial compression strength P_{ud}. Landet's tests (Landet and Johnsen, 1987) and Ueda's tests (Ueda and Rashed, 1985) indicated that both the dent directional angle β and the dent-to-thickness ratio dd/t have a significant effect on the behavior of dented beams. They can be determined by the following formulas (Duan, Loh, and Chen, 1990a).

$$\frac{P_{ud}}{P_u} = \exp\left(-0.08\,\frac{dd}{t}\right) \tag{6.23}$$

$$\frac{M_{ud}}{M_u} = \exp\left(-0.06\,\frac{dd}{t}\cos\beta\right) \tag{6.24}$$

in which $\cos\beta \geq 0$; P_u and M_u are mean compression yield strength and mean maximum moment capacity of an undented cylindrical section, including the effect of local buckling, respectively. They are predicted by API-ASD formulas (API-ASD 1989):

$$\frac{P_u}{P_y} = \begin{cases} 1.0 & \text{for} \quad D/t \leq 100 \\ 1.95 - 0.3(D/t)^{0.25} & \text{for} \quad D/t > 100 \end{cases} \tag{6.25}$$

$$\frac{M_u}{M_p} = \begin{cases} 1.0 & \text{for} \quad 0 < \dfrac{F_yD}{t} \leq 17{,}240 \\ 1.13 - 1.54\,\dfrac{F_yD}{Et} & \text{for} \quad 17{,}240 < \dfrac{F_yD}{t} \leq 44{,}820 \\ 0.96 - 0.77\,\dfrac{F_yD}{Et} & \text{for} \quad 44{,}280 < \dfrac{F_yD}{t} \leq 137{,}900 \end{cases} \tag{6.26}$$

in which M_p is the plastic bending moment capacity of an undented beam and (F_yD/t) is in SI units.

The peak curvature ϕ_{pud} corresponding to the moment capacity of a dented cylindrical section m_{pud} is shown in Fig. 6.7. Test data (Landet and Johnsen, 1987) show that the ϕ_{pud} value depends mainly on the axial compression ratio p and the dent directional angle β. It decreases with an increasing p value and with a decreasing β value. The following empirical formulas (Duan, Loh, and Chen, 1990a) can be used for the prediction of the peak curvature ϕ_{pud}.

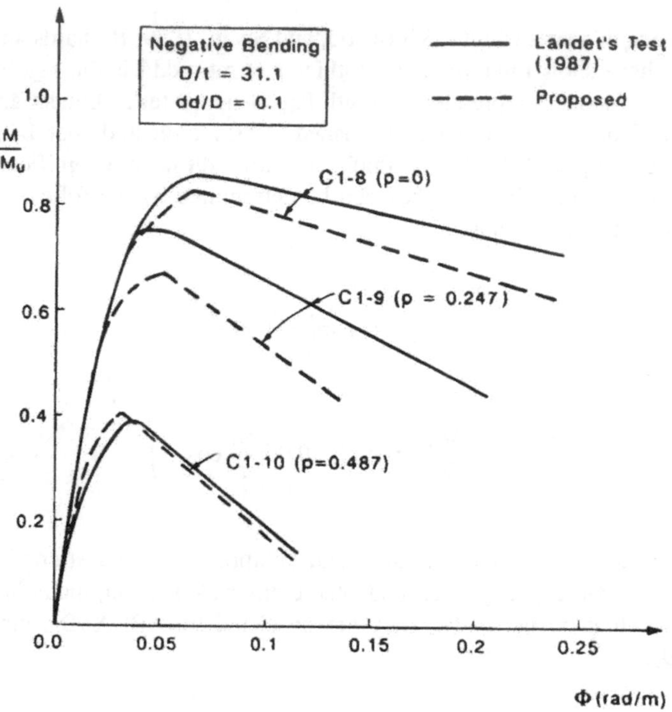

FIGURE 6.8 M-P-Φ curves for dented sections, dd/t = 3.11 and β = 0.

$$\phi_{pud} = \left[1 + \frac{2\beta}{\pi}\right]\phi_o^* \tag{6.27}$$

$$\phi_o^* = 2.8 - 0.1\frac{dd}{t} - 2.5p \geq 1.0 \tag{6.28}$$

Landet's tests indicate that the slope of the descending branch of the M-P-Φ curve for a dented cylindrical section has an almost constant value, especially in the case of negative bending, and this slope S_d increases linearly with an increasing axial compression ratio p. It increases sharply with an increase of the dent directional angle β. From a statistical regression analysis of the slopes based on Landet's tests, it is found that the S_d-p-β relationship can be approximated by the following formulas (Duan, Loh, and Chen, 1990a):

$$S_d = 0.1 + 0.4p + 0.6 \sin\beta + 0.8p \sin\beta \tag{6.29}$$

where $\beta \leq \pi/2$.

Figures 6.8 to 6.10 show comparisons of proposed M-P-Φ expressions for dented cylindrical sections subjected to bending with different directions with Landet's tests. Figure 6.8 is for negative bending ($\beta = 0$); Fig. 6.9 is for neutral bending ($\beta = \pi/2$);

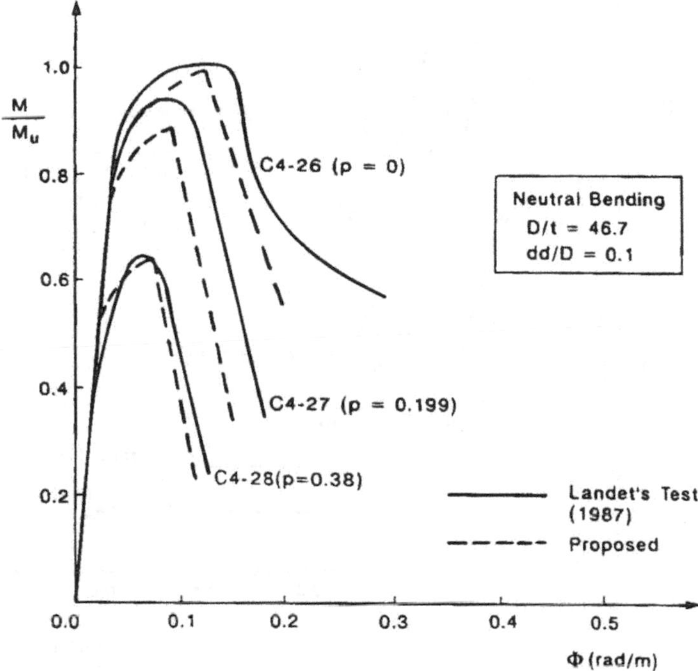

FIGURE 6.9 M-P-Φ curves for dented sections, $dd/t = 4.67$ and $\beta = \pi/2$.

and Fig. 6.10 is for positive bending ($\beta = \pi$). In these tests, the dimensionless ratios are $D/t = 31.1$ and 46.7, $dd/D = 0.1$, $p = 0$ to 0.487, and $M/M_u = M/M_p$. It is observed that the proposed M-P-Φ expressions fit reasonably well with the test curves in both ascending and descending branches; but the peak values of the M-P-Φ curves are off somewhat due to the approximate nature of the M_{pud} and ϕ_{pud} values.

6.2.3 M-P-ϵ_O Expressions

The moment-thrust-axial strain relation of a cross section is the basic quantity required for computing axial shortening of compression members. Based on the assumption that the M-P-ϵ_O relationship is not affected significantly by the local buckling of cross section and residual stresses, Sohal and Chen (1987) have successfully used the M-P-ϵ_O relationship of the undented cylindrical section in pre- and post-local buckling analysis. For simplicity, the improved M-P-ϵ_O expressions presented in this section will be used for the analysis of dented cylindrical members with and without residual stresses.

Exact M-P-ϵ_O Relations for Circular Cylindrical Sections

An undented circular cylindrical section subjected to an axial compression load P, combined with a bending moment M, is shown in Fig. 6.11. Ellis (1958) derived the following exact M-P-Φ-ϵ_O relations using the thin-walled theory.

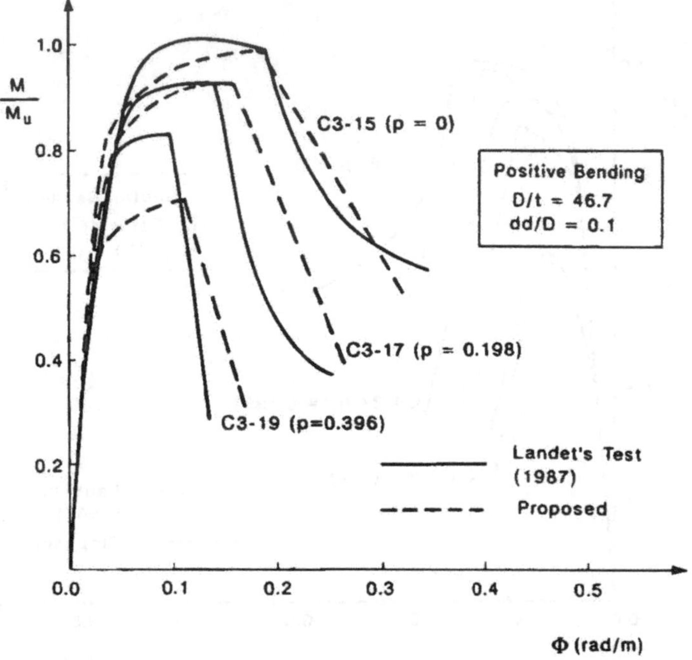

FIGURE 6.10 M-P-Φ curves for dented sections, dd/t = 4.67 and β = π.

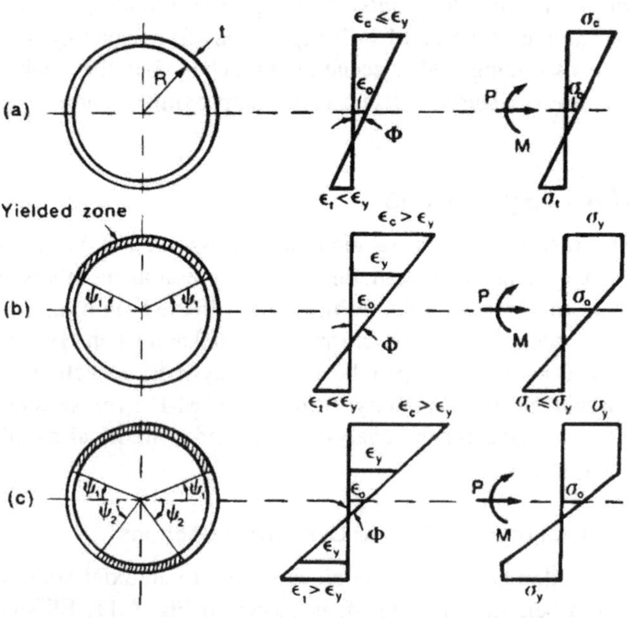

FIGURE 6.11 Stress-strain states of the cylindrical section.

Case 1: Elastic State (Fig. 6.11a)

$$p = \epsilon \tag{6.30}$$

$$\overline{m} = \frac{\pi\phi}{4} \tag{6.31}$$

Case 2: Yielded in the Compression Zone Only (Fig. 6.11b)

$$p = 1 - \frac{\phi}{\pi}\left[\left(\frac{\pi}{2} + \psi_1\right)\sin\psi_1 + \cos\psi_1\right] \tag{6.32}$$

$$\overline{m} = \phi\left(\frac{\phi}{8} + \frac{\psi_1}{4} + \frac{\sin\psi_1\cos\psi_1}{4}\right) \tag{6.33}$$

Case 3: Yielded in Both Zones (Fig. 6.11c)

$$p = \frac{2\psi_2}{\pi} - \frac{\phi}{\pi}[(\psi_1 + \psi_2)\sin\psi_1 + (\cos\psi_1 - \cos\psi_2)] \tag{6.34}$$

$$\overline{m} = \frac{\phi}{4}(\psi_1 + \psi_2 + \sin\psi_1\cos\psi_1 + \sin\psi_2\cos\psi_2) \tag{6.35}$$

where

$$\psi_1 = \sin^{-1}\left(\frac{1 - \epsilon}{\phi}\right) \tag{6.36}$$

$$\psi_2 = \sin^{-1}\left(\frac{1 + \epsilon}{\phi}\right) \tag{6.37}$$

$$p = \frac{P}{P_y}, \qquad \overline{m} = \frac{M}{M_p}, \qquad \phi = \frac{\Phi}{\Phi_y}, \qquad \epsilon = \frac{\epsilon_o}{\epsilon_y} \tag{6.38}$$

where ϵ_o is axial strain at the centroid of the cross section; ϵ_y is the initial yield axial strain.

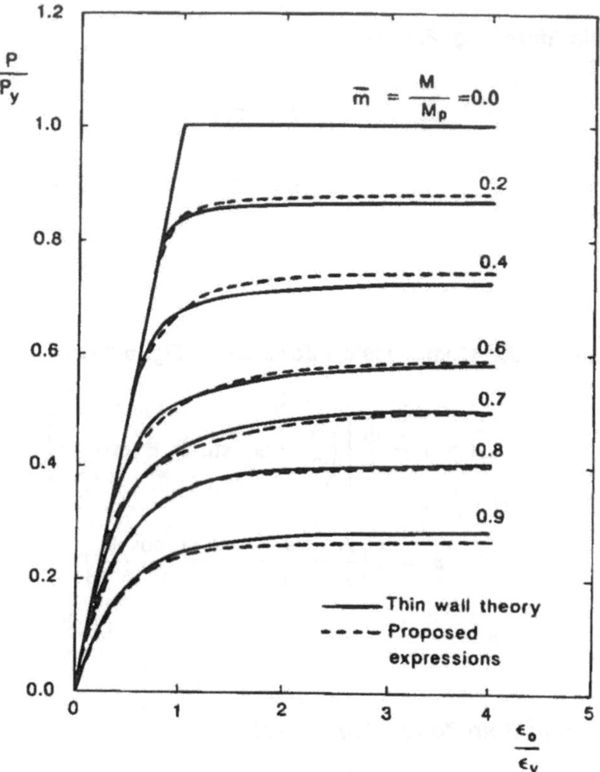

FIGURE 6.12 M-P-ϵ_o curves for cylindrical sections.

Approximate M-P-ϵ_O Expressions

Although the M-P-Φ-ϵ_O relations (Eqs. 6.30 to 6.35) are obtained in closed form, the relationships can only be obtained by an iterative numerical procedure. In the numerical analysis, the moment M and the axial force P are usually known, and the required quantities are axial strain ϵ_O. Thus, it is more convenient to express the relationship between the axial force P and the axial strain ϵ_O explicitly.

The nonlinear M-P-ϵ_O curves obtained by the thin-walled theory (Ellis 1958) are shown in Fig. 6.12 for the undented cylindrical section. A three-regime M-P-ϵ_O expression similar to that of M-P-Φ curves was proposed by Chen and Han (1985). However, Chen-Han expressions (1985) are too complicated. For simplicity, the following modifications are adopted:

$$\epsilon = \begin{cases} p & \text{for} \quad p \leq p_1, \overline{m} < 0.75 \\ p + \mathrm{k}_1 p^2 & \text{for} \quad p \leq p_1, \overline{m} \geq 0.75 \\ [c/(b - p)]^2 & \text{for} \quad p_1 < p \leq p_2 \\ \sqrt{f/(p_o - p)} & \text{for} \quad p > p_2 \end{cases} \tag{6.39}$$

in which

$$b = \frac{p_2\sqrt{\epsilon_2} - p_1\sqrt{\epsilon_1}}{\sqrt{\epsilon_2} - \sqrt{\epsilon_1}} \tag{6.40}$$

$$c = \frac{p_2 - p_1}{1/\sqrt{\epsilon_1} - 1/\sqrt{\epsilon_2}} \tag{6.41}$$

$$f = (p_o - p_2)\epsilon_2^2 \tag{6.42}$$

$$k_1 = \frac{\epsilon_1 - p_1}{p_1^2} \tag{6.43}$$

The boundaries of the three-regime curve ϵ_1, p_1, ϵ_2, p_2, and p_0 that will be used to determined the parameters, a, b, c, f, and k_1 of M-P-ϵ_O curves can be calculated by the following formulas:

$$p_1 = \begin{cases} (1 - \overline{m})^{1.2} & \text{for } \overline{m} < 0.75 \\ (1 - \overline{m})^{0.9} & \text{for } \overline{m} \geq 0.75 \end{cases} \tag{6.44}$$

$$\epsilon_1 = \begin{cases} (1 - \overline{m})^{1.2} & \text{for } \overline{m} < 0.75 \\ 1.2 - \overline{m} & \text{for } \overline{m} \geq 0.75 \end{cases} \tag{6.45}$$

$$p_2 = \begin{cases} (1 - 0.9\,\overline{m})^{0.8} & \text{for } \overline{m} < 0.75 \\ 1.3 - 1.18\,\overline{m} & \text{for } \overline{m} \geq 0.75 \end{cases} \tag{6.46}$$

$$\epsilon_2 = \begin{cases} 1 + \overline{m}^2 & \text{for } \overline{m} < 0.75 \\ 1.0 & \text{for } m \geq 0.75 \end{cases} \tag{6.47}$$

$$p_o = (1 - \overline{m})^{0.57} \tag{6.48}$$

Using Eq. (6.39) together with Eqs. (6.40) to (6.48), the approximate M-P-ϵ_O curves for an undented cylindrical cross-section with M/M_p = 0, 0.2, 0.4, 0.6, 0.7, 0.8, and 0.9 are plotted in Fig. 6.12. It is seen that the exact M-P-ϵ_O curves (Ellis 1958) can all be closely approximated.

6.3 MEMBER ANALYSIS CONSIDERING DENT DAMAGE EFFECT

The behavior and strength of damaged members may be carried out by a nonlinear shell finite element analysis (Padula and Ostapenko, 1989; MacIntyre and Mirkemoe, 1989). This type of analysis, however, usually results in high computing costs. A number of simplified methods were proposed (Smith, Somerville, and Swan, 1981; Ellinas, 1984; Ueda and Rashed, 1985; Richards and Andronicou, 1985; Yao, Taby, and Moan, 1986; Taby and Moan, 1985 and 1987) for predicting behavior and strength of damaged members subjected to axial compression combined with negative bending (compression at the dented side). Little attention has been paid to dented members subjected to different directional loads with respect to the dent, although local damage may occur in any locations along the members in actual offshore structures.

In this section, based on the M-P-Φ relationships achieved in Section 6.2, an analytical procedure and a computer program BCDENT are developed to simulate the behavior of a general dented cylindrical beam-column subjected to loads with different directions with respect to the multiple dents.

6.3.1 General Description

The cylindrical beam-column under consideration is treated as an individual member. An initial geometric imperfection w_i (out-of-straightness) is assumed, and its boundary conditions can be either pinned, elastically restrained, or a combination thereof. The dents can be multiple in different directions and locations. All loads are proportionally increased. The loading cases include:

1. Constant end axial loads, increasing end bending moments.
2. Constant end axial loads, increasing lateral loads either linearly distributed loads or two concentrated loads.
3. Constant end bending moments and/or lateral loads, increasing end axial loads.

In the analysis, the following assumptions are made:

1. Deformations are small.
2. Shear and torsional deformations are negligible.
3. No strain reversal occurs in the member, i.e., the deformations of the member are always increasing.
4. The member is divided into a number of segments. The properties of these segments are described by their corresponding M-P-Φ and M-P-ϵ_O relationships.
5. For an undented section ($dd/t < 1.0$), the nonlinear M-P-Φ relation is represented by Eq. (6.2) together with Eqs. (6.7) to (6.13) and Eq. (6.14) presented in Section 6.2. The adopted M-P-Φ relationships (see Fig. 6.5) include both ascending and descending branches due to local buckling for undented cylindrical sections with 1% out-of-roundness and with or without residual stresses.
6. For a dented section ($dd/t \geq 1.0$), the three-regime expressions, Eq. (6.15), together with Eqs. (6.16) to (6.29) described in Section 6.2 are used to approximate its M-P-Φ curve (Fig. 6.7).
7. M-P-ϵ_O expressions, Eq. (6.39), together with Eqs. (6.40) to (6.48) developed in Section 6.2 are used for both dented and undented sections with or without residual stresses.

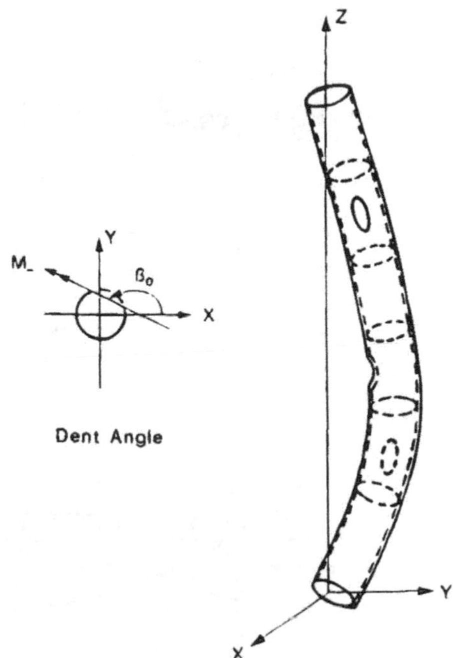

FIGURE 6.13 Coordinate system for a biaxially loaded dented cylindrical beam-column.

The coordinate system for a biaxially loaded, dented cylindrical beam-column is shown in Fig. 6.13. The right-hand rule for the sign convention is used in the present analysis (Fig. 6.14). Figure 6.15 shows typical beam-columns subjected to biaxial loadings.

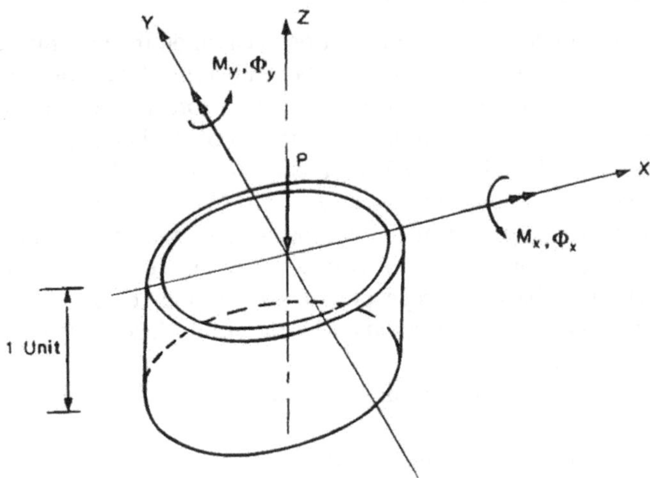

FIGURE 6.14 Moment, curvature, and axial force in cross section.

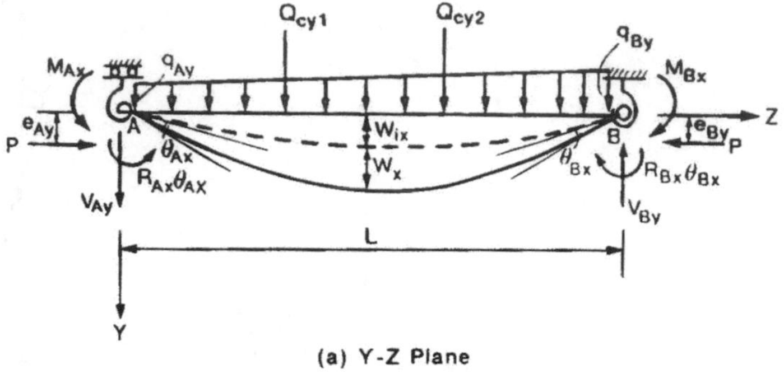

(a) Y-Z Plane

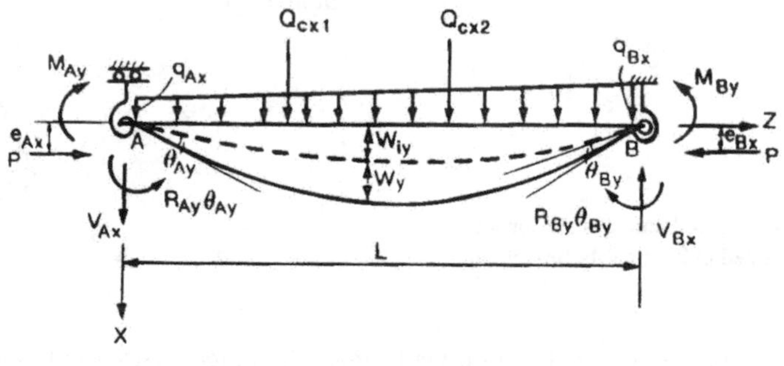

(b) X-Z Plane

FIGURE 6.15 Computing model for dented cylindrical members.

6.3.2 Numerical Procedure

The analytical procedure used in this section is based on the incremental deflection approach. The deflection shape of the member with a specified deflection at a control station is first assumed. The increment of the loads corresponding to this deflection is then computed, followed with the computation of the bending moments, taking into account the axial load and the lateral loads. The new deflections are calculated by integrating the curvature from the known M-P-Φ relationships along the length of the member. Comparisons of the computed new deflection with the assumed deflections are then made to check whether they are consistent. Using the most updated deflection shape and repeating this procedure for successive deflection increments at the control station would lead to the desired load-deflection relationship for the dented member.

The general calculation steps are summarized in the following (Fig. 6.16):

1. Divide the member into a sufficient number of segments (the length of each segment may be different). The division points are called stations or nodes.

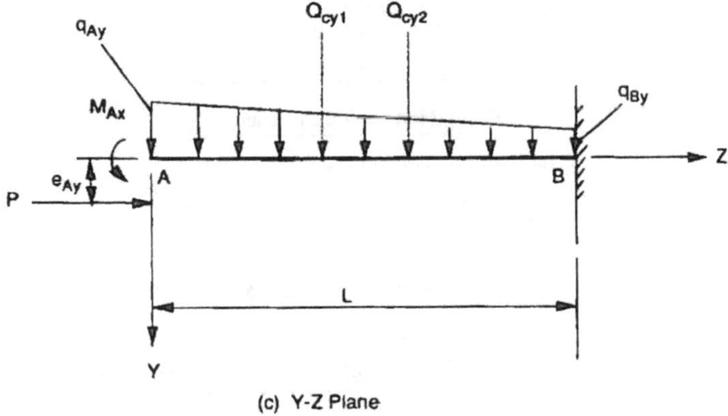

(c) Y-Z Plane

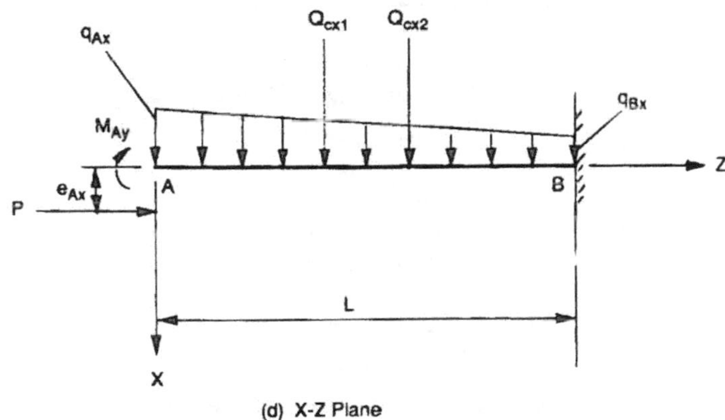

(d) X-Z Plane

FIGURE 6.15 *Continued.*

2. Assume the initially imperfect deflections along the member w_i.
3. Assume an additional deflection of the same shape Δw_i with the specified deflection δ at the control station.
4. Assume an increment of axial load, Δp or an increment of lateral load, Δq. (Increments can be negative or positive.)
5. Calculate the current peak moments and curvatures for all stations, Φ_{lb} and M_{max} for undented sections, Φ_{pud} and M_{pud} for dented sections.
6. Calculate the bending moment M_i at all stations considering the second-order effect (P-δ) due to the axial load.
7. Compare M_i with M_{pud} or M_{max}, and the updated Φ_i with Φ_{pud} or Φ_{lb}. If $M_i > M_{pud}$ (or M_{max}), repeat Step 4.
8. Calculate the curvature at all stations from a known M-P-Φ relationship.
   (a) If $M_i < M_{pud}$ (or M_{max}) and $\Phi_i < \Phi_{pud}$ (or Φ_{lb}), the curvature is on the ascending branch.

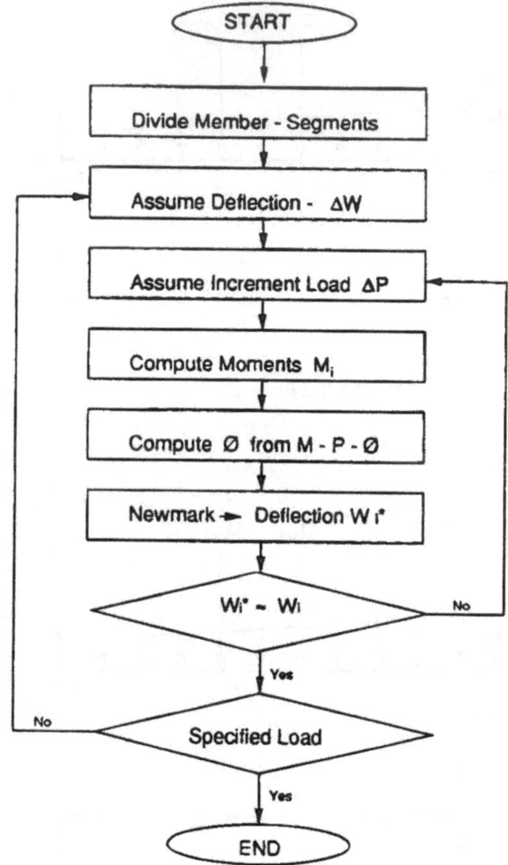

FIGURE 6.16 Flowchart of numerical procedure.

(b) If $M_i < M_{pud}$ (or M_{max}) and $\Phi_i > \Phi_{pud}$ (or Φ_{lb}), the curvature is on the descending branch.

9. Use Newmark's integration procedure (Newmark, 1943; Chen and Atsuta, 1976) to determine the deflections at all stations w_a*.
10. Compare w_a* with $w_a = w_i + \Sigma\, \Delta w$ at the control station. If the maximum difference is within the tolerance (say, $10^{-3}w_a$), do the next step. Otherwise, keep the deflection delta at the control station to be constant (Fig. 6.17), and scale the current deflection shape to the newly obtained w_a*. Repeat Step 4.
11. Compare w_a* with $w_a = w_i + \Sigma\, \Delta w$ at all other stations. If the maximum difference is within the tolerance (say, $10^{-3}w_a$), increase the next additional deflection at the control station with the same deflection shape newly obtained in the last comparison. Otherwise, update the current deflection shape to the newly obtained w_a*. Repeat Step 6.

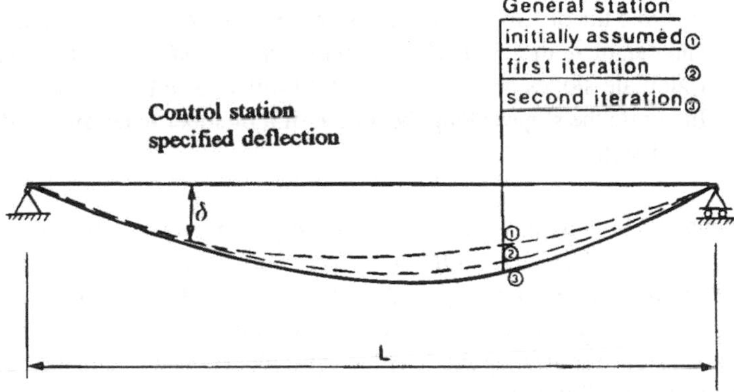

FIGURE 6.17 Iteration for member deflection.

12. Repeat Steps 4 to 11 until loads decrease to a specified percentage level of the peak loads.

Newmark's Integration Procedure

Newmark's integration method (Newmark, 1947; Chen and Atsuta, 1976) is a useful procedure to compute the deflection shape from a given curvature distribution. It is based on the conjugate beam concept. One of the characteristics of Newmark's method is that the curvature distribution on the conjugate beam is replaced by a series of *Equivalent Concentrated Loads* (Fig. 6.18). The magnitude of the *Equivalent Concentrated Loads* can be expressed in terms of the curvature diagram at the points of concentrated loading and the distance between the concentrated loads. They are obtained from equilibrium consideration. Herein, we shall only outline the key steps and derive the *Equivalent Concentrated Loads* for segments with different length. Newmark's procedure includes the following three steps:

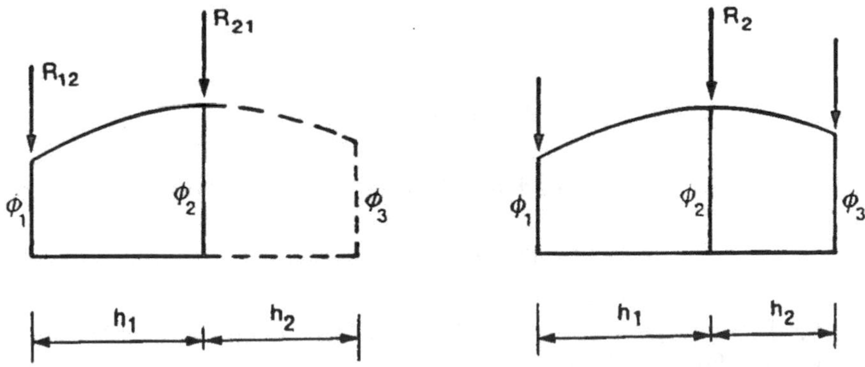

FIGURE 6.18 Equivalent concentrated loads.

1. Assume the distribution of curvature between two stations to be quadratic and calculate the *Equivalent Concentrated Loads* acting at these stations.
2. Calculate slopes at all stations by a numerical integration of curvatures.
3. Integrate the slopes along the length of a member to obtain the deflections at all stations.

In Newmark's method, the member is usually divided into several equal-length segments. For dented cylindrical members, the length of dented segments may be different from that of undented segments. The formulas of *Equivalent Concentrated Loads* for unequal-length segments are not available in the open literature. The general expressions for the magnitude of the equivalent concentrated loads, therefore, are derived in the following.

Equivalent Concentrated Loads

The representation of the curvature diagram between two stations by a second-order polynomial expression is shown in Fig. 6.18. The value of the curvature diagram at the stations are denoted as ϕ_1, ϕ_2, and ϕ_3. The lengths of neighbor segments are denoted as h_1 and h_2. The equivalent concentrated loads are denoted as R_{12}, R_{21}, and R_2. From simple static equilibrium, general expressions for *Equivalent Concentrated Loads* are obtained as follows:

$$R_{12} = \frac{\phi_1 h_1 h_2(3h_1 + 4h_2) + \phi_2 h_1(h_1^2 + 3h_1h_2 + 2h_2^2) - \phi_3 h_1^3}{12\,h_2(h_1 + h_2)} \tag{6.49}$$

$$R_{21} = \frac{\phi_1 h_1 h_2(h_1 + 2h_2) + \phi_2 h_1(h_1^2 + 5h_1h_2 + 4h_2^2) - \phi_3 h_1^3}{12\,h_2(h_1 + h_2)} \tag{6.50}$$

$$R_2 = \frac{\phi_1 A + \phi_2 B + \phi_3 C}{12\,h_1 h_2(h_1 + h_2)} \tag{6.51}$$

where

$$A = h_1^2 h_2(h_1 + 2h_2) - h_2^4 \tag{6.52}$$

$$B = (h_1^2 + h_2^2)^2 + 6h_1^2h_2^2 + 5h_1^3h_2 + 5h_1h_2^3 \tag{6.53}$$

$$C = h_1h_2^2(h_2 + 2h_1) - h_1^4 \tag{6.54}$$

For equal-length segments, $h_1 = h_2 = h$, the general formulas of equivalent concentrated loads, Eqs. (6.49), (6.50), and (6.51), reduce to the following conventional formulas, respectively.

$$R_{12} = \frac{h(7\phi_1 + 6\phi_2 - \phi_3)}{24} \tag{6.55}$$

$$R_{21} = \frac{h(3\phi_1 + 10\phi_2 - \phi_3)}{24} \tag{6.56}$$

$$R_2 = \frac{h(\phi_1 + 10\phi_2 - \phi_3)}{12} \tag{6.57}$$

6.3.3 Load-Shortening Relations

The total axial-shortening of a beam-column consists of two parts: the axial shortening due to axial strain and the axial shortening due to the geometry change of lateral deflection, i.e.,

$$\Delta = \Delta_s + \Delta_g \tag{6.58}$$

where Δ_s is axial-shortening due to axial strain and Δ_g is axial shortening due to geometric change.

The axial shortening due to axial strain is obtained from:

$$\Delta_s = \sum_{i=1}^{i=N} (\Delta L_i)\epsilon_{oi} \tag{6.59}$$

where N is number of segments; ΔL_i is length of segment i; and ϵ_{oi} is axial strain of segment i determined from M-P-ϵ_O expressions described in Section 6.2.

As long as the deflection of a member is obtained, the corresponding axial shortening due to geometric change of deflection is obtained from:

$$\Delta_g = \sum_{i=1}^{i=N} \Delta L_i - \sqrt{(\Delta L_i)^2 - (w_{1i} - w_{2i})^2} \tag{6.60}$$

where w_{1i} and w_{2i} are total resultant deflection at two ends of segment i.

6.4 COMPUTER IMPLEMENTATION

6.4.1 Program BCDENT

BCDENT is an analysis program to determine the behavior of multiply dented cylindrical members with different boundary conditions (pin-end, elastic restrained, or cantilever) subjected to axial compression combined with biaxial bending due to end moments and/or linearly distributed or concentrated lateral loads. The program traces the load-displacement curve of a dented tube up to its peak load and includes a portion of the post-peak behavior down to 20% of the peak load. The program BCDENT uses the M-P-Φ approach and the Newmark numerical procedure described in Section 6.3. The analysis results are valid for the range $D/t \leq 80$ and $dd/D \leq 0.2$, for which the experimental data have been used for the verification and development of the present program (Duan, Loh, and Chen, 1990b).

The computer program BCDENT is written in FORTRAN 77. It has been tested in two computing environments. The first is in a Gould-NP1 Main-Frame, Sun Workstation, and HP-Apollo Workstation using UNIX Fortran 77 compiler. The second is in an IBM or compatible Personal Computer using a MicroSoft Fortran 77 compiler, version 5.1.

6.4.2 Structure of Program BCDENT

The program BCDENT consists of a main program and 27 subroutines as listed in Fig. 6.19. The function of the main program and its subroutines are explained as follows:

- main: control all subroutines
- datapt: read and print all data required in the analysis
- capact: calculate the sectional load-carrying capacity of undented cylindrical sections including the effects of local buckling (based on modified API-RP-2A specification)
- respint: compute initial response for constant axial loads, or constant end-moments, or lateral loads
- respaxd: compute load-displacement response for increasing axial load only
- respcon: compute load-displacement response for increasing lateral concentrated load only
- respmed: compute load-displacement response for increasing end-moment only
- respdis: compute load-displacement response for increasing lateral distributed loads only
- bedm: compute bending moments at every station with respect to X-X and Y-Y axes, respectively, and magnitude and direction of resultant moments
- phimbr: compute curvatures at every station for initial constant axial load
- phimbs: compute curvatures at every station for constant axial load and increasing lateral loads
- mphiud: compute curvature for a given moment and axial load for undented circular cylindrical sections
- mphidt: compute curvature for a given moment and axial load for dented circular cylindrical sections
- abcf: determine the constants a, b, c, and f for given boundaries of the three-regime M-P-Φ curve
- mpud: determine boundaries of the three-regime M-P-Φ curve for a given axial load on an undented cylindrical section
- descen: determine the parameters of descending branch of M-P-Φ curve for undented cylindrical sections

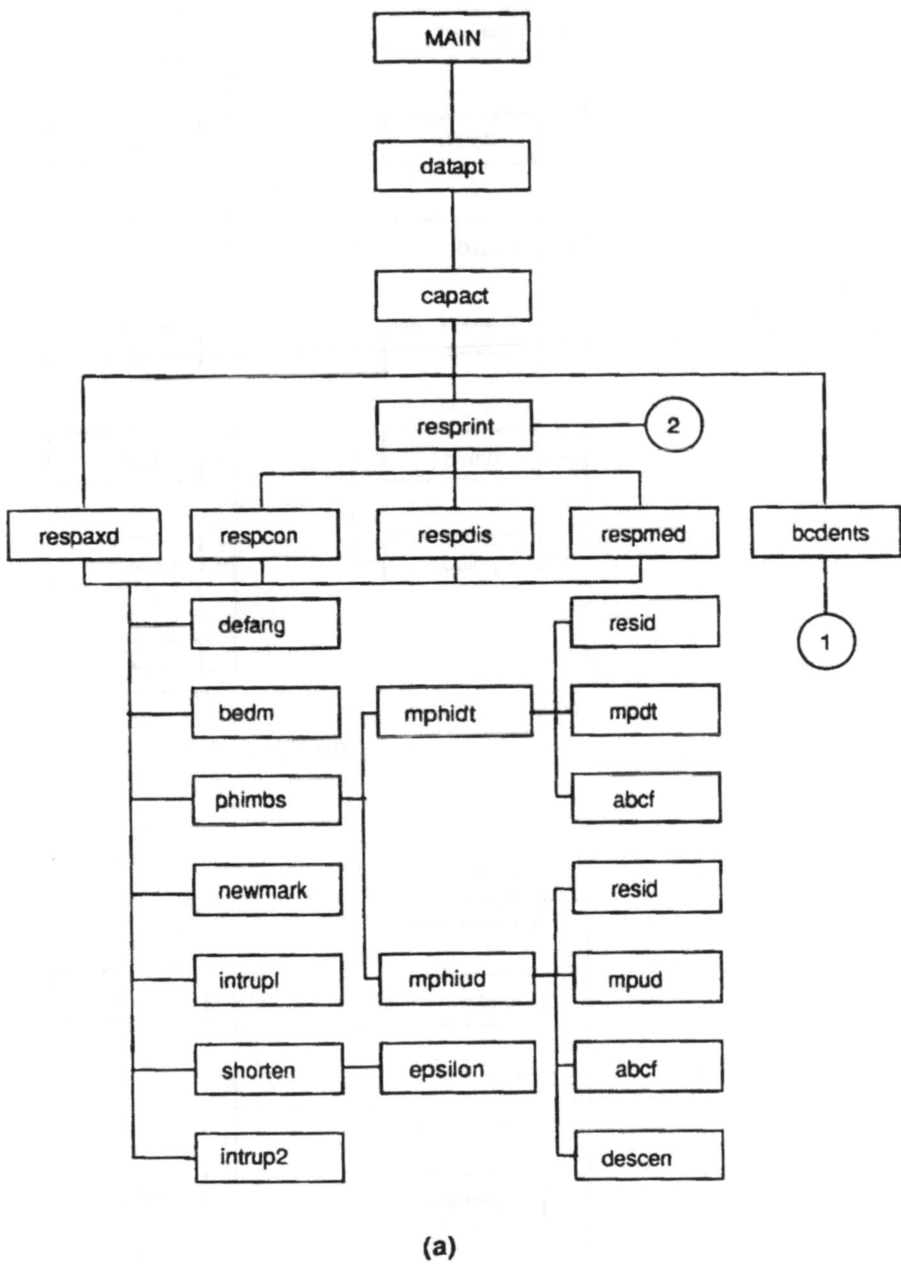

(a)

FIGURE 6.19 Structure of program BCDENT.

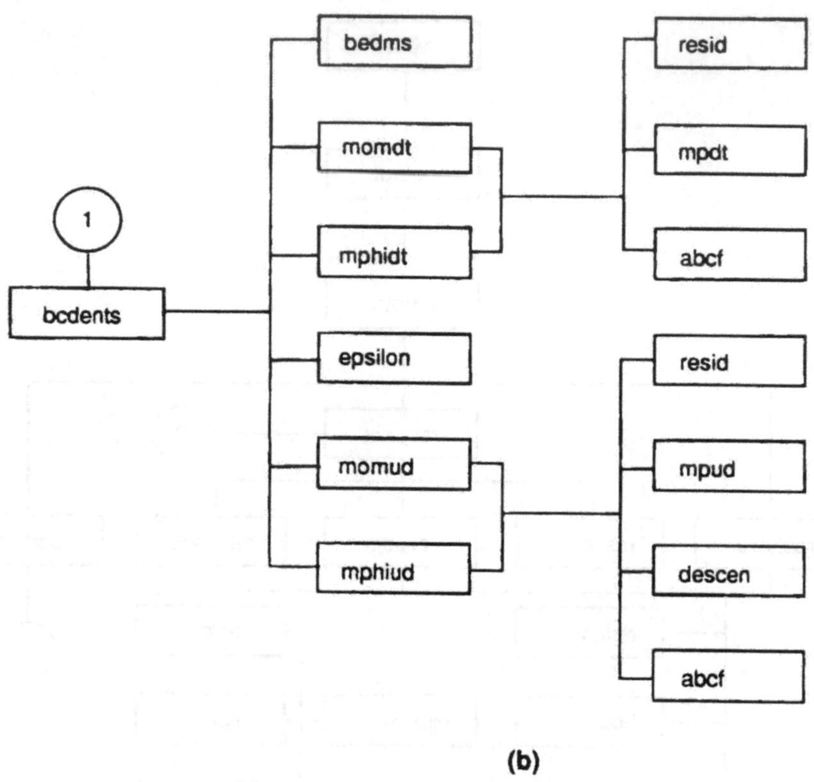

(b)

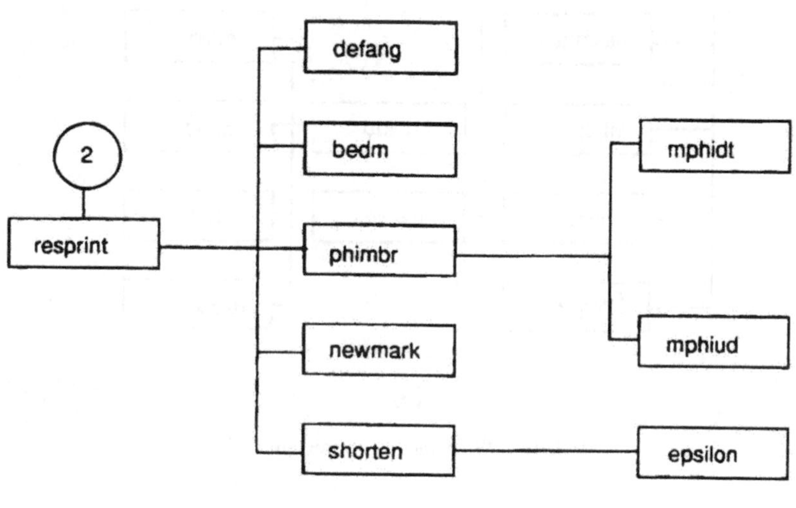

(c)

FIGURE 6.19 *Continued*

mpdt: determine boundary parameters of the three-regime M-P-Φ curve for a given axial load on dented cylindrical sections

resid: determine the residual stresses factors $f_1(R)$ and $f_2(R)$ considering the effects of residual stresses on M-P-Φ curve

shorten: compute axial-shortening of the member

epsilon: compute axial strain for a given moment and axial load for cylindrical sections

newmark: compute the deflections and rotations at all stations by the Newmark integration procedure

intrup1: print the interrupt information when the number of iterations are more than 500 times

intrup2: print the interrupt information when the currently computed deflection is less than the deflection obtained at last step

defang: compute magnitudes and directional angles of resultant deflections

bcdents: use the simple plastic hinge approach to analyze approximately the behavior of dented cylindrical members subjected to constant lateral loads with an increasing axial load

bedms: compute bending moments at the critical section with respect to X-X and Y-Y axes by the simple plastic hinge approach

momud: determine moments for a given curvature and axial load for undented cylindrical sections

momdt: determine moments for a given curvature and axial load for dented cylindrical sections

The complete source code listing for the program BCDENT (Version 1.2—for IBM PC) is provided in the attached diskette. Users are encouraged to modify or improve the program for their use. Program BCDENT is a research development; thus, some improvements are still required to enhance its capability for practical engineering use.

6.5 USER'S MANUAL FOR BCDENT

This section presents the necessary instruction for the user to execute BCDENT. The user is advised to read all the instructions carefully.

6.5.1 Input Data Organization

The input required by BCDENT is divided into six general classes. GEOMETRY AND MATERIAL cards give the geometry and material properties of the dented cylindrical members to be analyzed. LOAD DEFINITION cards give loading types, starting loads, and other information. INITIAL DEFLECTION cards allow a user to input initial deflections with respect to X-X and Y-Y axes at every station or node. SEGMENT DEFINITION cards are used to input segment length, dent depth, and angles. The CONTROL card gives deflection increment and computation stop point. SIMPLE HINGE cards allow a user to input data required by Simple Plastic-

Hinge Approach for analysis of pin-end and cantilever dented members subjected to an increasing axial load only .

6.5.2 Input Data Formats

Geometry and Material Cards

Card	Columns	Variable	Description	Format
1	1-5	nbond	indicator for boundary conditions of member analyzed 100 = simply supported or elastic rationally restrained (Fig. 6.16) 200 = cantilever member (Fig. 6.16)	I5
	6-10	nlocal	indicator for consideration of local buckling of undented tubular sections 1 = M-P-Φ with local buckling 2 = M-P-Φ without local buckling	I5
	11-15	nstant	number of controlling station	I5
2	1-10	d	outer diameter of tube (mm)	F10.0
	11-20	t	thickness of tube (mm)	F10.0
	21-30	e	elastic modulus of material of tube (Mpa)	F10.0
	31-40	fy	yield strength of material of tube (Mpa)	F10.0
	41-50	tl	length of member (mm)	F10.0
3	1-5	ires	indicator for consideration of effects of residual stresses 1 = M-P-Φ with residual stresses 2 = M-P-Φ without residual stresses	I5
4	1-10	rstax	linear rotational restraint at A-end about X-X axis (N-mm/Rad)	F10.0
	11-20	rstay	linear rotational restraint at A-end about Y-Y (N-mm/Rad)	F10.0
	21-30	rstbx	linear rotational restraint at B-end about X-X axis (N-mm/Rad)	F10.0
	31-40	rstby	linear rotational restraint at B-end about Y-Y axis (N-mm/Rad)	F10.0

Users Note: 1. For pin-ended members, all end-restraints be given zeroes.
2. For cantilever members, all end-restraints be given zeroes.

Load Definition Cards

Card	Columns	Variable	Description	Format
5	1-5	ip	indicator for loading cases: 1 = increasing axial load only 2 = increasing axial load combined with constant end-moments and lateral concentrated loads 3 = increasing axial load combined with constant end-moments and lateral distributed loads 4 = increasing lateral concentrated loads combined with constant axial load 5 = increasing lateral distributed load combined with constant axial load 6 = increasing end-moments combined with constant axial load	I5
		If ip = 1, Loading Case 1, input		
5.1	1-10	pmin	minimum axial load ratio (P_{min}/P_y)	F10.0
	11-20	pmax	maximum axial load ratio (P_{max}/P_y)	F10.0
5.2	1-10	eax	eccentricity (X-X axis) at A-end (mm)	F10.0
	11-20	eay	eccentricity (Y-Y axis) at A-end (mm)	F10.0
	21-30	ebx	eccentricity (X-X axis) at B-end (mm)	F10.0
	31-40	eby	eccentricity (Y-Y axis) at B-end (mm)	F10.0
		If ip = 2, Loading Case 2, input		
5.1	1-10	pmin	minimum axial load ratio (P_{min}/P_y)	F10.0
	11-20	pmax	maximum axial load ratio (P_{max}/P_y)	F10.0
5.2	1-10	eax	eccentricity (X-X axis) at A-end (mm)	F10.0
	11-20	eay	eccentricity (Y-Y axis) at A-end (mm)	F10.0
	21-30	ebx	eccentricity (X-X axis) at B-end (mm)	F10.0
	31-40	eby	eccentricity (Y-Y axis) at B-end (mm)	F10.0
5.3	1-10	emax	A-end moment (X-X axis) ratio (M_{ax}/M_p)	F10.0
	11-20	emay	A-end moment (Y-Y axis) ratio (M_{ay}/M_p)	F10.0
	21-30	embx	B-end moment (X-X axis) ratio (M_{bx}/M_p)	F10.0
	31-40	emby	B-end moment (Y-Y axis) ratio (M_{by}/M_p)	F10.0
5.4	1-5	iqc	number of lateral concentrated loads 1 = only one concentrated load 2 = two concentrated loads	I5
		If iqc = 1, input		
5.5	1-10	qcx1	concentrated load at X-X axis (N)	F10.0
	11-20	qcy1	concentrated load at Y-Y axis (N)	F10.0
	21-30	zc1	distance from A-end to concentrated load (mm)	F10.0

If iqc = 2, input				
5.5	1-10	qcx1	first concentrated load at X-X axis (N)	F10.0
	11-20	qcy1	first concentrated load at Y-Y axis (N)	F10.0
	21-30	zc1	distance from A-end to first concentrated load (mm)	F10.0
	31-40	qcx2	second concentrated load at X-X axis (N)	F10.0
	41-50	qcy2	second concentrated load at Y-Y axis (N)	F10.0
	51-60	zc2	distance from A-end to second concentrated load (mm)	F10.0
If ip = 3, Loading Case 3, input				
5.1	1-10	pmin	minimum axial load ratio (P_{min}/P_y)	F10.0
	11-20	pmax	maximum axial load ratio (P_{max}/P_y)	F10.0
5.2	1-10	eax	eccentricity (X-X axis) at A-end (mm)	F10.0
	11-20	eay	eccentricity (Y-Y axis) at A-end (mm)	F10.0
	21-30	ebx	eccentricity (X-X axis) at B-end (mm)	F10.0
	31-40	eby	eccentricity (Y-Y axis) at B-end (mm)	F10.0
5.3	1-10	emax	A-end moment (X-X axis) ratio (M_{ax}/M_p)	F10.0
	11-20	emay	A-end moment (Y-Y axis) ratio (M_{ay}/M_p)	F10.0
	21-30	embx	B-end moment (X-X axis) ratio (M_{bx}/M_p)	F10.0
	31-40	emby	B-end moment (Y-Y axis) ratio (M_{by}/M_p)	F10.0
5.4	1-10	quax	density of distributed load at A-end (X-X axis) (N/mm)	F10.0
	11-20	quay	density of distributed load at A-end (Y-Y axis) (N/mm)	F10.0
	21-30	qubx	density of distributed load at B-end (X-X axis) (N/mm)	F10.0
	31-40	quby	density of distributed load at B-end (Y-Y axis) (N/mm)	F10.0
If ip = 4, Loading Case 4, input				
5.1	1-10	pratio	constant axial load ratio (P/P_y)	F10.0
	11-20	zc1	distance from A-end to first concentrated load (mm)	F10.0
	21-30	zc2	distance from A-end to second concentrated load (mm)	F10.0
5.2	1-10	qymin	minimum concentrated load (Y-Y axis) (N)	F10.0
	11-20	qymax	maximum concentrated load (Y-Y axis) (N)	F10.0
	21-30	rqxy	concentrated load ratio (Q_x/Q_y)	F10.0

If ip = 5, Loading Case 5, input				
5.1	1-10	pratio	constant axial load ratio (P/P_y)	F10.0
	11-20	qrx	end-distributed load ratio (X-X axis) (Q_{ubx}/Q_{uax})	F10.0
	21-30	qry	end-distributed load ratio (Y-Y axis) (Q_{uby}/Q_{uby})	F10.0
5.2	1-10	quymin	minimum density of distributed load (Y-Y axis) (N/mm)	F10.0
	10-20	quymax	maximum density of distributed load (Y-Y axis) (N/mm)	F10.0
	21-30	rquxy	distributed load ratio (Q_{ux}/Q_{uy})	F10.0
If ip = 6, Loading Case 6, input				
5.1	1-10	pratio	constant axial load ratio (P/P_y)	F10.0
	11-20	emrx	end-moment ratio (X-X axis) (M_{bx}/M_{ax})	F10.0
	21-30	emry	end-moment ratio (Y-Y axis) (M_{by}/M_{by})	F10.0
	31-35	iemax	direction of A-end moment (X-X axis)	I5
	46-50	iemay	direction of A-end moment (Y-Y axis)	I5
5.2	1-10	emymin	minimum end-moment ratio (Y-Y axis) (M_{ymin}/M_p)	F10.0
	11-20	emymax	maximum end-moment ratio (Y-Y axis) (M_{ymax}/M_p)	F10.0
	21-30	redmxy	A-end-moment ratio (M_{ay}/M_{ax})	F10.0

User Notes: For cantilever members, only data at the A-end are needed, and data at the B-end are given zeroes (see Fig. 6.16).

Initial Deflection Cards

Card	Columns	Variable	Description	Format
6	1-5	n	number of segments	I5
7.1	1-10	wint(1,1)	initial deflection (X-X axis) at Station 1 (mm)	F10.0
	11-20	wint(2,1)	initial deflection (Y-Y axis) at Station 1	F10.0
Continue F10.0 input for all stations (n + 1), using additional cards if needed				

Segment Definition Cards

Card	Columns	Variable	Description	Format
8.1	1-10	detl(1)	length of Segment 1 (mm)	F10.0
	11-20	ndent(1)	indicator for dented segment 1 = dented segment 0 = undented segment	I10
	21-30	ddent(1)	dent depth of Segment 1 (mm)	F10.0
	31-40	dbeta(1)	dent direction angle of Segment 1 (degree, see Fig. 6.7)	F10.0
Continue F10.0 input for all segments (n), using additional cards if needed				

Control Card

Card	Columns	Variable	Description	Format
9	1-10	ddwr	deflection incremental ratio (δ/L)	F10.0
	11-20	comstop	stop point, computation will stop when load is down to (comstop)x(peak-load) (from 0.2 to 1.0)	F10.0

Simple Hinge Cards

Card	Columns	Variable	Description	Format
10	1-5	method	indicator for analysis methods 1 = M-Φ integration method 2 = simple-hinge approach	I5
If iqc = 2, input				
11	1-10	sl	distance from A-end to critical section (mm)	F10.0
	11-20	dl	length of critical segment (mm)	F10.0
	21-30	wintx	initial deflection (X-X axis) at the critical section (mm)	F10.0
	31-40	winty	initial deflection (y-y Axis) at the critical section (mm)	F10.0
12	1-10	beta	dent angle respect to X-X axis (degree, See Fig. 6.7)	F10.0
	11-20	dd	dent depth (mm)	F10.0
13	1-10	ppmin	minimum axial load ratio (P_{min}/P_y)	F10.0
	11-20	ppmax	maximum axial load ratio (P_{max}/P_y)	F10.0
	21-30	dphi	curvature incremental ratio ($\delta\Phi/\Phi_y$)	F10.0

6.5.3 Operation of BCDENT on an IBM-PC

To run BCDENT on an IBM or compatible personal computer, a hard disk and MicroSoft Fortran 77 Compiler are generally required. An executable program file BCDENT is given. This program is executed by issuing the command "BCDENT" at the terminals. After entering command BCDENT, the screen will print the following message:

```
BCDENT—ANALYSIS OF DENTED CYLINDRICAL MEMBERS, PLEASE
ENTER YOUR DATA FILE NAME (UP TO 10 CHARACTERS) YOUR
OUTPUT FILE IS DATAFILE.out
and then enter your data file name.
```

6.5.4 Examples

Three sample problems are selected to illustrated how to use Program BCDENT. Details of these dented members are given in their outputs, respectively.

Example 1: Landet and Johnsen's Test D1-32, for loading case 1, increasing axial load only
Example 2: Landet and Johnsen's Test D1-32, using approximate simple plastic-hinge approach
Example 3: A dented cantilever member under constant axial load and increasing laterally concentrated loads.

The input data and parts of output results for these examples are listed in this section.

Input Data of Example 1

```
100    1  4
   140.        4.51    208942.        350.       2527.
  0
    0.0         0.0         0.0         0.0
  1
    0.1       0.745
    0.0         0.0         0.0         0.0
  7
    0.0         0.0
    0.0        5.15
    0.0        8.92
    0.0        10.3
    0.0         9.0
    0.0       7.794
    0.0         4.5
    0.0         0.0
```

```
287.848         0       0.0       0.0
287.848         0       0.0       0.0
287.848         0       0.0       0.0
800.            1      14.0       0.0
287.848         0       0.0       0.0
287.848         0       0.0       0.0
287.848         0       0.0       0.0
  0.0008        0.2
  1
```

Output Results of Example 1

```
Boundary Indicator = 100
Analyzed Member is Simply Supported
Or Elastic-Rotational Restrained

Local Buckling Considered in M-P-Φ Curves
        For Undented Sections

ANALYSIS OF DENTED CYLINDRICAL MEMBER
################################
(Valid for Range of D/t <= 80 and dd/D < = 0.2)

INPUT DATA FOR DENTED MEMBER

Outer Diameter                              D =   140.0000  mm
Thickness                                   t =   4.5100  mm
Elastic Modulus                             E =   208942.0000  Mpa
Yield Strength                             Fy =   350.0000  Mpa
Member Length                               L =   2527.0000  mm
Residual Stress Indicator ires                =   0

Linear Rotational End-Restraint Stiffness Rst
   Rsax (A-end X-X axis) =  .0000E+00 N-mm/Rad
   Rsay (A-end Y-Y axis) =  .0000E+00 N-mm/Rad
   Rsbx (B-end X-X axis) =  .0000E+00 N-mm/Rad
   Rsby (B-end Y-Y axis) =  .0000E+00 N-mm/Rad

Loading Case Indicator ip = 1
   Axial Load Only

Eccentricity A-end X-X eax =  .000 mm
   Eccentricity A-end Y-Y eay =  .000 mm
   Eccentricity B-end X-X ebx =  .000 mm
   Eccentricity B-end Y-Y eby =  .000 mm

Limits of Axial Load Ratio P/Py
   Minimum Axial Load Ratio Pmin/Py =  .1000
   Maximum Axial Load Ratio Pmax/Py =  .7450
```

```
  Number of Segments n = 7

      INITIAL DEFLECTIONS

Stations   Wint (X-X)   Wint (Y-Y)
             (mm)         (mm)
   1        .0000        .0000
   2        .0000       5.1500
   3        .0000       8.9200
   4        .0000      10.3000
   5        .0000       9.0000
   6        .0000       7.7940
   7        .0000       4.5000
   8        .0000        .0000

      DENT INFORMATION

Segments  Delta      Dent        Dent     Dent
          Length   Indicator     Depth    Angle
           (mm)                  (mm)   (degree)

   1     287.848       0          .000     .000
   2     287.848       0          .000     .000
   3     287.848       0          .000     .000
   4     800.000       1        14.000     .000
   5     287.848       0          .000     .000
   6     287.848       0          .000     .000
   7     287.848       0          .000     .000

  Controlling Parameters
  Controlling Station Number Nstat  =      4
  Deflection Incremental Ratio Dw/L  =     .0008
  Deflection Incremental Dw  =    2.0216 mm
  Your Computation Will Stop Until The Load Down to 20.0% Peak Load

  Moment-Curvature Integration Approach is Used in the Analysis

  Sectional Properties
     Cross-section Area A     =   .1920E+04   mm2
     Moment of inertia I      =   .4410E+07   mm4
     Radius of Gyration r     =   47.9295    mm
     Slenderness Ratio L/r    =   52.7233

  Undented Sectional Load-Carrying Capacities
     Axial Yield Strength Py        =  .6719E+03 KN
     Axial Ultimate Strength Pu     =   .6719E + 03 KN
     Yield Moment My                =  .2276E + 02 KN-m
     Plastic Moment Mp              =  .2898E + 02 KN-m
     Ultimate Moment Mu             =  .2898E + 02 KN-m
     Yield Curvature Phiy           =  .23930086E-04 Rad/mm

  Iteration Number              =     14
  Relative Error                =    -.8180E-04
```

Axial Load Ratio = .3374
Axial Load = .2267E+06 N
Dent Segment = 4
Total Deflection −Y = 11.917 mm
Total Deflection −X = .000 mm
Local Shortening = .000 mm
Axial Shortening of Member = 1.614 mm
Axial Shortening/L = .6387E− 03

Station	Mom-X (Mx/Mp)	Def-Y (mm)	Slope-X (Rad)	Mom-Y (My/Mp)	Def-X (mm)	Slope-Y (Rad)
1	.000	.000	.285E−02	.000	.000	.697E−18
2	.047	5.969	.244E−02	.000	.000	.599E−18
3	.082	10.443	.174E−02	.000	.000	.426E−18
	.096	12.323	−.119E−04	.000	.000	−.291E−20
5	.086	11.013	−.179E−02	.000	.000	−.438E−18
6	.073	9.292	−.242E−02	.000	.000	−.594E−18
7	.041	5.300	−.278E−02	.000	.000	−.681E−18
8	.000	.000	.102E−06	.000	.000	.249E−22

Iteration Number = 10
Relative Error = .2824E−03
Axial Load Ratio = .5145
Axial Load = .3457E+06 N
Dent Segment = 4
Total Deflection −Y = 14.193 mm
Total Deflection −X = .000 mm
Local Shortening = .001 mm
Axial Shortening of Member = 2.427 mm
Axial Shortening/L = .9603E − 03

Station	Mom-X (Mx/Mp)	Def-Y (mm)	Slope-X (Rad)	Mom-Y (My/Mp)	Def-X (mm)	Slope-Y (Rad)
1	.000	.000	.556E−02	.000	.000	.136E−17
2	.081	6.751	.487E−02	.000	.000	.119E−17
3	.142	11.922	.362E−02	.000	.000	.886E−18
4	.171	14.344	−.183E−04	.000	.000	−.448E−20
5	.155	13.029	−.370E−02	.000	.000	−.905E−18
6	.128	10.759	−.484E−02	.000	.000	−.119E−17
7	.072	6.072	−.546E−02	.000	.000	−.134E−17
8	.000	.000	.198E−06	.000	.000	.485E−22

Iteration Number = 13
Relative Error = .2306E−03
Axial Load Ratio = .5919
Axial Load = .3977E+06 N
Dent Segment = 4
Total Deflection −Y = 16.471 mm
Total Deflection −X = .000 mm
Local Shortening = .001 mm
Axial Shortening of Member = 2.826 mm
Axial Shortening/L = .1118E−02

Station	Mom-X (Mx/Mp)	Def-Y (mm)	Slope-X (Rad)	Mom-Y (My/Mp)	Def-X (mm)	Slope-Y (Rad)
1	.000	.000	.815E−02	.000	.000	.200E−17
2	.103	7.497	.727E−02	.000	.000	.178E−17
3	.183	13.359	.562E−02	.000	.000	.138E−17
4	.225	6.357	−.210E−04	.000	.000	−.515E−20
5	.206	5.039	−.571E−02	.000	.000	−.140E−17
6	.167	12.190	−.723E−02	.000	.000	−.177E−17
7	.094	6.814	−.804E−02	.000	.000	−.197E−17
8	.000	.000	.289E−06	.000	.000	.709E−22

Peak Axial Load P = 397719.959 N

Iteration Number = 6
Relative Error = −.8420E−03
Axial Load Ratio = .4503
Axial Load = .3026E+06 N
Dent Segment = 4
Total Deflection −Y = 41.583 mm
Total Deflection −X = .000 mm
Local Shortening = .037 mm
Axial Shortening of Member = 4.238 mm
Axial Shortening/L = .1677E−02

Station	Mom-X (Mx/Mp)	Def-Y (mm)	Slope-X (Rad)	Mom-Y (My/Mp)	Def-X (mm)	Slope-Y (Rad)
1	.000	.000	.346E−01	.000	.000	.847E−17
2	.158	15.108	.332E−01	.000	.000	.814E−17
3	.297	28.443	.298E−01	.000	.000	.730E−17
4	.401	38.398	−.138E−04	.000	.000	−.339E−20
5	.387	37.085	−.299E−01	.000	.000	−.731E−17
6	.285	27.284	−.332E−01	.000	.000	−.813E−17
7	.151	14.434	−.345E−01	.000	.000	−.845E−17
8	.000	.000	.121E−05	.000	.000	.297E−21

Iteration Number = 7
Relative Error = −.5188E−03
Axial Load Ratio = .2486
Axial Load = .1670E+06 N
Dent Segment = 4
Total Deflection −Y = 87.356 mm
Total Deflection −X = .000 mm
Local Shortening = .226 mm
Axial Shortening of Member = 9.821 mm
Axial Shortening/L = .3886E−02

Station	Mom-X (Mx/Mp)	Def-Y (mm)	Slope-X (Rad)	Mom-Y (My/Mp)	Def-X (mm)	Slope-Y (Rad)
1	.000	.000	817E−01	.000	.000	200E−16
2	.165	28.670	.803E−01	.000	.000	197E−16
3	.320	55.545	.750E−01	.000	.000	.184E−16
4	.452	78.502	−.314E−05	.000	.000	−.769E−21
5	.445	77.195	−.750E−01	.000	.000	−.184E−16
6	.313	54.402	−.802E−01	.000	.000	−.197E−16

```
   7       .161     28.007  -.817E-01   .000      .000   -.200E-16
   8       .000       .000   .285E-05   .000      .000    .699E-21

  Iteration Number                  =   7
  Relative Error                    =
                                       .7870E-04
  Axial Load Ratio                  =   .1178
  Axial Load                        =   .7916E+05 N
  Dent Segment                      =   4
  Total Deflection -Y               =   159.973 mm
  Total Deflection -X               =   .000 mm
  Local Shortening                  =   .859 mm
  Axial Shortening of Member        =   26.762   mm
  Axial Shortening/L                =   .1059E-01

 Station   Mom-X     Def-Y     Slope-X     Mom-Y     Def-X    Slope-Y
          (Mx/Mp)    (mm)       (Rad)     (My/Mp)    (mm)      (Rad)
   1       .000       .000   .156E+00    .000      .000     .382E-16
   2       .137     50.042   .155E+00    .000      .000     .379E-16
   3       .269     98.359   .147E+00    .000      .000     .360E-16
   4       .388    142.093   .672E-05    .000      .000     .165E-20
   5       .385    140.790  -.147E+00    .000      .000    -.360E-16
   6       .266     97.225  -.155E+00    .000      .000    -.379E-16
   7       .135     49.385  -.156E+00    .000      .000    -.382E-16
   8       .000       .000   .544E-05    .000      .000     .133E-20

 Axial Load Less Than 20.0% Peak Load
 Computation Finish, Thank you for your effort
 Good Bye!
```

Input Data of Example 2

```
100    1  4
    140.       4.51    208942.       350.       2527.
  0
     0.0        0.0        0.0        0.0
  1
     0.1      0.724
     0.0        0.0        0.0        0.0
  7
     0.0        0.0
     0.0       5.15
     0.0       8.92
     0.0       10.3
     0.0        9.0
     0.0      7.794
     0.0        4.5
     0.0        0.0
287.848           0        0.0        0.0
287.848           0        0.0        0.0
287.848           0        0.0        0.0
```

```
800.            1        14.0       0.0
287.848         0         0.0       0.0
287.848         0         0.0       0.0
287.848         0         0.0       0.0
 0.0005       0.2
  2
  863.5       800.        0.0      10.3
    0.0       14.0
   0.01      0.724        0.1
```

Output Results of Example 2

```
Boundary Indicator = 100
Analyzed Member is Simply Supported
Or Elastic-Rotational Restrained

Local Buckling Considered in M-P-PHI Curves
        For Undented Sections

ANALYSIS OF DENTED CYLINDRICAL MEMBER
#########################################
(Valid for Range of D/t <= 80 and dd/D <= 0.2)

INPUT DATA FOR DENTED MEMBER
  Outer Diameter  D  =   140.0000  mm
  Thickness     t  =   4.5100  mm
  Elastic Modulus E  =  208942.0000  Mpa
  Yield Strength Fy  =   350.0000  Mpa
  Member Length   L   =  2527.0000  mm
  Residual Stress Indicator ires =   0

Linear Rotational End-Restraint Stiffness Rst
  Rsax (A-end X-X axis) =    .0000E+00  N-mm/Rad
  Rsay (A-end Y-Y axis) =    .0000E+00  N-mm/Rad
  Rsbx (B-end X-X axis) =    .0000E+00  N-mm/Rad
  Rsby (B-end Y-Y axis) =    .0000E+00  N-mm/Rad

Loading Case Indicator ip =  1
  Axial Load Only

  Eccentricity A-end X-X eax =    .000  mm
  Eccentricity A-end Y-Y eay =    .000  mm
  Eccentricity B-end X-X ebx =    .000  mm
  Eccentricity B-end Y-Y eby =    .000  mm

Limits of Axial Load Ratio P/Py
  Minimum Axial Load Ratio Pmin/Py  =  .1000
  Maximum Axial Load Ratio Pmax/Py  =  .7240

Number of Segments  n =  7
```

INITIAL DEFLECTIONS

Stations	Wint (X-X) (mm)	Wint (Y-Y) (mm)
1	.0000	.0000
2	.0000	5.1500
3	.0000	8.9200
4	.0000	10.3000
5	.0000	9.0000
6	.0000	7.7940
7	.0000	4.5000
8	.0000	.0000

DENT INFORMATION

Segments	Delta Length (mm)	Dent Indicator	Dent Depth (mm)	Dent Angle (degree)
1	287.848	0	.000	.000
2	287.848	0	.000	.000
3	287.848	0	.000	.000
4	800.000	1	14.000	.000
5	287.848	0	.000	.000
6	287.848	0	.000	.000
7	287.848	0	.000	.000

Controlling Parameters
Controlling Station Number Nstat = 4
Deflection Incremental Ratio Dw/L = .0005
Deflection Incremental Dw = 1.2635 mm

Your Computation Will Stop Until The Load Down to 20.0% Peak Load

Simple Plastic-Hinge Approach is Used in the Analysis
(Approximated Estimation)

Controlling Parameters
Distance of Critical Section L1 = 863.5000 mm
Length of Critical Segment DL = 800.0000 mm
Initial Deflection (X-X) Wintx = .0000 mm
Initial Deflection (Y-Y) Winty = 10.3000 mm
Dent Angle Respect to X-X Beta = .000 Degree
Dent Depth dd = 14.000 mm

Minimum Axial Load Ratio Pmin/Py = .0100
Maximum Axial Load Ratio Pmax/Py = .7240
Curvature Incremental dPhi/Phiy = .1000 Rad/mm

Sectional Properties
Cross-section Area A = .1920E+04 mm2
Moment of inertia I = .4410E+07 mm4

Radius of Gyration r = 47.9295 mm
Slenderness Ratio L/r = 52.7233

Undented Sectional Load-Carrying Capacities

Axial Yield Strength	Py	=	.6719E+03 KN
Axial Ultimate Strength	Pu	=	.6719E+03 KN
Yield Moment	My	=	.2276E+02 KN-m
Plastic Moment	Mp	=	.2898E+02 KN-m
Ultimate Moment	Mu	=	.2898E+02 KN-m
Yield Curvature	Phiy	=	.23930086E−04 Rad/mm

P/Py	P (N)	M/My	Phi/Phiy (mm)	Def-X (mm)	Def-Y (mm)	Short
.000	.0000E+00	.000	.000	.000	10.300	.000
.262	.1762E+06	.091	.100	.000	11.721	1.114
.438	.2941E+06	.170	.200	.000	13.143	1.866
.533	.3582E+06	.229	.300	.000	14.564	2.288
.585	.3933E+06	.276	.400	.000	15.985	2.532
.587	.3945E+06	.302	.500	.000	17.407	2.571
.582	.3912E+06	.324	.600	.000	18.828	2.608

Peak Axial Load P = 394465.211 N

.576	.3871E+06	.344	.700	.000	20.249	2.653
.524	.3523E+06	.446	1.300	.000	28.777	3.064
.501	.3363E+06	.467	1.500	.000	31.620	3.255
.478	.3211E+06	.486	1.700	.000	34.462	3.475
.456	.3067E+06	.503	1.900	.000	37.305	3.721
.426	.2865E+06	.523	2.200	.000	41.569	4.143
.399	.2680E+06	.540	2.500	.000	45.833	4.626
.374	.2511E+06	.553	2.800	.000	50.097	5.170
.351	.2356E+06	.563	3.100	.000	54.361	5.777
.323	.2169E+06	.572	3.500	.000	60.046	6.682
.298	.2001E+06	.578	3.900	.000	65.731	7.697
.276	.1852E+06	.581	4.300	.000	71.416	8.823
.251	.1686E+06	.582	4.800	.000	78.523	10.385
.225	.1513E+06	.579	5.400	.000	87.051	12.487
.199	.1340E+06	.571	6.100	.000	97.000	15.255
.175	.1175E+06	.559	6.900	.000	108.371	18.836
.149	.1003E+06	.540	7.900	.000	122.584	23.940
.125	.8377E+05	.514	9.100	.000	139.640	30.993
.123	.8254E+05	.512	9.200	.000	141.061	31.627
.121	.8135E+05	.509	9.300	.000	142.482	32.268
.119	.8016E+05	.507	9.400	.000	143.903	32.916
.118	.7901E+05	.504	9.500	.000	145.325	33.571
.116	.7787E+05	.502	9.600	.000	146.746	34.233

Axial Load Less Than 20.0% Peak Load
Computation Finish, Thank you for your effort
Good Bye!

Input Data of Example 2

```
100    1  1
   140.        3.01   214524.        388.  2628.812
 0
    0.0         0.0         0.0         0.0
 4
  0.375     1126.65 1502.1905
     1.       1200.         3.
 7
   2.63        2.63
  2.045       2.045
  1.489       1.489
  0.990       0.990
  0.574       0.574
  0.260       0.260
  0.066       0.066
    0.0         0.0
 375.55           0         0.0         0.0
 375.55           0         0.0         0.0
 375.55           0         0.0         0.0
375.5405          1        14.0       180.0
375.5405          0         0.0         0.0
375.5405          0         0.0         0.0
375.5405          0         0.0         0.0
 0.0005         0.2
 1
```

Output Results of Example 3

```
Boundary Indicator = 101
Analyzed Member is Cantilever with Free-A-End

Local Buckling Considered in M-P-PHI Curves
      For Undented Sections

ANALYSIS OF DENTED CYLINDRICAL MEMBER
####################################
(Valid for Range of D/t < = 80 and dd/D < = 0.2)

INPUT DATA FOR DENTED MEMBER
   Outer Diameter D =       140.0000 mm
   Thickness    t =     3.0100 mm
   Elastic Modulus  E =   214524.0000 Mpa
   Yield Strength Fy =    388.0000 Mpa
   Member Length   L =   2628.8120 mm
   Residual Stress Indicator ires  =   0
 Linear Rotational End-Restraint Stiffness Rst
   Rsax (A-end X-X axis) =  .0000E+00 N-mm/Rad
   Rsay (A-end Y-Y axis) =  .0000E+00 N-mm/Rad
   Rsbx (B-end X-X axis) =  .0000E+00 N-mm/Rad
   Rsby (B-end Y-Y axis) =  .0000E+00 N-mm/Rad
```

```
Loading Case Indicator ip = 4
   Constant Axial Load Ratio P/Py =    .3750
   Increasing Lateral Concentrated Loads

   Load #1—Distance from A-end zc1 = 1126.6500 mm
   Load #2—Distance from A-end zc2 = 1502.1905 mm

Limits of Lateral Concentrated Load

   Qymin (Y-Y axis) = .1000E + 01 N
   Qymax (Y-Y axis) = .1200E + 04 N
   Load Ratio Qx/Qy = .3000E + 01

Number of Segments n = 7
```

INITIAL DEFLECTIONS

Stations	Wint (X-X) (mm)	Wint (Y-Y) (mm)
1	2.6300	2.6300
2	2.0450	2.0450
3	1.4890	1.4890
4	.9900	.9900
5	.5740	.5740
6	.2600	.2600
7	.0660	.0660
8	.0000	.0000

DENT INFORMATION

Segments	Delta Length (mm)	Dent Indicator	Dent Depth (mm)	Dent Angle (degree)
1	375.550	0	.000	.000
2	375.550	0	.000	.000
3	375.550	0	.000	.000
4	375.541	1	14.000	180.000
5	375.541	0	.000	.000
6	375.541	0	.000	.000
7	375.541	0	.000	.000

```
Controlling Parameters
   Controlling Station Number Nstat  = 1
   Deflection Incremental Ratio Dw/L = .0005
   Deflection Incremental  Dw    =     1.3144 mm

Your Computation Will Stop Until The Load Down to 20.0% Peak Load

Moment-Curvature Integration Approach is Used
in the Analysis

Sectional Properties
   Cross-section Area    A =   .1295E + 04  mm2
   Moment of inertia     I =   .3040E + 07  mm4
```

```
Radius of Gyration      r =   48.4450 mm
Slenderness Ratio     L/r =   54.2639

Undented Sectional Load-Carrying Capacities
  Axial Yield Strength         Py =   .5026E+03 KN
  Axial Ultimate Strength      Pu =   .5026E+03 KN
  Yield Moment                 My =   .1721E+02 KN-m
  Plastic Moment               Mp =   .2192E+02 KN-m
  Ultimate Moment              Mu =   .2193E+02 KN-m
  Yield Curvature            Phiy =   .25837935E-04 Rad/mm

  Iteration Number              =  27
  Relative Error                =  .0000E+00
  Axial Load Ratio              =  .3750
  Axial Load                    =  .1885E+06 N
  Dent Segment                  =  4
    Total Deflection -Y         =  4.027 mm
    Total Deflection -X         =  4.027 mm
    Local Shortening            =  .001 mm
  Axial Shortening of Member    =  1.783  mm
  Axial Shortening/L            =  .6783E-03
```

Station	Mom-X (Mx/Mp)	Def-Y (mm)	Slope-X (Rad)	Mom-Y (My/Mp)	Def-X (mm)	Slope-Y (Rad)
1	.000	14.199	.699E−02	.000	14.199	.699E−02
2	−.028	10.987	.665E−02	.028	10.987	.665E−02
3	−.054	7.932	.597E−02	.054	7.932	.597E−02
4	−.077	5.191	.478E−02	.077	5.191	.478E−02
5	−.096	2.980	.353E−02	.096	2.980	.353E−02
6	−.111	1.339	.217E−02	.111	1.339	.217E−02
7	−.119	.330	.703E−03	.119	.330	.703E−03
8	−.122	.000	.165E−17	.122	.000	.101E−17

```
Iteration Number              =  10
Relative Error                =  -.4605E-03
Axial Load Ratio              =  .3750
Axial Load                    =  .1885E+06 N
Concentrated Load Qy          =  .1127E+03 N
Concentrated Load Qx          =  .3382E+03 N
Dent Segment                  =  4
Total Deflection -Y           =  5.349 mm
Total Deflection -X           =  6.071 mm
Local Shortening              =  .001 mm
Axial Shortening of Member    =  1.966  mm
Axial Shortening/L            =  .7480E-03
```

Station	Mom-X (Mx/Mp)	Def-Y (mm)	Slope-X (Rad)	Mom-Y (My/Mp)	Def-X (mm)	Slope-Y (Rad)
1	.000	18.729	.969E−02	.000	20.735	.107E−01
2	−.040	14.503	.921E−02	.036	16.148	.102E−01
3	−.074	10.489	.827E−02	.074	11.759	.926E−02
4	−.105	6.885	.665E−02	.109	7.783	.758E−02
5	−.132	3.972	.498E−02	.143	4.522	.577E−02

6	−.154	1.788	.308E−02	.175	2.042	.361E−02
7	−.170	.437	.989E−03	.200	.494	.114E−02
8	−.177	.000	.734E−18	.216	.000	−.147E−17

```
Iteration Number              =    10
Relative Error                = −.4655E−03
Axial Load Ratio              =   .3750
Axial Load                    =   .1885E+06 N
Concentrated Load Qy          =   .3101E+03 N
Concentrated Load Qx          =   .9304E+03 N
Dent Segment                  =    4
Total Deflection −Y           =    7.076 mm
Total Deflection −X           =    9.870 mm
Local Shortening              =   .000 mm
Axial Shortening of Member    =   2.173  mm
Axial Shortening/L            =   .8268E−03
```

Station	Mom-X (Mx/Mp)	Def-Y (mm)	Slope-X (Rad)	Mom-Y (My/Mp)	Def-X (mm)	Slope-Y (Rad)
1	.000	24.409	.130E−01	.000	32.878	.175E−01
2	−.054	18.942	.123E−01	.057	25.719	.168E−01
3	−.098	13.749	.111E−01	.116	18.852	.153E−01
4	−.138	9.080	.902E−02	.170	12.597	.127E−01
5	−.176	5.277	.688E−02	.231	7.395	.993E−02
6	−.211	2.381	.428E−02	.297	3.352	.628E−02
7	−.237	.578	.136E−02	.351	.800	.196E−02
8	−.253	.000	−.147E−17	.390	.000	−.220E−17

```
Iteration Number              =    15
Relative Error                = −.3612E−03
Axial Load Ratio              =   .3750
Axial Load                    =   .1885E+06 N
Concentrated Load Qy          =   .4027E+03 N
Concentrated Load Qx          =   .1208E+03 N
Dent Segment                  =    4
Total Deflection −Y           =    8.022 mm
Total Deflection −X           =    12.162 mm
Local Shortening              =    .000 mm
Axial Shortening of Member    =    2.336  mm
Axial Shortening/L            =   .8884E−03
```

Station	Mom-X (Mx/Mp)	Def-Y (mm)	Slope-X (Rad)	Mom-Y (My/Mp)	Def-X (mm)	Slope-Y (Rad)
1	.000	27.391	.147E−01	.000	40.171	.216E−01
2	−.059	21.294	.140E−01	.070	31.463	.208E−01
3	−.109	15.495	.126E−01	.142	23.112	.190E−01
4	−.154	10.273	.103E−01	.208	15.495	.158E−01
5	−.197	5.997	.791E−02	.283	9.132	.124E−01
6	−.240	2.711	.497E−02	.367	4.147	.792E−02
7	−.271	.650	.155E−02	.436	.977	.243E−02
8	−.290	.000	.220E−17	.486	.000	.367E−18

```
Iteration Number                  =   10
Relative Error                    = -.9587E-03
Axial Load Ratio                  =  .3750
Axial Load                        =  .1885E+06 N
Concentrated Load Qy              =  .4084E+03 N
Concentrated Load Qx              =  .1225E+04 N
Dent Segment                      =   4
Total Deflection -Y               =   8.022 mm
Total Deflection -X               =   12.545 mm
Local Shortening                  =  .000 mm
Axial Shortening of Member        =   2.372   mm
Axial Shortening/L                =   .9025E-03
```

Station	Mom-X (Mx/Mp)	Def-Y (mm)	Slope-X (Rad)	Mom-Y (My/Mp)	Def-X (mm)	Slope-Y (Rad)
1	.000	27.939	.150E-01	.000	41.353	.223E-01
2	-.061	21.725	.143E-01	.072	32.401	.214E-01
3	-.111	15.818	.128E-01	.146	23.812	.195E-01
4	-.156	10.496	.105E-01	.213	15.976	.163E-01
5	-.201	6.134	.811E-02	.291	9.424	.129E-01
6	-.244	2.776	.511E-02	.377	4.284	.821E-02
7	-.276	.663	.159E-02	.447	1.006	.250E-02
8	-.296	.000	.129E-17	.498	.000	.367E-18

```
Iteration Number                  =   8
Relative Error                    = -.7938E-03
Axial Load Ratio                  =  .3750
Axial Load                        =   .1885E+06 N
Concentrated Load Qy              = .4105E+03 N
Concentrated Load Qx              = .1231E+04 N
Dent Segment                      =  4
Total Deflection -Y               =   8.385 mm
Total Deflection -X               =  12.914 mm
Local Shortening                  =   .000 mm
Axial Shortening of Member        =  2.410   mm
Axial Shortening/L                =   .9167E-03
```

Station	Mom-X (Mx/Mp)	Def-Y (mm)	Slope-X (Rad)	Mom-Y (My/Mp)	Def-X (mm)	Slope-Y (Rad)
1	.000	28.502	.153E-01	.000	42.512	.229E-01
2	-.062	22.172	.145E-01	.074	33.315	.220E-01
3	-.113	16.152	.131E-01	.150	24.492	.201E-01
4	-.159	10.727	.107E-01	.220	16.441	.168E-01
5	-.204	6.275	.830E-02	.299	9.707	.133E-01
6	-.248	2.843	.525E-02	.386	4.416	.849E-02
7	-.281	.677	.163E-02	.458	1.033	.258E-02
8	-.301	.000	-.184E-18	.509	.000	-.257E-17

Peak Concentrated Load Qy = 410.461 N

Iteration Number = 4
Elative Error = −.2479E−03
Axial Load Ratio = .3750
Axial Load = .1885E+06 N
Concentrated Load Qy = .3430E+03 N
Concentrated Load Qx = .1029E+04 N
Dent Segment = 4
Total Deflection −Y = 10.297 mm
Total Deflection −X = 15.986 mm
Local Shortening = .000 mm
Axial Shortening of Member = 2.820 mm
Axial Shortening/L = .1073E−02

Station	Mom-X (Mx/Mp)	Def-Y (mm)	Slope-X (Rad)	Mom-Y (My/Mp)	Def-X (mm)	Slope-Y (Rad)
1	.000	34.357	.185E−01	.000	52.185	.284E−01
2	−.070	26.819	.177E−01	.093	40.938	.272E−01
3	−.132	19.632	.160E−01	.186	30.154	.249E−01
4	−.187	13.126	.132E−01	.271	20.312	.209E−01
5	−.239	7.736	.104E−01	.360	12.048	.166E−01
6	−.287	3.524	.667E−02	.451	5.506	.108E−01
7	−.322	.825	.202E−02	.523	1.271	.321E−02
8	−.341	.000	−.367E−18	.569	.000	.367E−18

Iteration Number = 11
Relative Error = −.7917E−03
Axial Load Ratio = .3750
Axial Load = .1885E+06 N
Concentrated Load Qy = .1569E+03 N
Concentrated Load Qx = .4708E+03 N
Dent Segment = 4
Total Deflection −Y = 12.104 mm
Total Deflection −X = 17.498 mm
Local Shortening = .000 mm
Axial Shortening of Member = 3.191 mm
Axial Shortening/L = .1214E−02

Station	Mom-X (Mx/Mp)	Def-Y (mm)	Slope-X (Rad)	Mom-Y (My/Mp)	Def-X (mm)	Slope-Y (Rad)
1	.000	40.851	.223E−01	.000	58.532	.325E−01
2	−.078	31.873	.214E−01	.109	45.752	.311E−01
3	−.152	23.284	.194E−01	.214	33.506	.284E−01
4	−.219	15.482	.160E−01	.310	22.350	.234E−01
5	−.277	9.066	.123E−01	.398	13.128	.182E−01
6	−.324	4.127	.788E−02	.475	5.986	.117E−01
7	−.357	.973	.241E−02	.531	1.402	.356E−02
8	−.371	.000	.257E−17	.559	.000	.000E+00

Iteration Number = 17
Relative Error = −.5727E−03

```
Axial Load Ratio                 = .3750
Axial Load                       = .1885E+06 N
Concentrated Load Qy             = .5646E+03 N
Concentrated Load Qx             = .1694E+03 N
Dent Segment                     = 4
Total Deflection -Y              = 12.295 mm
Total Deflection -X              = 17.179 mm
Local Shortening                 = .000 mm
Axial Shortening of Member       = 3.246  mm
Axial Shortening/L               = .1235E-02

Station    Mom-X     Def-Y     Slope-X     Mom-Y     Def-X     Slope-Y
          (Mx/Mp)    (mm)       (Rad)     (My/Mp)    (mm)       (Rad)

   1        .000    42.341    .234E-01    .000     59.072    .333E-01
   2      -.079     32.959    .225E-01    .114     45.991    .319E-01
   3      -.156     23.972    .204E-01    .221     33.465    .290E-01
   4      -.226     15.802    .166E-01    .319     22.075    .235E-01
   5      -.284      9.168    .125E-01    .402     12.820    .178E-01
   6      -.329      4.169    .795E-02    .468      5.832    .113E-01
   7      -.358       .990    .246E-02    .512      1.381    .350E-02
   8      -.369       .000   -.147E-17    .529       .000   -.514E-17

 Lateral Concentrated Load Less Than 20.0% Peak Load
      Computation Finish, Thank you for your effort
      Good Bye!
```

6.6 SOLUTIONS OF UNDENTED CYLINDRICAL MEMBER BEHAVIOR

In this section, the numerical comparisons are made for undented members with and without local buckling to confirm the validity of the proposed M-P-Φ-based analysis method and the program BCDENT for these special cases.

6.6.1 Undented Columns Without Local Buckling

Axial load-deflection and axial load-shortening curves by the program BCDENT, including the post-buckling branch of simply supported cylindrical columns with the slenderness ratios (L/r) of 80, 120, and 160, and the initial imperfection of 0.001L, are shown in Figs. 6.20 and 6.21. Features of the member are: outer diameter D = 114.3 mm (4.5 in.), thickness t = 2.38 mm (0.0938 in.), elastic modulus E = 206,700 MPa (30,000 ksi), and yield strength F_y = 248 Mpa (36 ksi). In these figures, the results obtained using the Finite Segment Method by Sugimoto and Chen (1985) are also plotted for comparisons. It is observed that the present results agree well with those reported previously by Sugimoto and Chen (1985).

6.6.2 Undented Columns with Local Buckling

The local buckling behavior of cross sectional distortion can also be considered in the program BCDENT. To further evaluate the present program, the column with

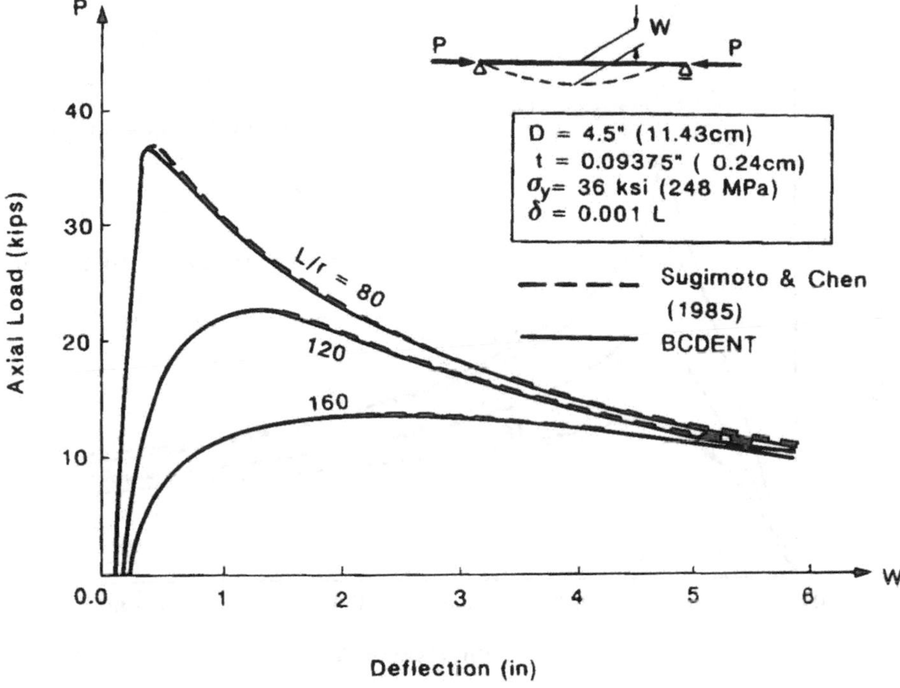

FIGURE 6.20 Axial load-deflection curves for undented cylindrical members without local buckling.

the slenderness ratio equal to 80 and the diameter-to-thickness ratio (D/t) equal to 48 and 100 is chosen for the present analysis. The initial out-of-straightness is taken as 0.001L.

Figure 6.22 shows the comparison of the computed load-deflection and load-shortening curves of a pin-ended member under axial load. After local buckling, the present results are slightly lower than those obtained using the Assumed Deflection Method by Sohal and Chen (1988). This difference is due to the fact that the equilibrium between external loads and internal resistance is enforced only at one critical section in the Assumed Deflection Method, while the equilibrium equation is satisfied at every station (ends of every segments) in the present analysis. Therefore, it is not surprising that the Assumed Deflection Method gives slightly higher curves than the present curves, although the same M-P-Φ curve for descending branch is used in both methods.

6.7 SOLUTIONS OF DENTED CYLINDRICAL MEMBER BEHAVIOR

This section compares the results of dented cylindrical members predicted by the computer program BCDENT with the available experimental tests, providing the confirmation of the validity of the proposed method and the computer program BCDENT.

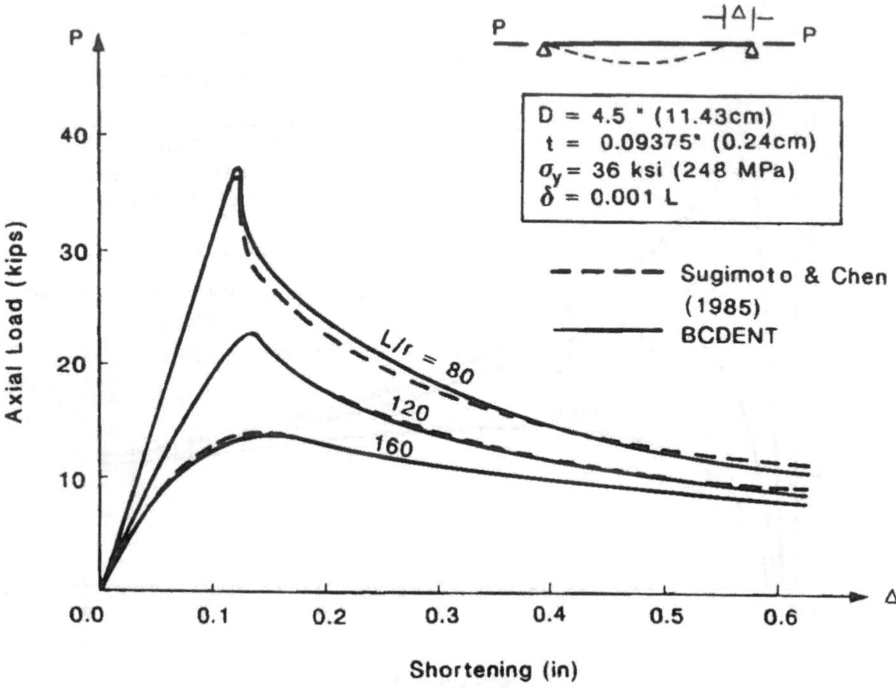

FIGURE 6.21 Axial load-shortening curves for undented cylindrical members without local buckling.

6.7.1 Pin-Ended Dented Columns

In the numerical analysis by the program BCDENT for the Landet and Johnsen (1987) tests, the member is divided into seven segments with eight stations. Since the adopted M-P-Φ relationships for dented cylindrical sections are based on an average moment and curvature along an 800 mm segment, the length of the dented segment is taken here as 800 mm and the dented M-P-Φ curves are adopted at two stations of the dented segment, as shown in Fig. 6.23. The axial load-shortening curves computed by BCDENT are compared with test results for dent depth ratio (dd/D) of 0.1 (D1-35) and 0.2 (D1-36) in Figs. 6.24 and 6.25. The predicted maximum strength is quite close to the test result, but the present curves are generally lower than the tested curves. There are several factors contributing to this difference. In the analytical studies, the ends of members are assumed to be perfectly pin-ended. In the experiments, some end restraints always exist. Other contributing factors are that the M-P-Φ curves used in the BCDENT are developed on the basis of constant axial load tests, and that the present M-P-Φ curves at the pre-maximum region are softer than those tested (Duan, Loh, and Chen, 1990a), especially for the case of dd/D = 0.2.

Figure 6.26 shows a comparison of axial load-deflection curves for Test D1-32 (D/t = 31 and dd/D = 0.1). The analytical result is carried out by using ABAQUS, a commercial finite element program, as reported by MacIntyre and Birkemoe (1989) and that is also shown in Fig. 6.26. It is obvious that the present curve is much closer to the test results than those obtained by ABAQUS, except that the computed maximum load is 7% higher than the test value. Note that a typical analysis (denting

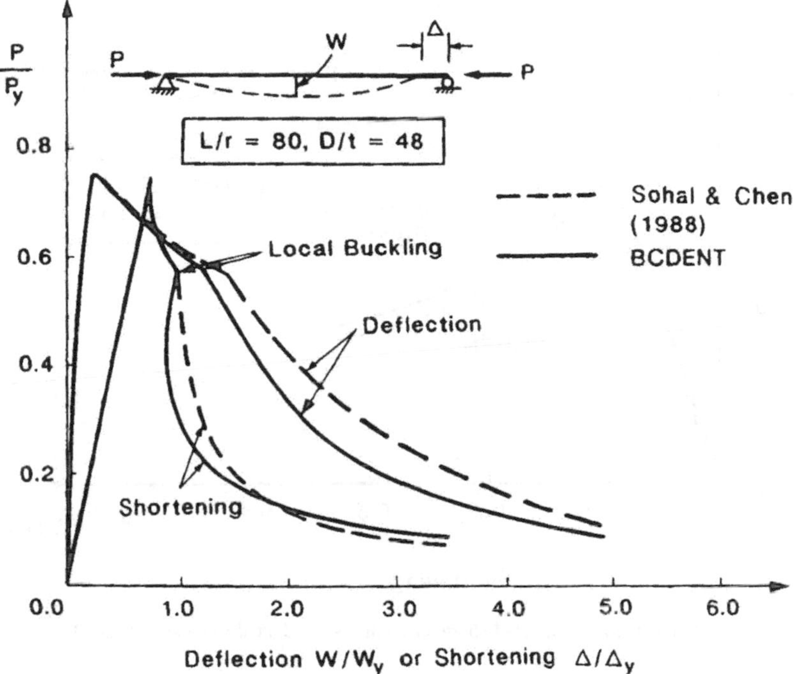

FIGURE 6.22 Axial load-deflection, shortening curves for undented cylindrical members with local buckling (D/t = 48).

and subsequent loading) by ABAQUS requires approximately 35 minutes of CPU time on a Cray X-MP/24 supercomputer (MacIntyre and Birkemoe, 1989).

The comparisons of axial load-deflection and shortening curves for Taby (1986) test 2BCBC are shown in Fig. 6.27. Test 2BCBC is subjected to an eccentric axial load with double curvature, and dent is located at L/4 point. In the analysis, the dented segment is taken approximately as 2D. A good agreement is observed for Test 2BCBC in Fig. 6.27.

6.7.2 Dented Beam-Columns

The load-deflection curves computed by BCDENT for three Landet and Johnsen (1987) beam-column tests are shown in Figs. 6.28 to 6.30. These three tests are

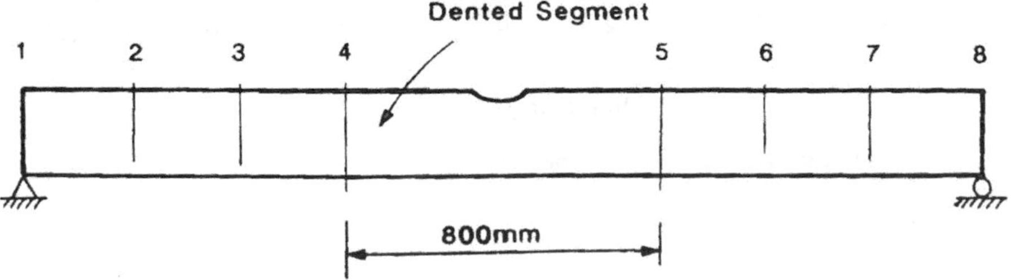

FIGURE 6.23 Divided segments and stations for analysis of Landet and Johnsen's Tests (1987).

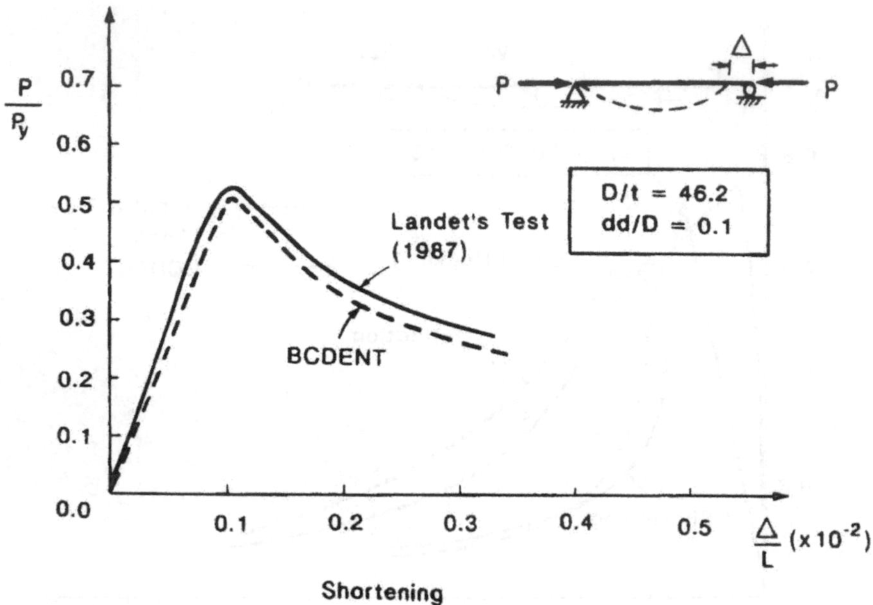

FIGURE 6.24 Axial load-shortening curves for dented member test D1-35.

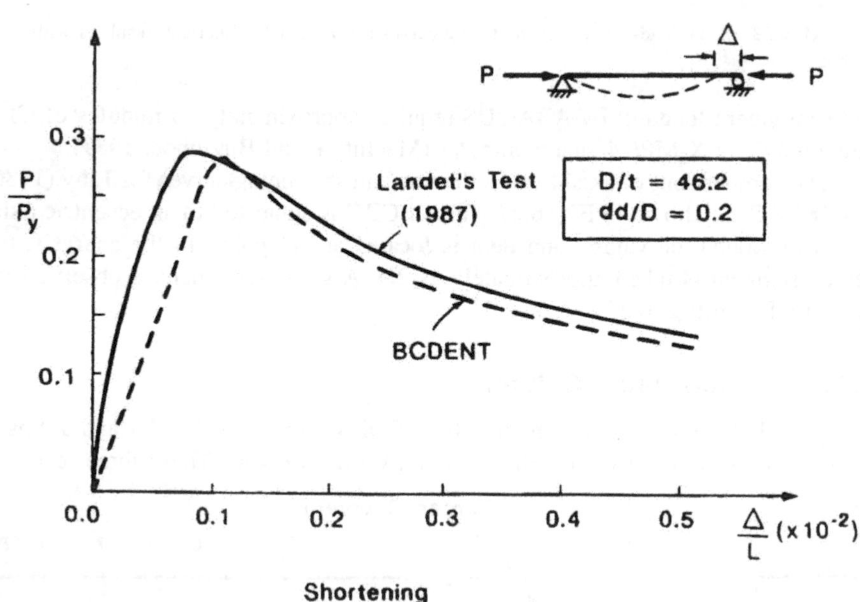

FIGURE 6.25 Axial load-shortening curves for dented member test D1-36.

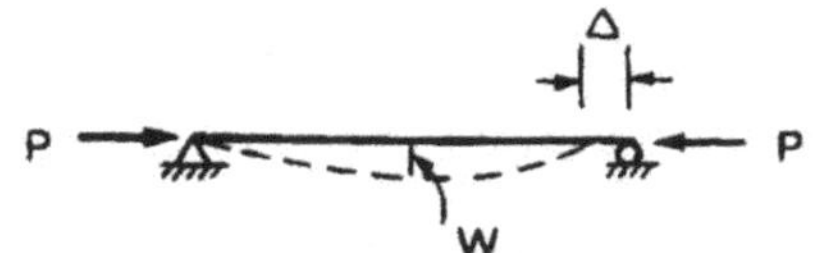

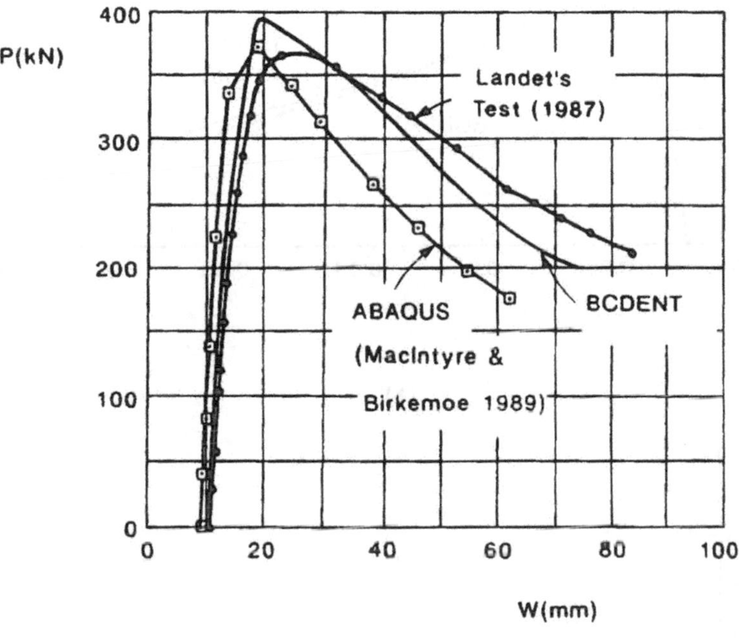

FIGURE 6.26 Axial load-deflection curves for dented member test D1-32.

subjected to two lateral concentrated loads under a constant axial compression. Figure 6.28 is for negative bending, Fig. 6.29 for positive bending, and Fig. 6.30 for neutral bending. In these figures, crosses are the computed maximum lateral concentrated loads Q_{max}, and circles represent the computed maximum bending moments at the critical sections M_{max}. It can be seen that the M_{max} point usually lies after the Q_{max} point. A typical stability behavior of the dented beam-column is predicted by the program BCDENT.

For the present comparison, different M-P-Φ curves are used for analysis of Tests D2-37. In Fig. 6.28, the dashed line is based on the present M-P-Φ curve, while the dotted-dashed line is based on the present M-P-Φ formula using tested peak moment and curvature. It is obvious that the M-P-Φ relationships significantly affect the behavior and strength of dented beam-columns. From Fig. 6.28 it may be concluded that the proposed M-P-Φ-based model can simulate accurately the actual behavior of damaged tubular members if more accurate M-P-Φ relationships are used.

A typical analysis for a dented member subjected to an increasing axial load requires approximately 5 to 10 minutes of CPU time on a Gould NP1 computer.

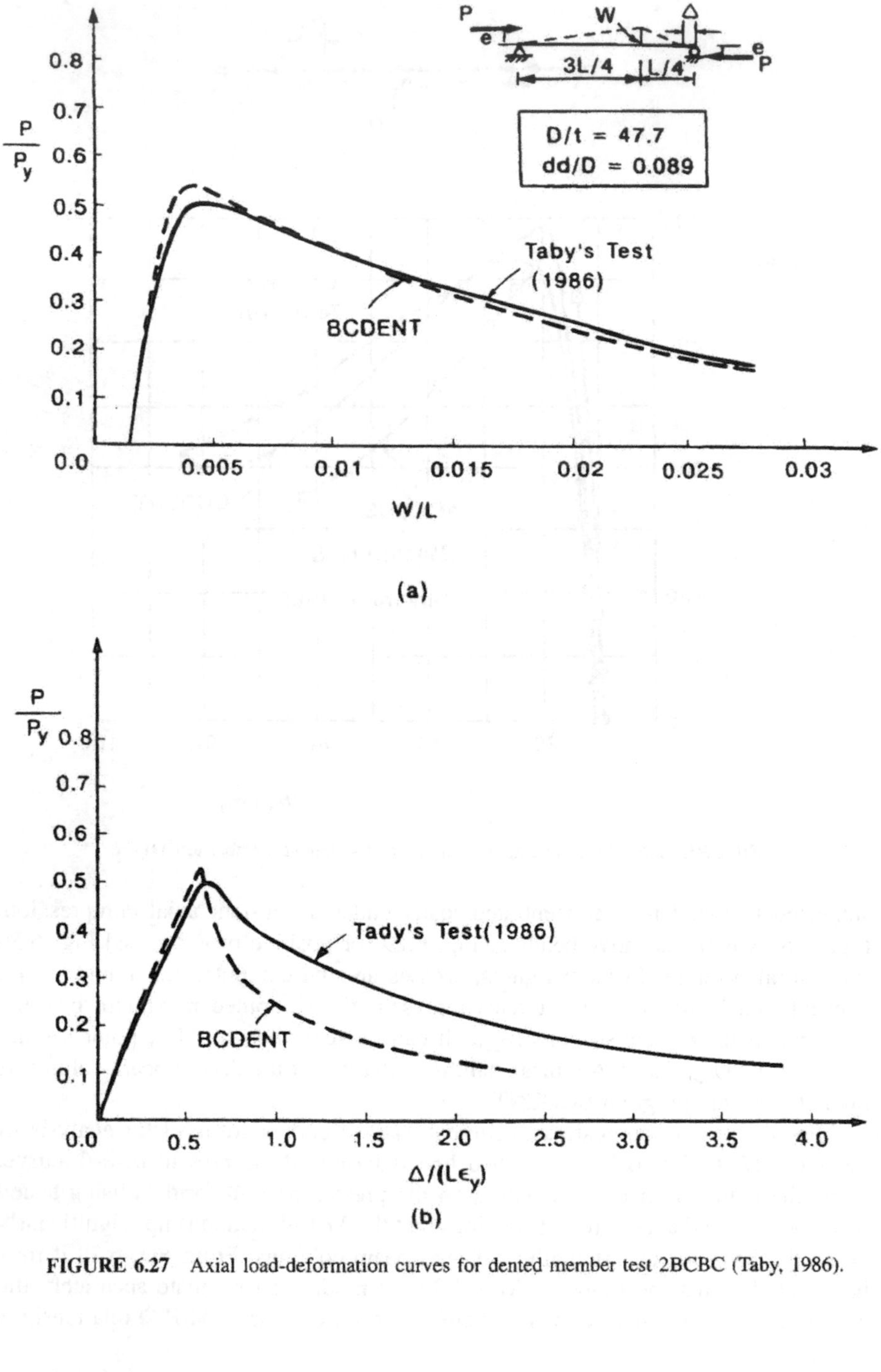

(a)

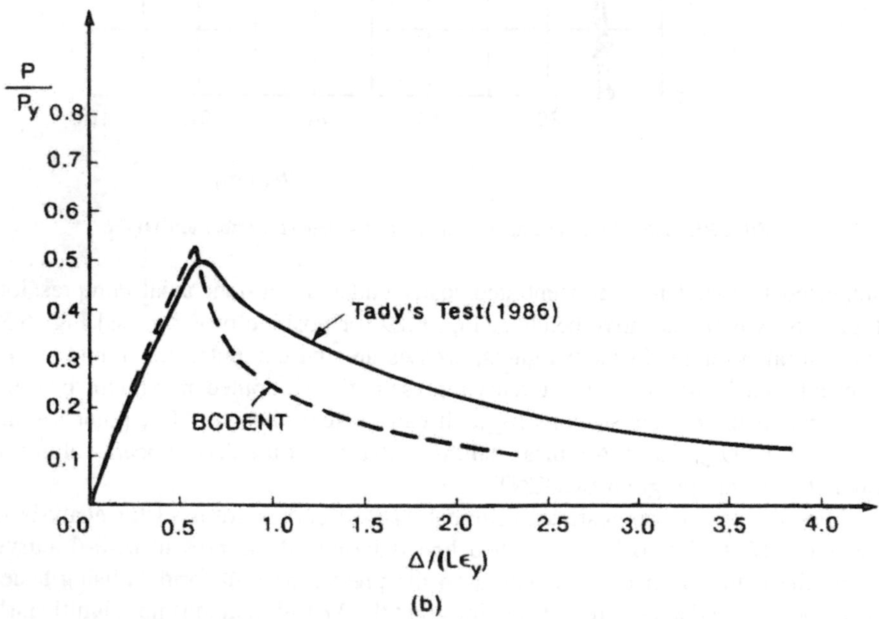

(b)

FIGURE 6.27 Axial load-deformation curves for dented member test 2BCBC (Taby, 1986).

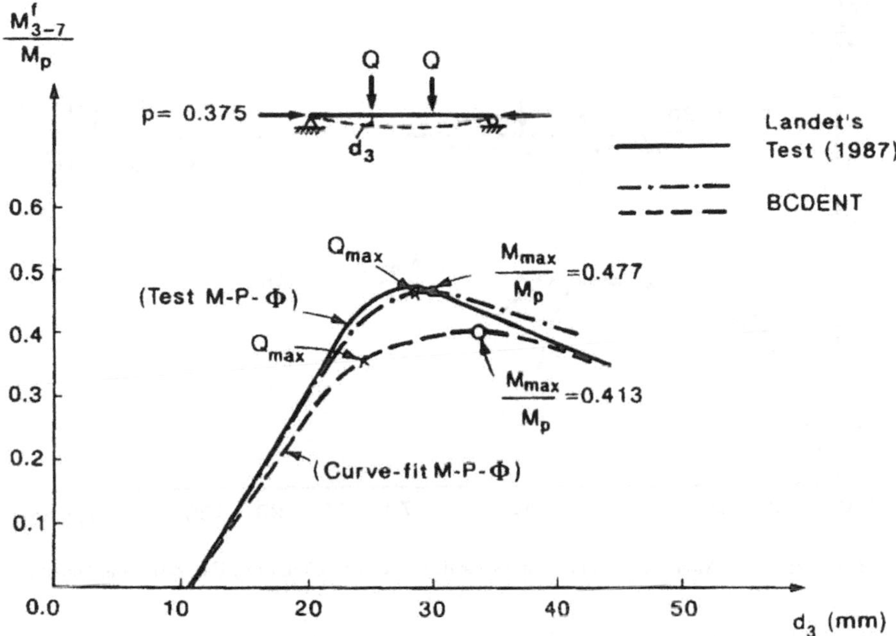

FIGURE 6.28 Moment-deflection curves for dented beam-column test D2-37 (negative bending).

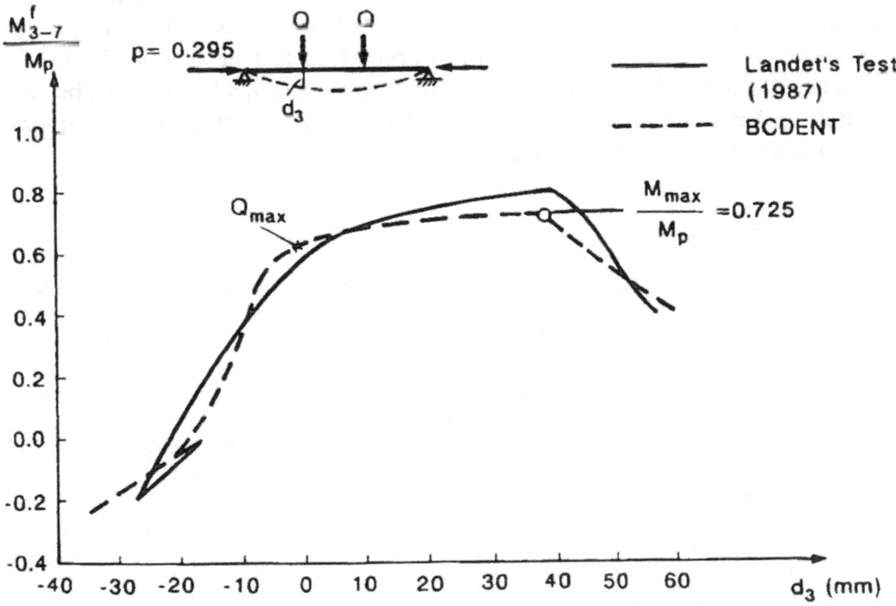

FIGURE 6.29 Moment-deflection curves for dented beam-column test C3-25 (positive bending).

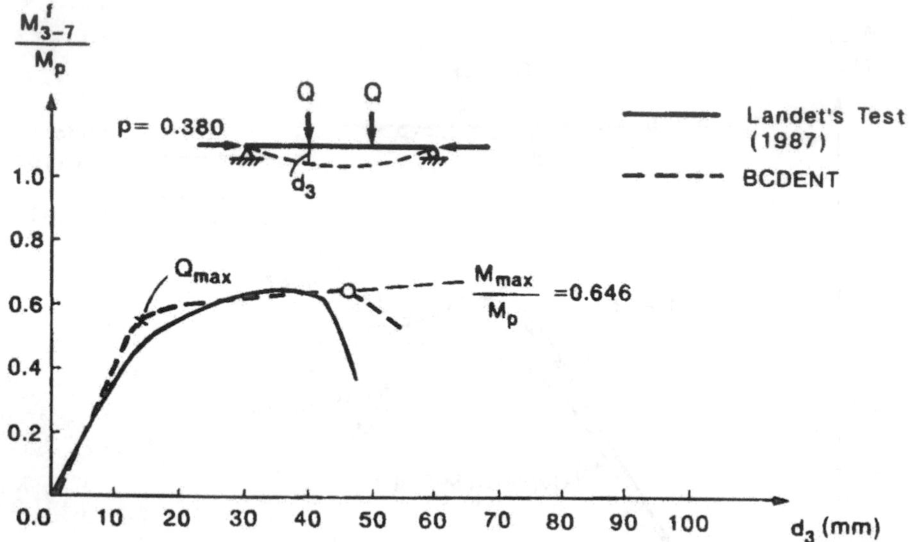

FIGURE 6.30 Moment-deflection curves for dented beam-column test C4-28 (neutral bending).

For dented members subjected to an increasing lateral load and a constant axial compression this requires approximately 1 minute of CPU time.

The M-P-Φ relationships have a significant effect on the behavior and strength of dented cylindrical members. The M-P-Φ expressions implemented in BCDENT are an average representation of all existing M-P-Φ test results. Refinement of these expressions would be possible when additional test data become available. It is concluded that the M-P-Φ-based model can simulate accurately the actual behavior of damaged cylindrical members, provided that a more accurate M-P-Φ relationship is available and is used in the analysis.

REFERENCES

AISC (1993) *Load and Resistance Factor Design Specification for Structural Steel Buildings,* American Institute of Steel Construction, Chicago, IL.

AISC (1989) *Specification for Structural Steel Buildings—Allowable Stress and Plastic Design,* American Institute of Steel Construction, Chicago, IL.

API (1989) *Recommended Practice for Planning, Designing and Constructing Fixed Offshore Platform,* 18th Ed., API-RP-2A, American Petroleum Institute, Washington, DC.

API (1989) *Draft Recommended Practice for Planning, Designing and Constructing fixed Offshore Platforms-Load and Factor Design,* API-RP-2A-LRFD, American Petroleum Institute, Washington, DC.

Chen, W. F. (1971) Further studies of inelastic beam-column problems, *Journal of the Structural Division,* 97, ST2; 529–44.

Chen, W. F. and Atsuta, F. (1976) *Theory of Beam-Columns,* Vol. I: *In-Plane Behavior and Design,* McGraw-Hill, New York, NY.

Chen, W. F. and Han, D. J. (1985) *Tubular Members in Offshore Structures,* Pitman, London.

Chen, W. F. and Ross, D. A. (1977) Test of Fabricated Tubular Columns, *Journal of the Structural Division,* ASCE, 103(ST3), 619–34.

Duan, L., and Chen, W. F. (1990) Design Interaction Equation for Cylindrical Tubular Beam-Columns, *Journal of Structural Engineering,* ASCE, 116, 7; 1794–812.

Duan, L., Loh, J. T., and Chen, W. F. (1990a) *Moment-Thrust-Curvature Relationships for Dented Tubular Section,* Structural Engineering Report No. CE-STR-90-26, School of Civil Engineering, Purdue University, West Lafayette, IN.

Duan, L., Loh, J. T., and Chen, W. F. (1990b) *M-P-Φ–Based Analysis of Dented Tubular Members,* Structural Engineering Report No. CE-STR-90-27, School of Civil Engineering, Purdue University, West Lafayette, IN.

Ellinas, C. P. (1984) Ultimate Strength of Damaged Tubular Bracing Members, *Journal of the Structural Engineering,* ASCE, 110, 2; 245–59.

Ellis, J. S. (1958) Plastic Behavior of Compression Members, *Transactions,* Engineering Institute of Canada, 2, 2; 49–60.

Gu, Y. N. and Li, R. P. (1992) On the Assessment of Strength of Platform with Damaged Members, *Proceedings of the Second International Offshore and Polar Engineering Conference,* ISOPE, San Francisco, CA. June 14–19, IV, 408–14.

Landet, E. and Johnsen, R. H. (1987) Investigation on Ultimate Strength of Dented Pipes, Technical Report No. 87-3278, VERITEC, May 20.

Loh, J. T. (1990) A Unified Design Procedure for Tubular Members, *Proceedings, 22th Annual Offshore Technology Conference,* OTC 6310, Houston, TX, May 7–10, 365–79.

MacIntyre, J. and Birkemoe, P. C. (1989) *Damage of Steel Tubular Members in Offshore Structures: A Nonlinear Finite Element Analysis,* Department of Civil Engineering, University of Toronto, Toronto, Ontario, Canada, M5S 1A4.

Marshall, P. W. (1970) Stability Problems in Offshore Structures, *Proceedings, Annual Technical Meeting of the Column Research Council,* St. Louis, MO.

Newmark, N. M. (1943) Numerical Procedure for Computing Deflection, Moments, and Buckling Loads, *Transactions,* ASCE, 108, 1161.

Padula, J. A. and Ostapenko, A. (1989) *Axial Behavior of Damaged Tubular Columns,* Fritz Engineering Laboratory Report No. 508.11, Lehigh University, Bethlehem, PA.

Richards, D. M. and Andronicou, A (1985) Residual Strength of Dented Tubulars: Impact Energy Correlation, *Proceedings, Fourth International Offshore Mechanics and Arctic Engineering Symposium,* Dallas, Texas, February 17–21, 522–27.

Saleeb, A. F. (1979) Near-Bottom Bend of Flowlines, M. S. thesis, School of Civil Engineering, Purdue University, West Lafayette, IN.

Sherman, D. R. (1976) Tests of Circular Steel Tubes in Bending, *Journal of the Structural Division,* ASCE, 102(ST11), 2181–95.

Sherman, D. R. (1982) Research in North America on the Stability of Circular Tubes, *Proceedings, SSRC Annual Meeting,* Structural Stability Research Council, New Orleans, Louisiana.

Smith, C. S. and Dow, R. S. (1981) Residual Strength of Damaged Steel Ships and Offshore Structures, *Journal of Constructional Steel Research,* 1(4), 2–15.

Smith, C. S., Kirkwood, W., and Swan, J. W. (1979) Buckling Strength and Post-Collapse Behavior of Tubular Bracing Members Including Damage Effects, *Proceedings, Second International Conference on Behavior of Offshore Structures (BOSS '79),* Imperial College, London, England, August 28–31, 303–26.

Smith, C. S., Somerville, J. W., and Swan, J. W. (1981) Residual Strength and Stiffness of Damaged Steel Bracing Members, *Proceedings, 13th Annual Offshore Technology Conference,* Houston, TX, May 4–7, 273–91.

Sohal, I. S. and Chen, W. F. (1984) Moment-Curvature Expressions for Fabricated Tubes, *Journal of Structural Engineering,* ASCE, 110(11), 2738–57

Sohal, I. S. and Chen, W. F. (1987) Local Buckling and Sectional Behavior of Fabricated Tubes, *Journal of Structural Engineering,* ASCE, 113(3), 519–33.

Sohal, I. S. and Chen, W. F. (1988) Local and Post-Buckling Behavior of Tubular Beam-Columns, *Journal of Structural Engineering,* ASCE, 114, 5; 1073–90.

Sugimoto, H. and Chen, W. F. (1985) Inelastic Post-Buckling Behavior of Tubular Members, *Journal of Structural Engineering,* ASCE, 111, 9; 1965–78.

Taby, J. (1986) *Experiments with Damaged Tubulars,* Report No. 6.07, SINTEF, October.

Taby, J. and Moan, T. (1985) Collapse and Residual Strength of Damaged Tubular Members, *Proceedings, Fourth International Conference on Behavior of Offshore Structures,* Delft, The Netherlands, July 1–5, 395–408.

Taby, J. and Moan, T. (1987) Ultimate Behavior of Circular Tubular Members with Large initial Imperfections, *Proceedings, SSRC Annual Technical Session,* Structural Stability Research Council, Houston, Texas, March 24–25, 79–104.

Toma, S. (1980) *Analysis and Design of Fabricated Tubular Beam-Columns,* Ph.D. Dissertation, School of Civil Engineering, Purdue University, West Lafayette, IN.

Toma, S. and Chen, W. F. (1979) Analysis of Fabricated Tubular Columns, *Journal of the Structural Division,* ASCE, 105, ST11; 2343–66.

Toma, S. and Chen, W. F. (1980) *Elastic-Plastic Behavior of Beam-Columns and Struts,* Report Submitted to Exxon Production Research, Houston, Texas, February. (Structural Engineering Report No. CE-STR-80-1, School of Civil Engineering, Purdue University, West Lafayette. IN. 47907)

Ueda, Y. and Rashed, S. M. H. (1985) Behavior of Damaged Tubular Structural Members, *Proceedings, Fourth International Offshore Mechanics and Arctic Engineering Symposium,* ASME, Dallas, Texas, February 17–21, 528–36.

Yao, T., Taby, J., and Moan, T. (1986) Ultimate Strength and Post-Ultimate Strength Behavior of Damaged Tubular Members in Offshore Structures, *Proceedings, Fifth International Offshore Mechanics and Arctic Engineering Symposium,* III, ASME, 301–08.

7: Analysis of Internally Grout-Repaired Damaged Members

J.M. Ricles,
Department of Civil and Environmental Engineering, Lehigh University, Bethlehem, Pennsylvania

7.1 INTRODUCTION

Damage to offshore platforms is a topic of increasing concern because of the extraordinarily high stakes involved in terms of lost lives, environmental devastation, and wasted resources. There are currently over 6000 fixed platforms throughout the world, a number of which are known to have suffered some form of damage. Damage records from a decade of recorded international events indicate that a majority of accidents result from ship-platform collisions and impact with debris, including dropped objects. These accidents account for nearly 17% and 12%, respectively, of all recorded accidents (Ellinas and Valsgard, 1985).

The steel jacket of an offshore fixed platform is a truss system of cylindrical steel members used to support gravity and lateral loads (Fig. 7.1). Under lateral impact to a cylindrical member, its circular cross section is susceptible to localized denting of dent depth d_d, ovalization, and out-of-straightness of δ_p (Fig. 7.2). Tests on tubular bracing (Ricles et al., 1992, 1994a, 1994b; Bruin et al., 1995) indicate that a significant reduction in a member's axial load capacity occurs due to the dent damage. As shown in Fig. 7.3, the residual strength P_{res} of a dented cylindrical member decreases with dent depth d_d, where for $d_d = 20\%$ of the member's diameter D approximately a 50% reduction in member capacity occurs. The results shown in Fig. 7.3 are associated with a minimal out-of-straightness damage δ_p, where a maximum value of $\delta_p = 0.0096L$ existed in the specimens of length L. Because of the dramatic loss in capacity, it is critical that the residual strength of a damaged member be accurately evaluated. In addition, because the cost to replace entire platforms is prohibitive, there are strong incentives to develop reliable and economical ways to repair damaged platform cylindrical members.

Grout has been used extensively in offshore applications because of its high compressive strength, ease of placement, and low cost. Grout is readily used in the installation of most steel jackets to complete the connection between the foundation and the jacket itself. Grout has also been incorporated effectively into the repair and rehabilitation of damaged members. Traditional repair techniques available to operators have included the use of bolted sections, welded sections, and clamps. Grout repair offers a viable alternative to these repair methods, for it does not involve underwater welding or the highly precise fabrication of repair components.

0-8493-8282-3/96/$0.00+$.50

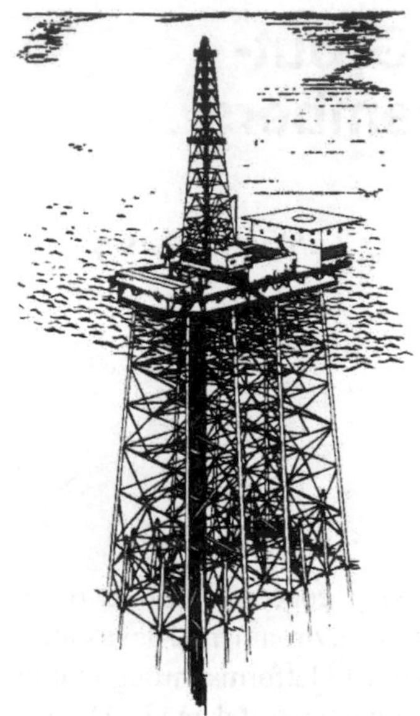

FIGURE 7.1 Typical fixed offshore platform steel jacket.

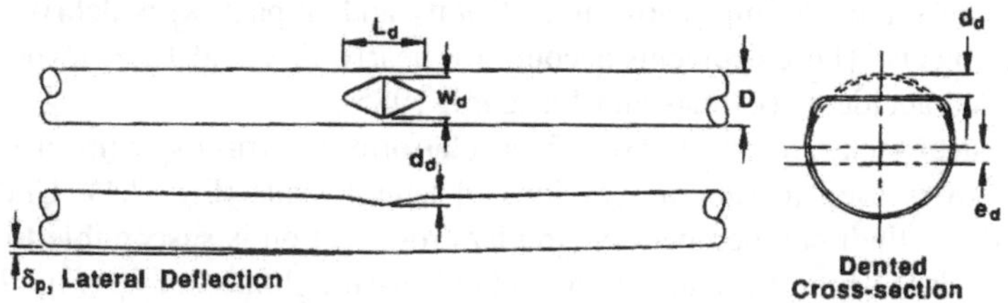

FIGURE 7.2 Dent-damaged cylindrical brace member.

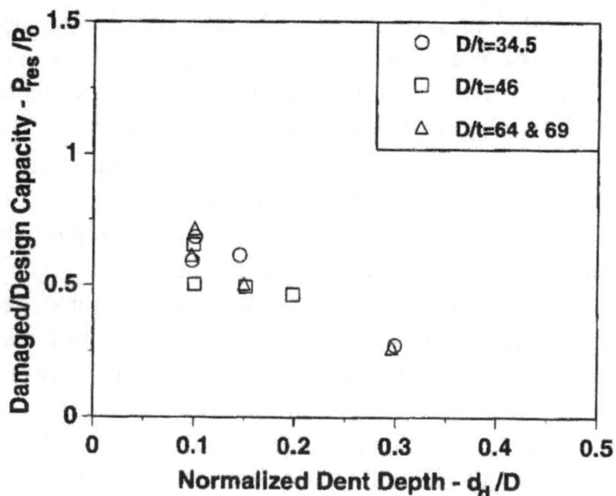

FIGURE 7.3 Effect of dent depth on member capacity.

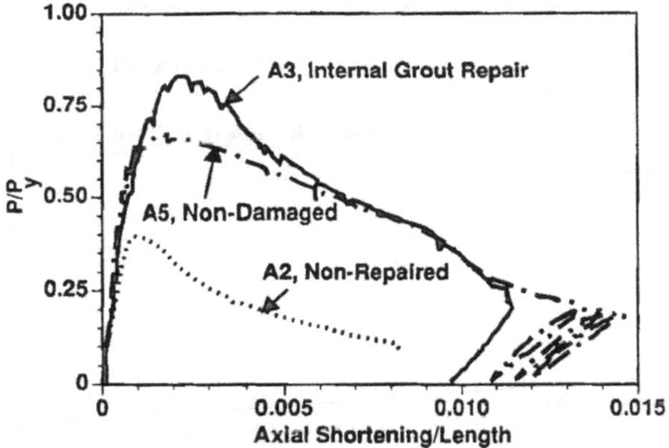

FIGURE 7.4 Normalized axial load-shortening relationship for undamaged, dent-damaged, and repaired specimens.

Furthermore, it effectively utilizes the already well-established knowledge base of underwater grouting developed from pile-jacket connections.

Basically, there are two types of grout repairs used for dented cylinders. These include: (1) internal grouting, either full or partial and (2) grouted steel clamps. This chapter is devoted to the analysis of the first type of repair, considering full grouting along the entire length of the dented member.

The method of grout repair of dented cylindrical bracing has been demonstrated experimentally by a number of test programs. Parsanejad (1987) tested 10 small-scale, completely grout-filled dented cylinders and concluded that full grouting of the members compensated for the loss of strength due to dent damage. Each test involved applying axial load to pin-ended specimens. The dent depth of the specimens ranged from 0 to about 15% of the member's diameter. Parsanejad and Gusheh (1988) also demonstrated that partial grouting of the dented region could improve the load-carrying capacity of damaged members. Under a United Kingdom project that addressed the technology of grout use for offshore platform construction and repair (U.K. Department of Energy, 1988) over 37 tests were conducted by Boswell and D'Mello (1990) on completely grout-filled dented cylindrical bracing. The test program consisted of concentric axial-loaded specimens with dent depths of less than 0.15D. Boswell and D'Mello also found the internal grouting repair method to be highly successful.

More recently Ricles et al. (1992, 1994a, 1994b), Bruin et al. (1995), and Fan (1994) conducted research studies on the experimental and analytical assessment of the effect of dent depth on the ability of a complete internal-grout repair to restore a dent-damaged brace member's axial load capacity. Dent depth among the test specimens ranged from 0.1D to 0.3D, with an out-of-straightness δ_p of less than 0.01L. Typical results in terms of normalized axial load P and shortening Δ are given in Fig. 7.4, involving specimens with $d_d = 0.1D$ and a diameter-to-thickness (D/t) ratio of 34.5. The internal grout repair was found to inhibit the growth of the dent, thereby enabling the dent-repaired member to achieve a higher strength than a

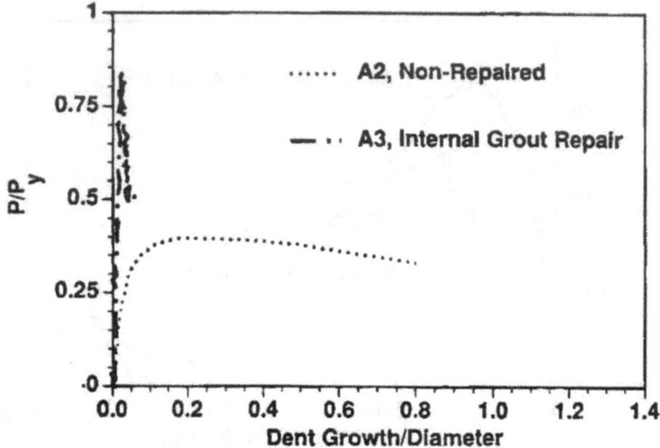

FIGURE 7.5 Normalized axial load–dent growth relationship for dent-damaged and repaired specimens.

similar non-repaired member (Fig. 7.5). However, the repaired capacity P_r was found to diminish with increasing dent depth, as shown in Fig. 7.6.

Because of the need to know a repaired member's performance, the analysis of grout-repaired cylinders is of importance and of a practical nature to engineers. There are three basic levels of analysis for grout-filled cylinders: (1) beam-column analysis; (2) moment-thrust-curvature integration analysis, and (3) finite element analysis. Among these, the first is the simplest to perform. Parsanejad (1987) took such an approach, in which he expressed the equilibrium of an internal grout-repaired dented brace in the form of a quadratic equation, based on beam-column theory, and transformed section properties in order to estimate a repaired member's capacity. While the method provides an estimate of capacity, the load-deformation behavior of the member is not predicted by the method. The finite element method, alternatively, provides as part of the solution the entire load-deformation behavior from

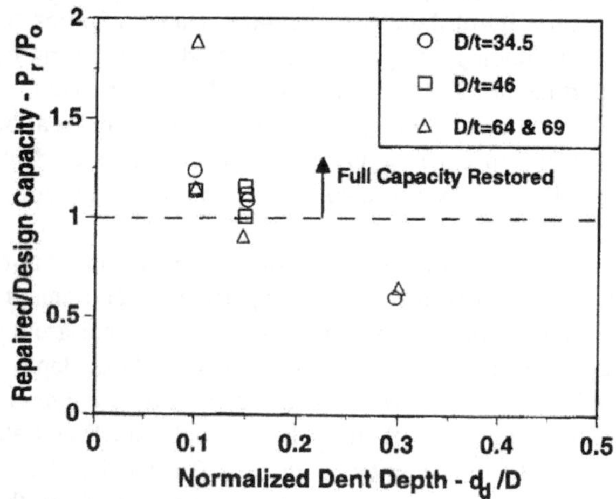

FIGURE 7.6 Grout repair effectiveness.

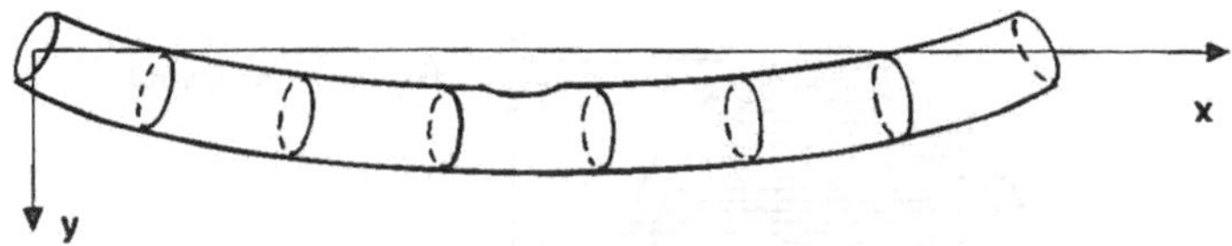

FIGURE 7.7 Dent-damaged cylindrical member with defined segments for M-P-Φ integration analysis.

which member capacity can be determined. The method, however, is computationally intensive and requires several hundred degrees of freedom to define the model, in addition to the use of plasticity theory and a nonlinear grout-constitutive relationship in conjunction with large deformation theory to arrive at a reasonably accurate solution. The moment-thrust-curvature (M-P-ϕ) integration method also provides a solution for the load-deformation behavior of a dented cylinder; however, it is less computationally intensive than the nonlinear finite element method.

Furthermore, it can readily account for the effects of out-of-straightness δ_p, dent depth d_d, single or multiple dents, boundary conditions, combined loading, steel stress-strain properties, and grout stress-strain properties. These attributes make the M-P-ϕ integration method appealing and suitable for use in engineering design offices for developing and assessing internal grout repairs.

The M-P-ϕ numerical integration analysis of internal grout-repaired cylinders involves dividing the member into segments, as shown in Fig. 7.7. The M-P-ϕ relationship for the cross section of an undented segment and dented segment is used to develop the force-deformation response of each segment and thus the entire member. This is accomplished through numerically integrating each segment along the member and enforcing both compatibility and equilibrium to arrive at a solution.

In the development of the M-P-ϕ integration procedure the following assumptions are made: (1) plane sections of the cross section remain plane; (2) deformations are small, i.e., small-displacement beam theory is used; (3) shear deformations are negligible; (4) strain reversal does not occur in the member; (5) residual stresses in the steel tube are neglected; and (6) the ovalization of the cross section due to dent growth is prevented by the grout.

7.2 M-P-ϕ ANALYSIS

The M-P-ϕ relationship that is used in conjunction with the numerical integration must be established first in order to analyze the behavior of a grout-filled tubular. This relationship must be numerically derived, unless a closed-form expression exists that is an approximation based on a regression analysis of data associated with the section's M-P-ϕ behavior. Some examples of these approximations were provided in Chapters 3–6 for both a dented and undented cylindrical cross section. To date, limited data is available to derive an approximation for a grout-filled dented steel cylinder, although closed-form approximations for concrete-filled steel cylinders have been derived by Chen and Atsuta (1976). Presented herein is the theory for the numerically derived solution for M-P-ϕ relationships of internally

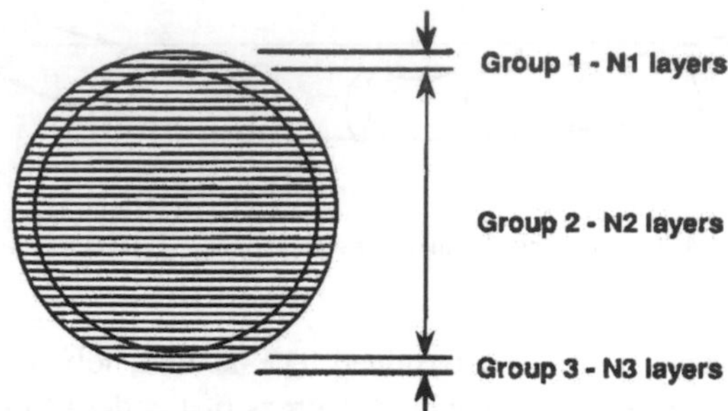

(a) Internal Grout Filled Undamaged Section

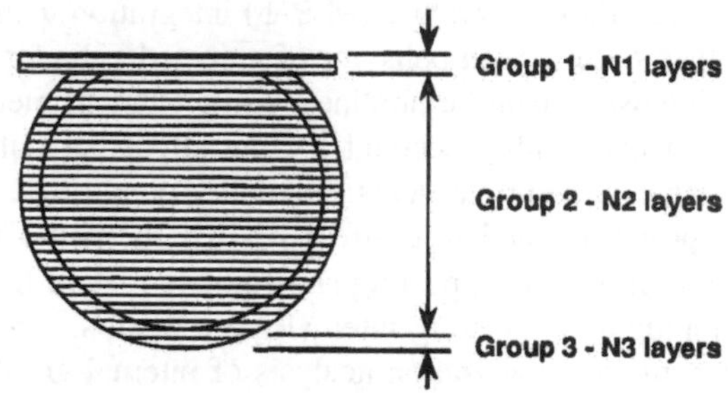

(b) Internal Grout Filled Damaged Section

FIGURE 7.8 Discretization of grouted cross section into layers.

grout-filled dented and undented steel cylinders. Also presented is the theory for axial strain–axial load-curvature (ϵ-P-ϕ) relationships that are numerically derived.

The numerical approach used to develop the M-P-ϕ and ϵ-P-ϕ relationships is based on the so-called "fiber model." This model idealizes the internally grouted section, either undented or dented, by dividing it into an appropriate number of discrete layers, as shown in Fig. 7.8. To preserve the area of the steel tube in the dented section of the same radius, an approximation is made where the width of the steel layer is extended in the dent saddle, as shown in Fig. 7.8b. Uniaxial stress-strain relationships of each of these layers are then used in the computations along with compatibility and equilibrium requirements to generate the M-P-ϕ and ϵ-P-ϕ.

Figure 7.9 shows two types of idealized stress-strain relationships, either of which could be used for the steel. Type 1 is an elastic–perfectly plastic stress-strain relationship. Type 2 takes into account the effect of local buckling, having a descending branch in the compression region that occurs at a strain of $\epsilon_{cr} = (1 + \gamma)\ \epsilon_y$, where ϵ_y is the yield strain and γ is an empirical parameter. An empirical expression for the descending branch of the steel's stress-strain relationship was adopted, based on the work by Sato and Suzuki (1992) on concrete-filled steel tubes, where the expression for stress σ as a function of strain ϵ is:

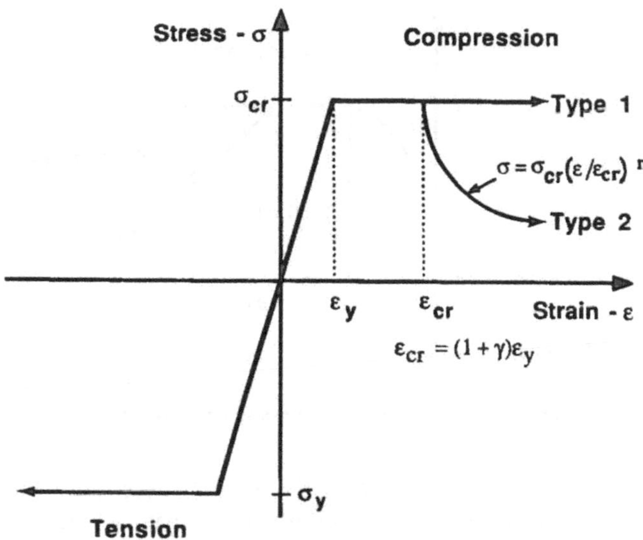

FIGURE 7.9 Steel fiber stress-strain relationship.

$$\sigma = \sigma_{cr}\left(\frac{\epsilon}{\epsilon_{cr}}\right)^n, \tag{7.1}$$

where $\epsilon \geq \epsilon_{cr}$, σ_{cr} is the stress when local buckling occurs, and n is an empirical parameter.

Based on observations from experiments on internal, grout-repaired dented cylinders (Ricles et al., 1992, 1994a) involving member D/t ratios of 34 to 69, the yield stress σ_y and a value of -0.2, respectively, are recommended for σ_{cr} and n. The value of γ ranges from 2 to 6, with smaller values used for cross sections possessing higher D/t ratios.

The stress-strain relationship for grout must consider the effects of confinement. It is well known that the compressive strength and ductility of concrete is significantly enhanced by confinement (Mander and Priestley, 1988), and that the same effect occurs in grout. In a grout-filled cylindrical brace, the steel tube will be loaded through force transfer from adjacent jacket members connected to the ends of the brace. The Poisson effect causes the steel wall of an axial-compressed cylindrical member to expand radially outward from the grout, due to the fact that grout has a smaller Poisson ratio than steel. This phenomenon, and that associated with the possibility of local buckling, is likely to lead to a condition where the grout is not fully confined by the steel cylinder.

The stress-strain relationships shown in Fig. 7.10 were thereby selected, consisting of no tension and Kent and Park's parabolic expression (Kent and Park, 1971) for the compressive ascending curve to the unconfined grout strength f_g, with either a sustained strength of f_g (Type 1 stress-strain curve) or a decrease in stress at a rate of β (Type 2 stress-strain curve) beyond a compressive strain of ϵ_{br}. The

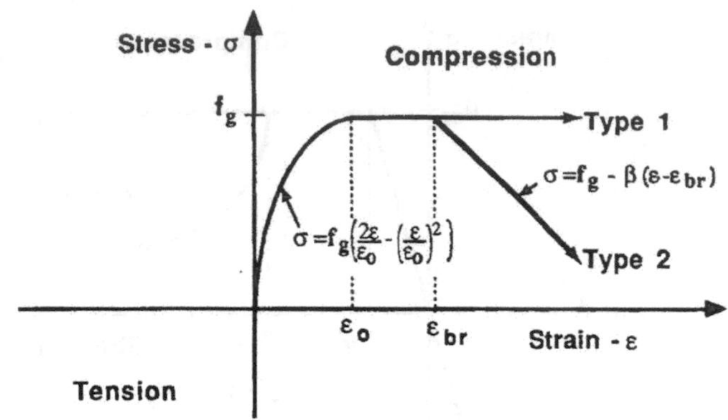

FIGURE 7.10 Grout fiber stress-strain relationship.

grout stress-strain relationship defined by the Type 2 stress-strain curve accounts for the effects of local buckling in the steel tube fibers that are adjacent to the grout fiber, where the ascending stress σ as a function of strain ϵ is

$$\sigma = f_g\left[\frac{2\epsilon}{\epsilon_0} - \left(\frac{\epsilon}{\epsilon_0}\right)^2\right], \qquad \text{where } 0 \le \epsilon \le \epsilon_0 \tag{7.2}$$

The initial value of Young's modulus for the grout (E_g) is equal to

$$E_g = \frac{2f_g}{\epsilon_0} \tag{7.3}$$

The strain ϵ_o is defined in Fig. 7.10, and the strain ϵ_{br} corresponding to the onset of local buckling of the steel fiber that is adjacent to the grout is determined by

$$\epsilon_{br} = \epsilon_o\,(1 + \alpha) \tag{7.4}$$

where α is an empirical parameter. A typical comparison between the above relation-

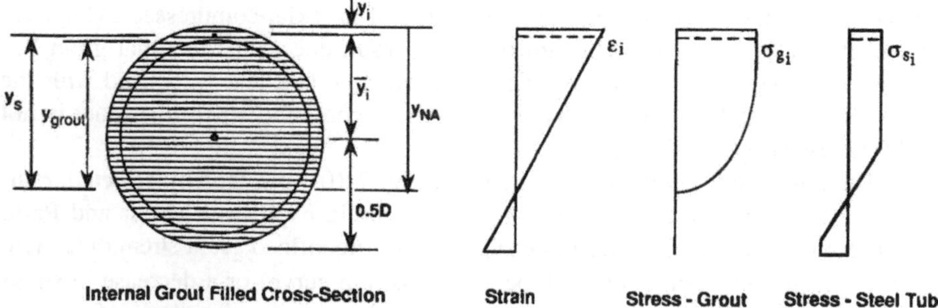

FIGURE 7.11 Stress and strain distribution on grouted cross section.

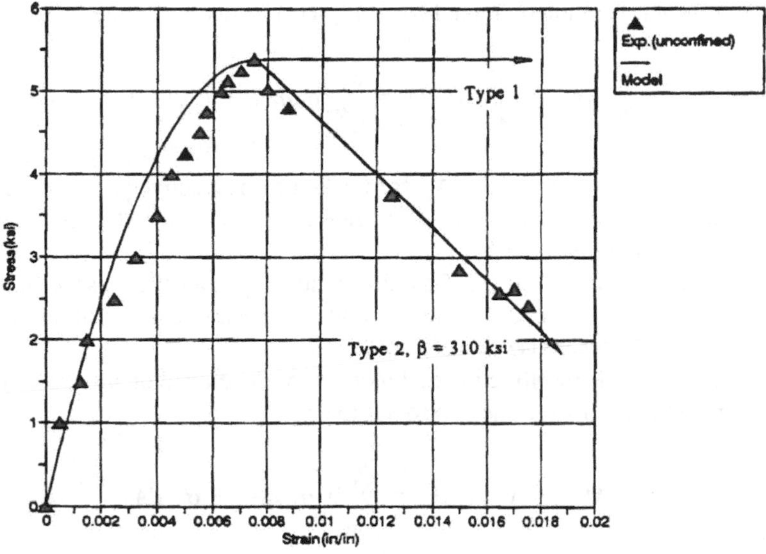

FIGURE 7.12 Comparison of grout model with a typical experimental stress-strain curve.

ships for Types 1 and 2 stress-strain curves and an experimental stress-strain curve for unconfined grout is shown in Fig. 7.12.

Because of the nonlinear stress-strain relationships of the materials, an iterative computational procedure is used to generate the M-P-ϕ and ϵ-P-ϕ relationships. First, the cross section to be analyzed is divided into N layers, where each layer may contain either a steel fiber or both a steel and grout fiber, and a Type 1 or 2 stress-strain curve is specified for the fibers. The subsequent 10 steps for generating the M-P-ϕ and ϵ-P-ϕ relationships for a specified axial compressive load P_{spec} is as follows:

Step 1. Specify an initial trial curvature ϕ.
Step 2. Assume a value y_{NA} for the position of the neutral axis of the cross-section.
Step 3. Calculate the corresponding axial strain in the fibers of each layer (positive for compression, negative for tension), where for layer i the strain ϵ_i is

$$\epsilon_i = \phi(y_{NA} - y_i) \tag{7.5}$$

where y_{NA} and y_i are, respectively, the distance from the top of the cross-section to the neutral axis, and the distance from the top of the cross-section to layer i (see Fig. 7.11).

Step 4. For each fiber in layer i determine the stress σ_i corresponding to the strain ϵ_i from Step 3.

Step 5. Calculate the internal axial force P_{cal} corresponding to the stress

$$P_{cal} = \int_A \sigma(y)dA = \sum_{i=1}^{N} (\sigma_{s_i} dA_{S_i} + \sigma_{g_i} dA_{g_i}) \tag{7.6}$$

where σ_{s_i}, σ_{g_i}, dA_{s_i} and dA_{g_i} are for layer i, respectively, the stress in the steel fiber, the stress in the grout fiber, the area of the steel fiber, and the area of the grout fiber.

Step 6. Compare P_{cal} with P_{spec}. If the difference is greater than a specified tolerance, go to Step 2 (where an updated neutral axis position must be assumed), otherwise proceed to Step 7.

Step 7. Calculate the resulting internal moment M developed on the cross-section, referencing it from the centroidal axis:

$$M = \int_A y \cdot \sigma(y)dA = \sum_{i=1}^{N} \bar{y}_i(\sigma_{s_i} dA_{s_i} + \sigma_{g_i} dA_{g_i}) \tag{7.7}$$

where $\bar{y}_i$ is the distance to layer i referenced 0.5D from above the bottom of the cross section.

Step 8. Calculate the resulting axial strain ϵ at the centroid of the cross-section:

$$\epsilon = \phi(y_{NA} - 0.5D) \tag{7.8}$$

where, as defined previously, y_{NA} is the distance from the top of the cross-section to the location of the neutral axis.

Step 9. Record the values of P_{cal}, ϕ, M, and ϵ.

Step 10. Increment the current curvature by an amount of $\Delta\phi$, where the updated curvature becomes $\phi = \phi + \Delta\phi$. Proceed to Step 2, unless the desired range in curvature has been achieved.

7.3 MEMBER ANALYSIS

In the analysis of an internal grout-repaired member the numerically derived M-P-ϕ and ϵ-P-ϕ relationships are utilized in conjunction with numerical integration. The numerical integration procedure is conceptually similar to Newmark's method (1943). Initially the method proceeds under incremental load control in order that the axial load applied to the member corresponds to that for a set of already computed M-P-ϕ and ϵ-P-ϕ relationships, thereby avoiding the need to iterate on both the computation of an M-P-ϕ relationship for a cross section as well as the load-displacement member response. As the ultimate load P is reached, the procedure switches to displacement control in order to compute the descending branch of the load-displacement response.

The procedure presented herein is for a simply supported member that may be subjected to end moments and transverse load, and may possess an out-of-

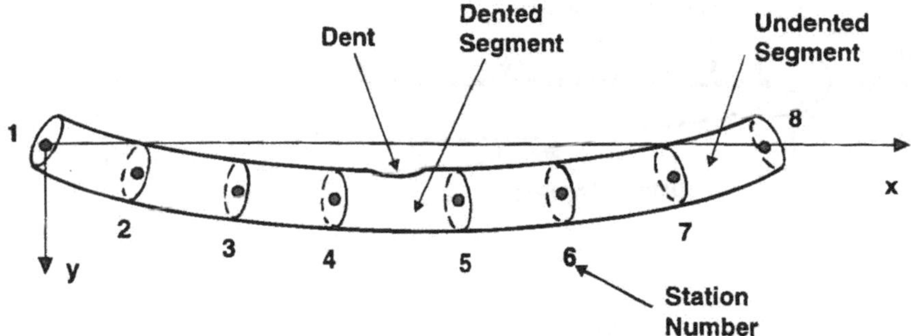

FIGURE 7.13 Dent-damaged cylindrical member with defined segments and stations for M-P-ϕ integration analysis.

straightness δ_p due to dent damage. The numerical procedure requires that the member is first divided into a specified number of segments, as shown in Fig. 7.13, resulting in stations. The behavior of a segment is described by the curvatures and forces developed at its end stations under applied moment and transverse and axial loading. The basic steps involved in the integration procedure to obtain the member load-displacement response is as follows:

Step 1. Divide the cylindrical member into a specified number of NSEG segments along its length, where the length Δx of each segment becomes:

$$\Delta x = \frac{L}{NSEG} \tag{7.9}$$

Step 2. Set the axial load P equal to zero and select a value for the axial load increment ΔP.

Step 3. Initialize the axial load, $P = P + \Delta P$.

Step 4. At each station i, assume the initial lateral deflection y_{0i}.

Step 5. Include the initial out-of-straightness at each station i:

$$y_{0i}^{*} = y_{0i} + \delta_{pi} \tag{7.10}$$

where δ_{pi} is the initial out-of-straightness at each station.

Step 6. Calculate the internal member moment M_i at all stations considering second-order effects, where at station i (beginning with station 2):

$$M_i = Py_{0i}^{*} + M_L + R_L x_i + \sum_{j=1}^{i-1} (x_i - \bar{x}_j)w(x_j)\Delta x \tag{7.11}$$

where M_L, R_L, x_i, $\bar{x}_j$, $w(x_j)$, are, respectively, equal to the left-end member moment, left-end member reaction, distance from the left support (e.g., station 1) to station i, distance from the left-end support to midlength of segment j, and the averaged distributed transverse load acting on segment j (Fig. 7.14).

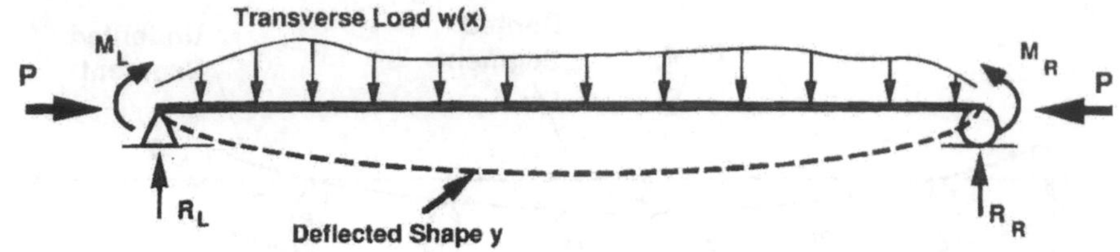

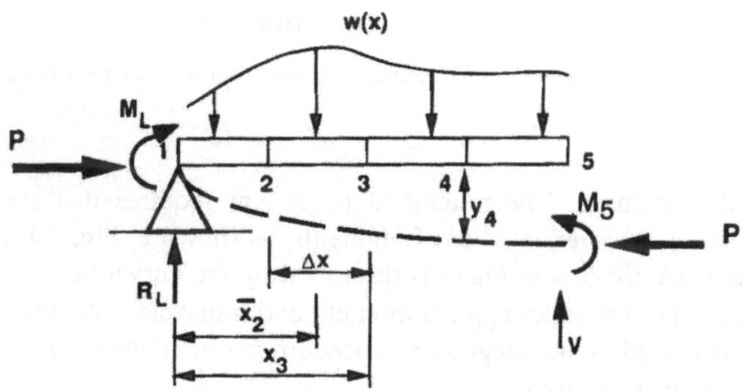

FIGURE 7.14 Beam-column divided into segments for analysis.

Step 7. At each station determine the curvature ϕ_i corresponding to the current axial load P and moment M_i, using the M-P-ϕ relationship for either an undented or dented tubular section with internal grout. To avoid divergence compare M_i with M_{peak}, where M_{peak} is the flexural capacity for the section at station i under the axial load P. If M_i is smaller than M_{peak}, then ϕ_i is obtained through interpolation of the numerical data points of the M-P-ϕ relationship on its ascending branch, and one then proceeds to Step 9. Otherwise, if M_i is greater than M_{peak} then ϕ_i must lie on the descending branch of the M-P-ϕ relationship. Such a branch exists only if local buckling effects are considered (e.g., the Type 2 stress-strain curve is utilized). If local buckling effects are not considered (e.g., stress-strain curve Type 1 is used to generate the M-P-ϕ relationship), then the analysis is terminated and the last value for P at which convergence was achieved under Step 11 is considered the last converged solution. If local buckling effects are considered and M_i exceeds M_{peak}, then the current station i is referred to as the control station, and proceed to Step 8.

Step 8. When local buckling effects are considered, the remaining part of the computation involves having the curvatures for all stations, except for the controlling station, on the ascending branch of the M-P-ϕ relationship. The curvature for the controlling station will always lie on the descending branch of the cross-section's M-P-ϕ relationship. The load P is reduced due to load shedding by the member after achieving M_{peak}. The numerical integration procedure with local buckling effects thereby continues by assigning a target

(a) Uncorrected Computed Displacements

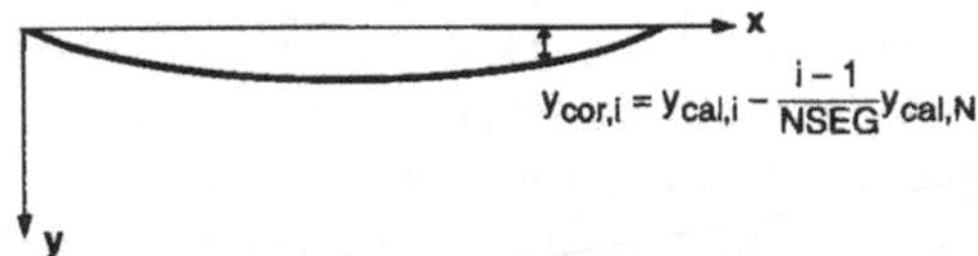

(b) Corrected Computed Displacements

FIGURE 7.15 Linear correction to computed deflections.

displacement to the control station. The axial load to cause this displacement to develop in the member is to be subsequently determined in Steps 9 to 11. Prior to proceeding to Step 9, the curvature ϕ_i must be determined that is associated with the axial load P and moment M_i. If M_i exceeds M_{peak}, then the axial load is decreased, where $P = P - \Delta P$, until M_i does not exceed M_{peak} in the control station and ϕ_i can be determined.

Step 9. Calculate the deflection $y_{cal,i}$ for each station, based on the station curvatures computed in Step 7 or 8, where θ_1 and $y_{cal,1}$ are at the left support (station 1) and are both equal to zero, whereas for other stations i (i = 2, . . . , NSEG + 1):

$$\theta_i = \theta_{i-1} - \left(\frac{\phi_{i-1} + \phi_i}{2}\right)\Delta x \tag{7.12}$$

and

$$y_{cal,i} = y_{cal,i-1} + \theta_{i-1}\Delta x - \int_{x_{i-1}}^{x_i} x\phi(x)dx \tag{7.13}$$

Assuming a linear variation in curvature between each station, Eqn. (7.13) can be simplified to the following:

$$y_{cal,i} = y_{cal,i-1} + \theta_{i-1}\Delta x - (2\phi_{i-1} + \phi_i)\,\frac{(\Delta x)^2}{6} \tag{7.14}$$

Step 10. Apply a linear correction to satisfy the right-end member boundary condition of zero deflection (Fig. 7.15), where at station i the corrected displacement $y_{cor,i}$ is equal to

$$y_{cor,i} = y_{cal,i} - \frac{i-1}{NSEG} y_{cal,N} \tag{7.15}$$

Step 11. Compare the computed deflection $y_{cor,i}$ with the assumed deflection y_{0i} at each station. If the difference between the two values at a station is greater than a specified tolerance and the analysis is on the ascending branch of the load-displacement response (e.g., M_{peak} has not been reached in Step 7), then for all stations, set y_{0i} equal to $y_{cor,i}$ and proceed to Step 5. If the difference between the two values at a station is greater than the specified tolerance and the analysis is on the descending branch of the load-displacement response (e.g., M_{peak} has been reached in Step 7), then the targeted value of displacement is kept at the control station, the axial load P and other station's computed displacements are adjusted, the moment M_i developed at each station, i, is computed using Eq. (7.11), and proceed to Step 8. If the difference between $y_{cor,i}$ and y_{0i} for all stations is within the specified tolerance, then convergence is achieved. The converged displacements for the current axial load P at station i is thus noted as y_i. The axial shortening Δ is then computed by Eq. (7.16) before proceeding to Step 12.

Step 12. If the analysis is on the ascending branch of the load-displacement response, go to Step 3. Otherwise, unless the target displacement has reached a specified limit, specify a new (larger) target displacement at the control station, assume the displacements y_{0i} at the other stations, compute M_i for each station using Eq. (7.11), and proceed to Step 8.

The axial shortening Δ of an internal grouted beam-column results from the effects of shortening Δ_s due to axial strain ϵ and shortening Δ_g due to geometric effects caused by the lateral deflection y_i at each station, where

$$\Delta = \Delta_s + \Delta_g \tag{7.16}$$

The axial shortening Δ_s is obtained from

$$\Delta_s = \Delta x \sum_{i=2}^{NSEG+1} \epsilon_i \tag{7.17}$$

where ϵ_i is the axial strain at the centroid of station i, which is obtained from the ϵ-P-ϕ relationship discussed previously. The shortening Δ_g is obtained from

$$\Delta_g = \sum_{i=2}^{NSEG+1} [\Delta x - \sqrt{\Delta x^2 - (y_i - y_{i-1})^2}] \tag{7.18}$$

Eq. (7.18) is based on summing the geometric shortening that occurs in each segment

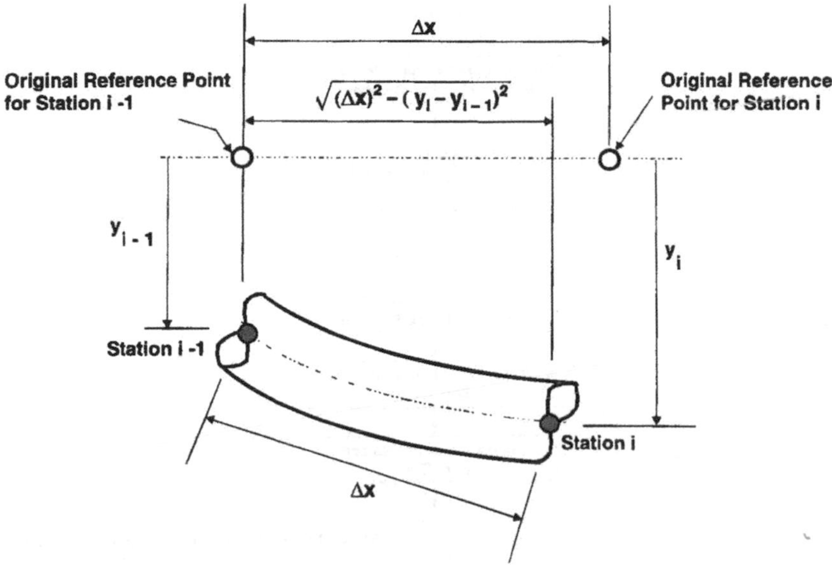

FIGURE 7.16 Determination of axial shortening due to geometric effects.

due to its relative transverse displacement y_i and y_{i-1} at its end stations i and $i - 1$, as shown in Fig. 7.16.

7.4 COMPUTER IMPLEMENTATION

The M-P-ϕ numerical integration procedure was programmed in FORTRAN 77 to develop the computer code DGROUT. A flow chart that summarizes the procedure is given in Fig. 7.17.

In the procedure, load control is used on the ascending branch of the member's force-deformation response in order to minimize the number of calls to subroutines DMPHI and UMPHI, which compute the moment-curvature relationship for a specified axial load for a dented and undented grout-filled cylindrical cross section. On the descending branch, the algorithm switches to displacement control, using specified values of the transverse displacement $y_{cntrl,\,i}$ at the control station. As noted previously, the control station first develops response on the descending branch of the M-P-ϕ relationship. Values of the displacement $y_{cntrl,\,i}$ are specified by the user-defined parameter κ, where the increment in displacement for each step of the analysis is κy_i and y_i is the converged transverse displacement that was computed at the control station corresponding to the peak axial load.

The convergence of the iteration process to compute the moment-curvature relationship for a specified axial load P is based on comparing P with the axial load P_{cal} determined from Eq. (7.6). Convergence occurs if:

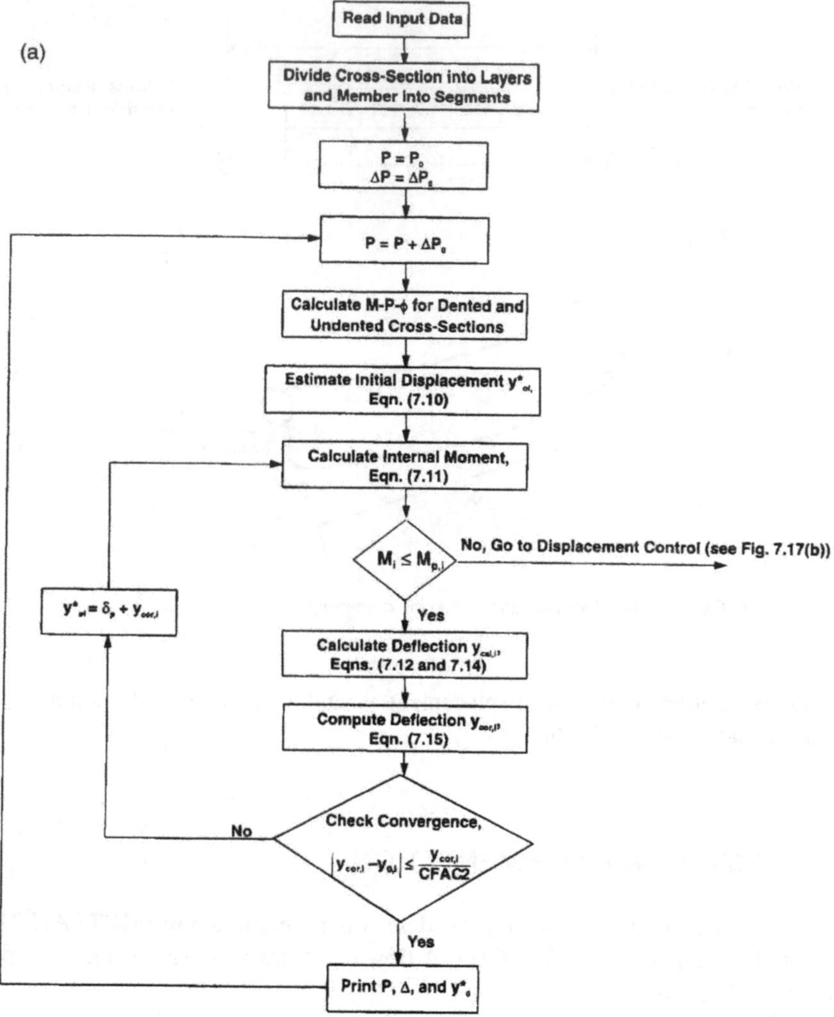

FIGURE 7.17 (a) M-P-ϕ integration analysis algorithm—load control branch.

$$|P - P_{cal}| \le \frac{P}{CFAC1} \tag{7.19}$$

where CFAC1 is a user-defined convergence factor. The position of the cross-section's neutral axis from previous cycles of iteration is used in conjunction with the axial load P and the bisection method to assume the position y_{NA} for the next cycle k + 1, as illustrated in Fig. 7.18. The iterative procedure is started by assigning bounds to the position of the neutral axis, between which DGROUT will seek the position of the neutral axis. DGROUT assigns the upper bound to be the top of the cross section, with the user defining the lower bound with respect to the distance below the top of the cross section.

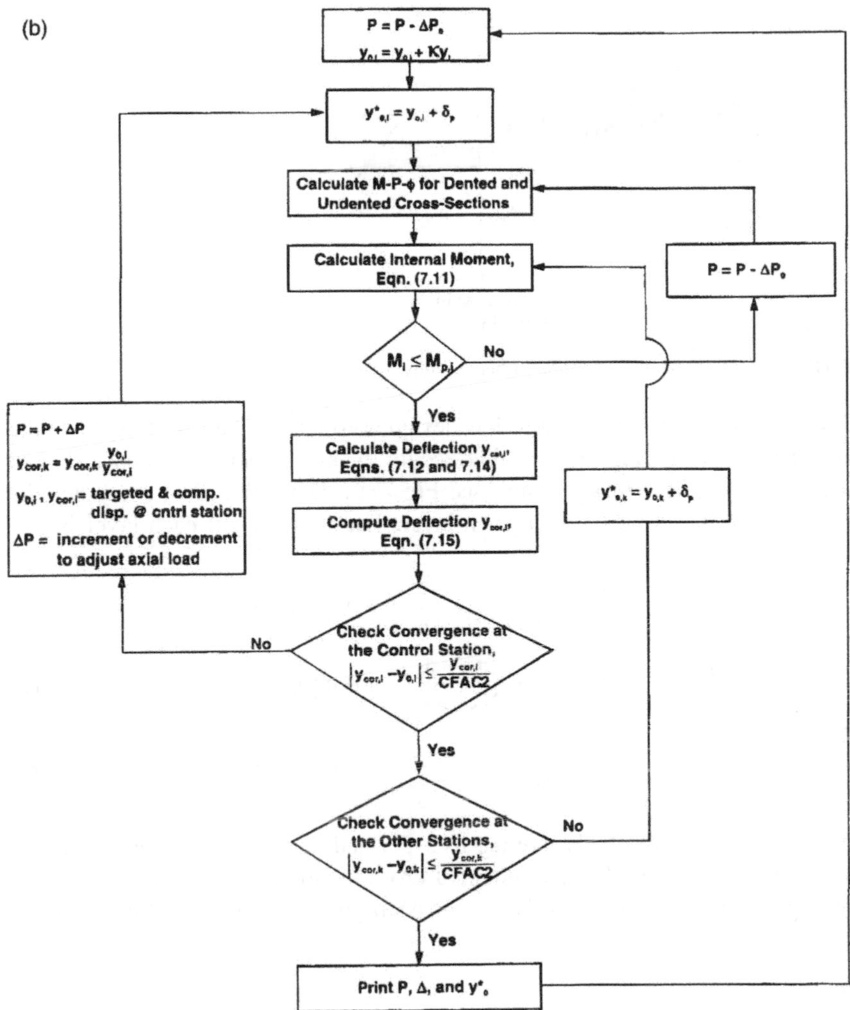

FIGURE 7.17 (b) M-P-φ integration analysis algorithm—displacement control branch.

The convergence of the integration procedure to determine the member's transverse displacement for a specified level of axial load, or axial load for a specified transverse displacement at the controlling station, is based on comparing values of each station's transverse displacements computed during the current and previous cycle of iteration. Convergence occurs if at each station j

$$|y_{cor,j} - y_{o,j}| \leq \frac{y_{cor,j}}{CFAC2} \tag{7.20}$$

where CFAC2 is a user-defined convergence factor.

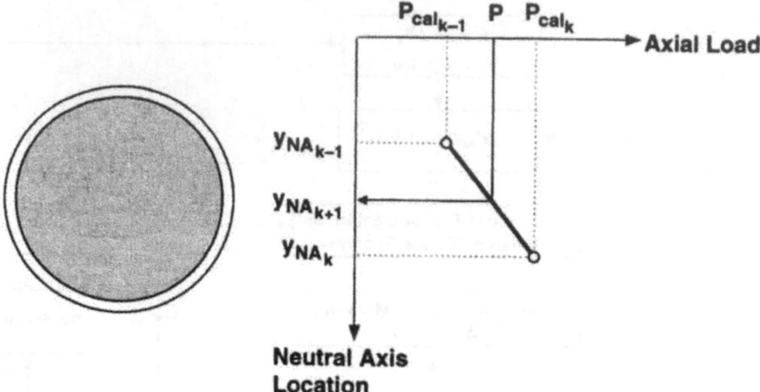

FIGURE 7.18 Estimating location of neutral axis for next iteration by bisection method.

The computer program DGROUT is designed to enable the user to specify the number of layers (N1, N2, and N3) that occurs in three groups (Groups 1, 2, and 3) used to describe the cross section, (see Fig. 7.8). DGROUT then computes the area of the steel and grout fibers in each layer. The thickness of each layer is

$$d_1 = \frac{t}{N1} \quad \text{(Group 1)} \tag{7.21a}$$

$$d_2 = \frac{D - 2t - d_d}{N2} \quad \text{(Group 2)} \tag{7.21b}$$

$$d_3 = \frac{t}{N3} \quad \text{(Group 3)} \tag{7.21c}$$

where D, t, and d_d, respectively, are the user-defined diameter, thickness, and dent depth of the cross section. For an undented cross section $d_d = 0$. The areas of the steel and grout fibers within each layer j of each group are computed as follows:

Group 1:

Steel fiber $A_{s1_j} =$

$$(1 + \Omega)\, 2d_1\sqrt{(0.5D)^2 - (0.5D - d_d - jd_1 + 0.5d_1)^2} \tag{7.22a}$$

Grout fiber $A_{g1_j} = 0.0$ (7.22b)

Group 2:

Steel fiber $A_{s2_j} =$

$$2d_2\sqrt{(0.5D)^2 - (0.5D - d_d - N1\, d_1 - jd_2 + 0.5d_2)^2} - A_{g2_j} \tag{7.23a}$$

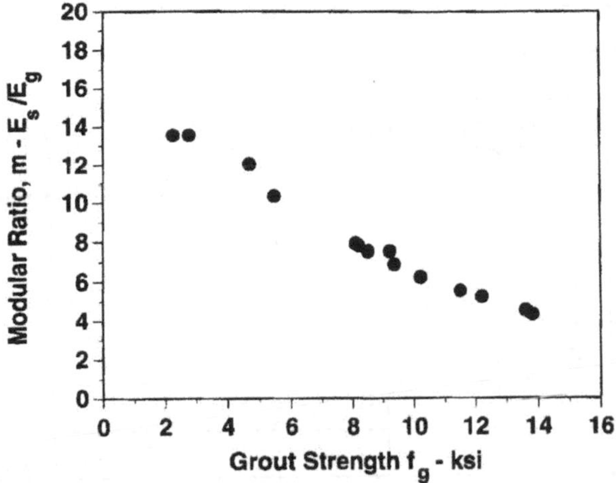

FIGURE 7.19 Typical modular ratio–grout strength relationship.

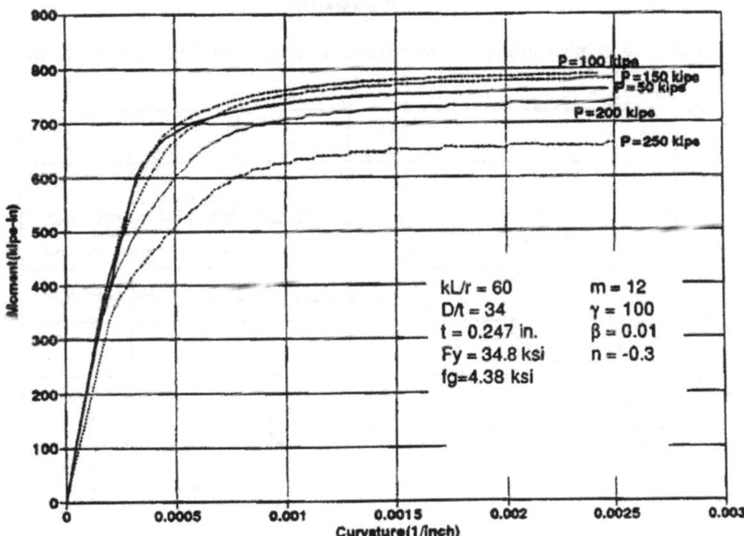

FIGURE 7.20 M-P-ϕ relationships for undamaged, internal, grout-filled cylindrical cross section.

Grout fiber $A_{g2_j} =$

$$2d_2\sqrt{(0.5D-t)^2-(0.5D-d_d-N1\,d_1-jd_2+0.5d_2)^2} \quad (7.23b)$$

Group 3:

Steel fiber $A_{s3_j} =$

$$2d_3\sqrt{(0.5D)^2-(0.5D-d_d-N1\,d_1-N2\,d_2-jd_3+0.5d_3)^2} \quad (7.24a)$$

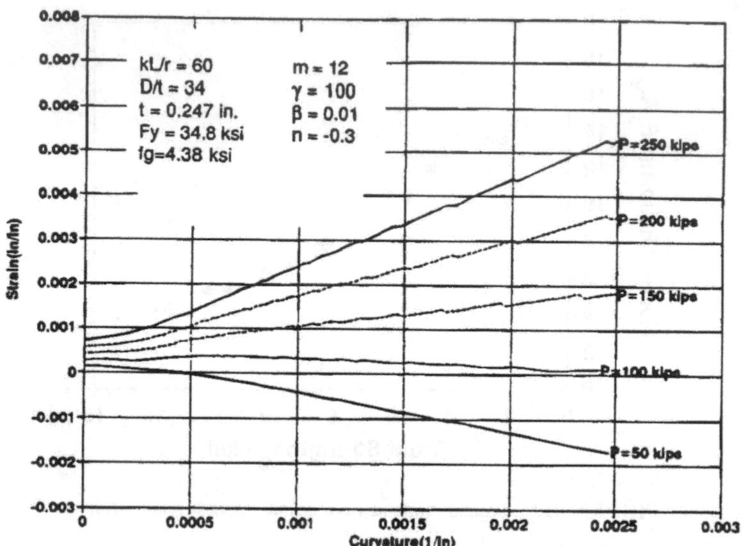

FIGURE 7.21 ϵ-P-ϕ relationships for undamaged, internal, grout-filled cylindrical cross section.

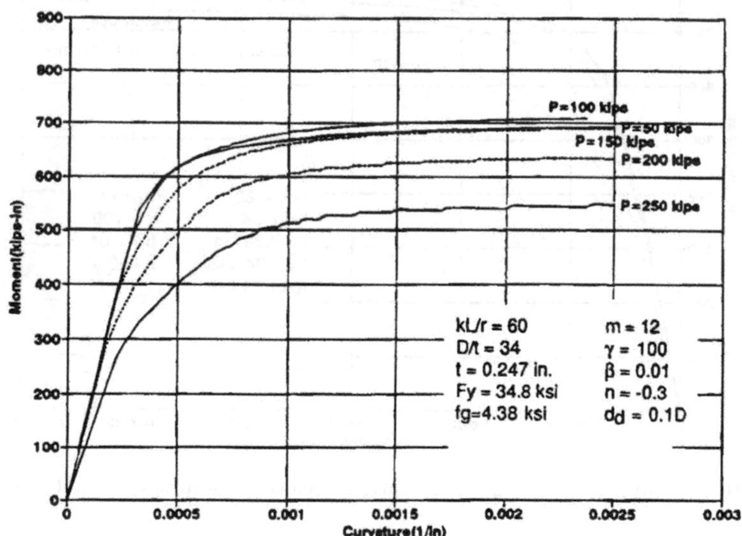

FIGURE 7.22 M-P-ϕ relationships for dent-damaged, internal, grout-repaired cylindrical cross section.

$$\text{Grout fiber } A_{g3_i} = 0.0 \tag{7.24b}$$

In Eq. (7.22a), Ω represents a correction factor in order to preserve the total steel area of the dented cross section, which was referenced in Fig. 7.8b. For an undented cross section, a value of $\Omega = 0$ should be used, whereas for a dented section

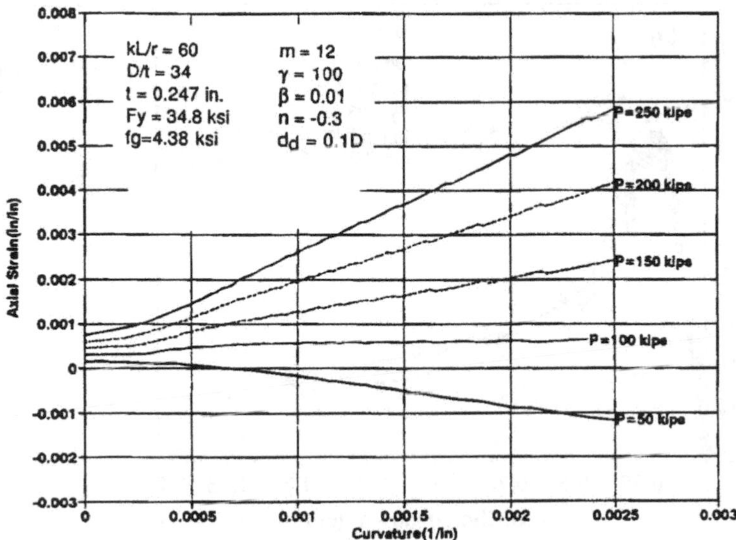

FIGURE 7.23 ϵ-P-ϕ relationships for dent-damaged, internal, grout-repaired cylindrical cross section.

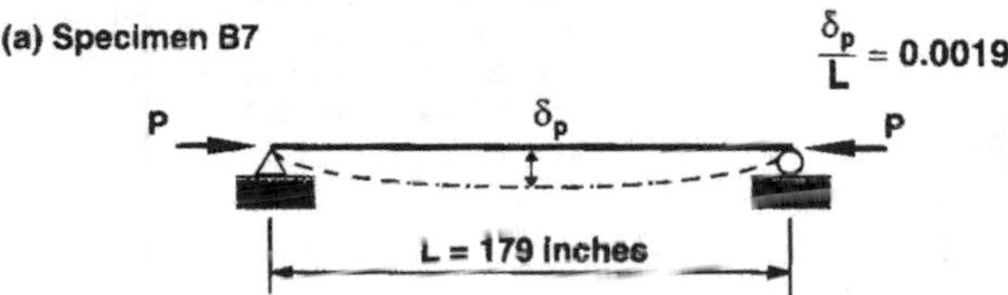

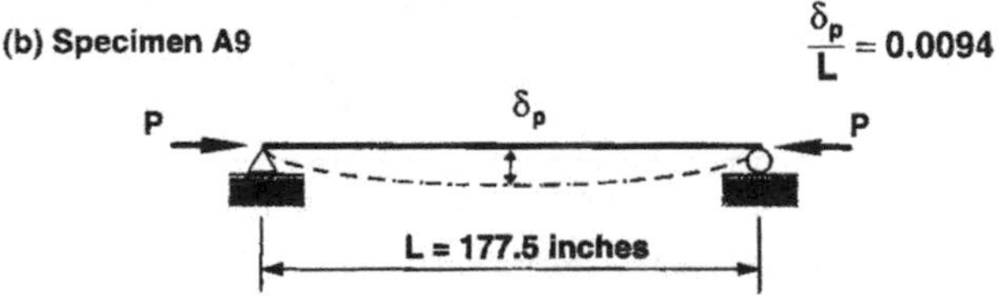

FIGURE 7.24 Experimental specimens: (a) B7 and (b) A9.

$$\Omega = \frac{\pi Dt - \sum_{j=1}^{N1} A_{s1_j} - \sum_{j=1}^{N1} A_{s2_j} - \sum_{j=1}^{N3} A_{s3_j}}{2d_1 \sum_{j=1}^{N1} \left(\sqrt{(0.5D)^2 - (0.5D - d_d - jd_1 + 0.5d_1)^2} \right)} \tag{7.25}$$

The accuracy of a given analysis is related to the number of layers used in the discretization of the cross-section. It is advisable to concentrate a sufficient number

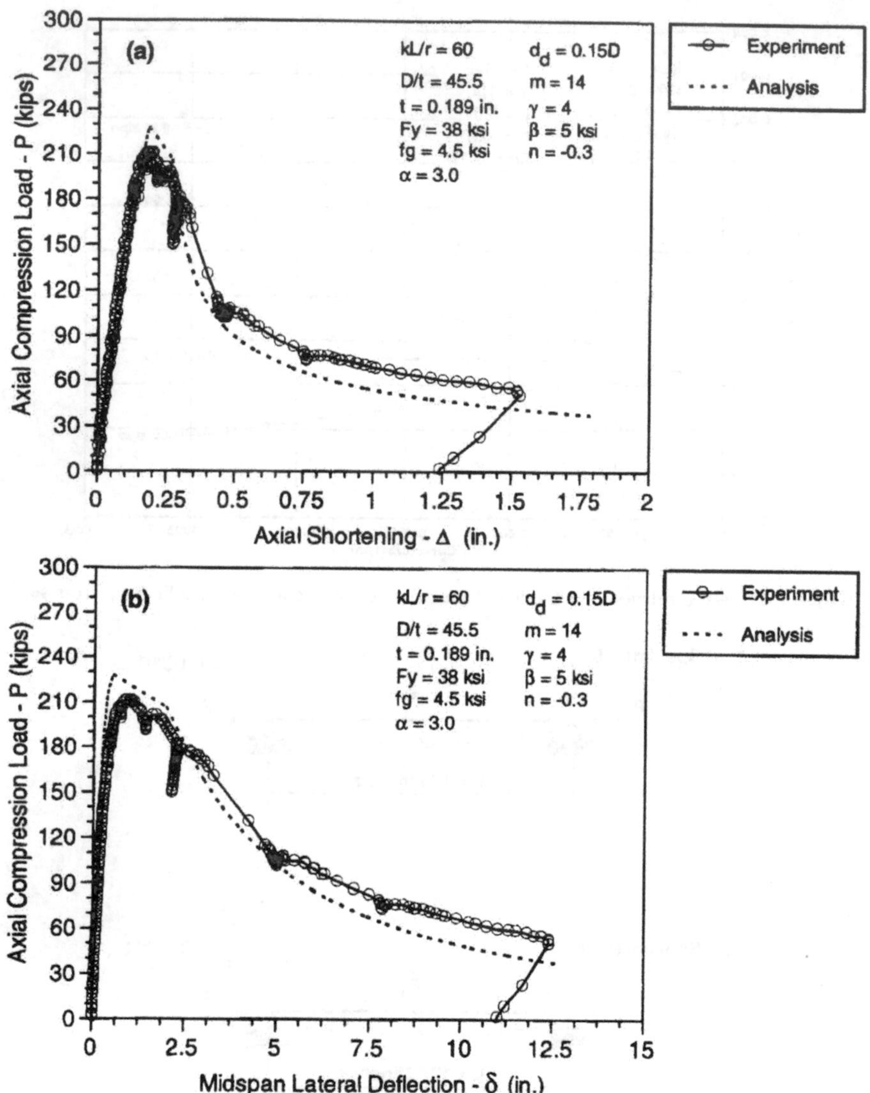

FIGURE 7.25 Comparison of predicted and experimental axial-load shortening (a) and axial load–midspan deflection response for specimen B7 (b).

of layers in Groups 1 and 3 in order to capture the spread of plasticity and local buckling through the thickness of the steel tube.

The current version of DGROUT has equally spaced stations and allows for only one dented segment. Furthermore, the effect of transverse loading is excluded; however, it can be easily programmed into DGROUT. End-moment effects are accounted for by specifying an end eccentricity, e, at the left end of the member,

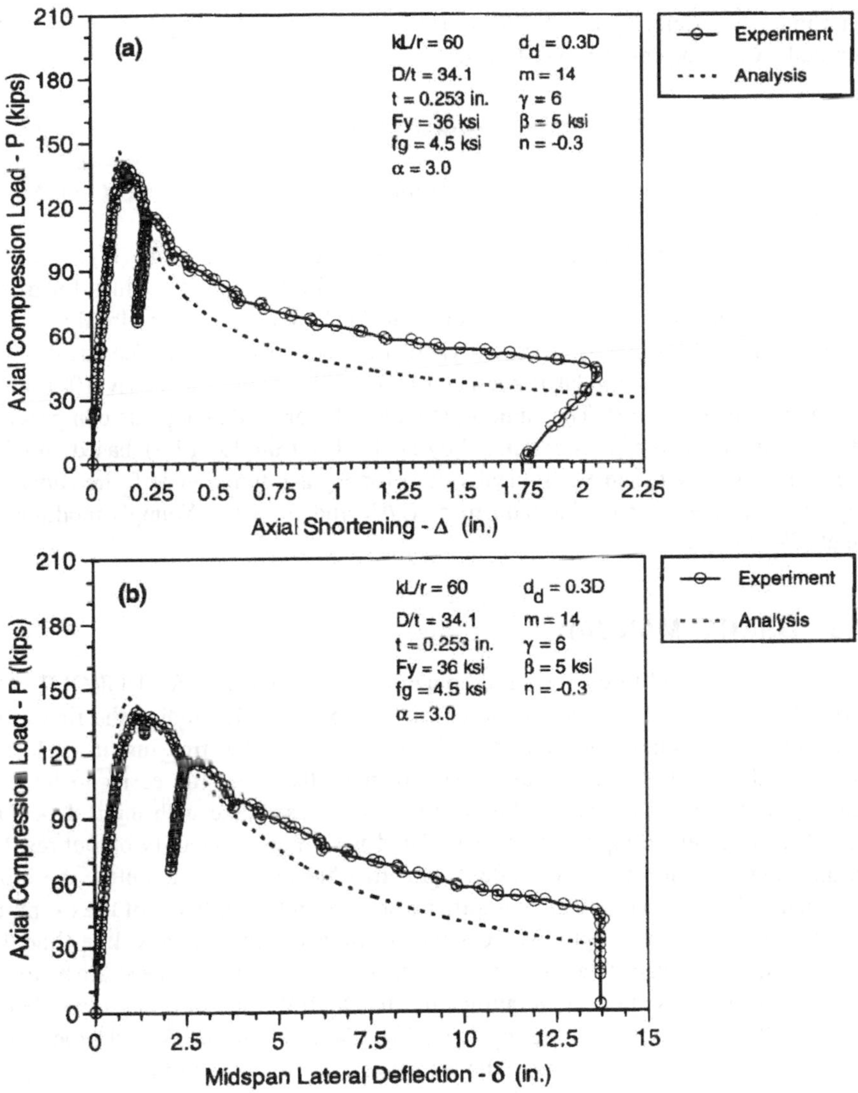

FIGURE 7.26 Comparison of predicted and experimental axial load shortening (a) and axial load–midspan deflection response for specimen A9 (b).

and the value for R representing the ratio of the left-end to right-end moment, which is:

$$R = \frac{M_L}{M_R} \tag{7.26}$$

where the value of R is positive for double curvature and negative for single curvature. In specifying a value for e, it is assumed that M_L is proportional to the applied axial compressive load P, where

$$M_L = P \cdot e \tag{7.27}$$

A positive value for eccentricity, e, results in a clockwise moment for M_L in DGROUT.

The stress-strain Type 2 curves for the steel (Fig. 7.9) and grout (Fig. 7.10) are implemented in DGROUT, where the user is required to specify values for α and β for the grout and n and γ for the steel. DGROUT assumes that the local buckling stress σ_{cr} is equal to the user-defined yield stress σ_y for the steel. The stress-strain Type 1 curve can be simulated by specifying a large value for γ (say 100) and a value close to zero for β. The strain ϵ_o at which the grout develops its compressive strength at the stress of f_g is computed by DGROUT from Eq. (7.3), based on user-defined values for f_g and E_g. Typical values for E_g as a function of f_g are shown in Fig. 7.19, where the modular ratio m = E_s/E_g and E_s is the Young's modulus of steel (29000 ksi).

7.5 USER'S MANUAL

The user is required to create a file containing the input data for DGROUT. The input data is in free-field format and consists of 51 lines. Excluding the first three lines, the subsequent 48 are related to 24 pairs of lines: the first line in each pair is intended to be utilized as a descriptor to make the input file easier to be read and edited by the user, whereas the second line is associated with input data. The first three lines in the input file are associated with titles to identify output results. In the example that follows, each descriptor line has been used to define the input data that follows on the next line. In other words, as noted on line 4 of the example, line 5 is the member's length, whereas line 46 indicates that the next line (line 47) should contain the identification numbers of three layers for which the strain histories will be saved and written to an output file. In DGROUT the layers are numbered sequentially, beginning with the top layer (first layer in Group 1), and proceeding throughout the cross section where the last layer is at the bottom of the cross section and whose identifying layer number is the sum of N1, N2, and N3.

Example Input Data File—Specimen A9

```
(1) 2 title lines for identifying output file (line 1)
Specimen A9 M-P-phi numerical integration analysis
8/6/95
(4) Member Length - L (line 4)
177.5
(6) Tube Outer Diameter - D (line 6)
```

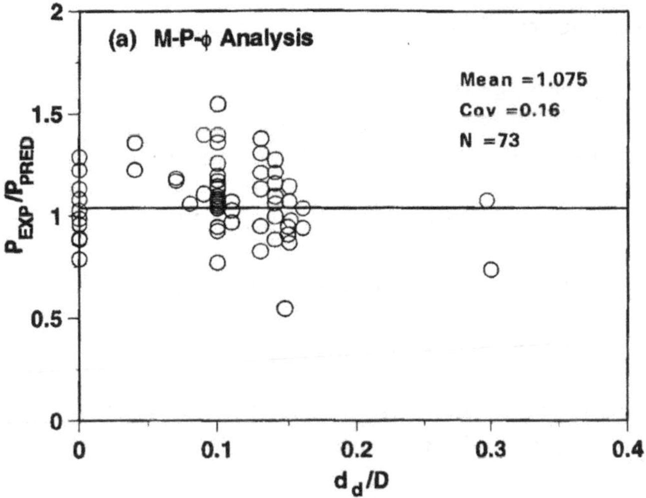

FIGURE 7.27 Comparison of experimental and predicted repaired member strength based on M-P-ϕ integration analysis.

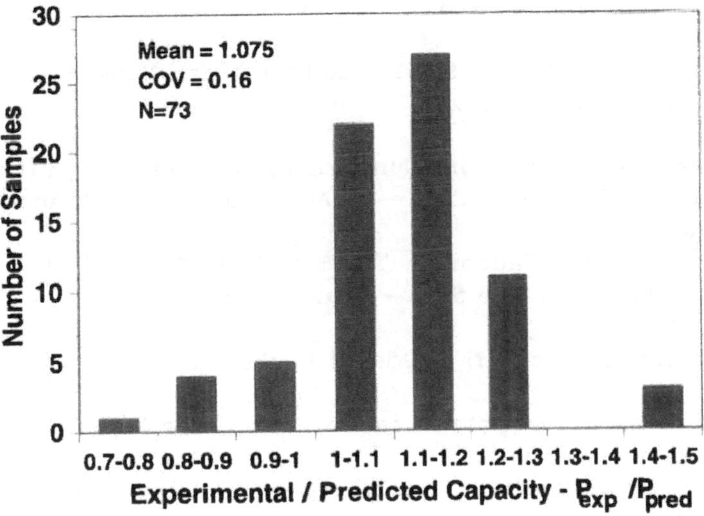

FIGURE 7.28 Histogram of experimental-to-predicted axial load capacity, internal, grout-repaired members.

```
8.6325
(8) Tube Thickness - t (line 8)
0.2533
(10) Dent Depth - dd (line 10)
2.563
(12) Steel Yield Stress - σy (line 12)
36
```

```
(14) Steel Young's Modulus - Es (line 14)
29000.
(16) Grout Compressive Strength - fg (line 16)
4.5
(18) Grout Tensile Strength - ft (line 18)
0.0
(20) Grout Young's Modulus - Eg (line 20)
2071
(22) Number of Segments - NSEG (line 22)
21
(24) Dent Segment Number With Dent (line 24)
11
(26) Number of Fibers in Groups 1, 2, and 3 of Dented
Section - N1, N2, and N3 (line 26)
10,60,10
(28) Number of Fibers in Groups 1, 2, and 3 of Undented
Section - N1, N2, and N3 (line 28)
10,60,10
(30) Left End Eccentricity - e (line 30)
0.
(32) Ratio of Left-End Moment to Right-End Moment - R
{Negative = Single Curv.} ( line 32)
-1.
(34) Convergence Factor counters for m-p-phi relationship
and numerical integration - ICFAC1 and ICFAC2 (line 34)
200,100
(36) Convergence Factors - CFAC1 and CFAC2; limit for
target displacement @ dent - ymax (line 36)
50,50,14
(38) load step size - N . (Pdel = (fyAs + fgAg)/N ) (line 38)
100
(40) Displacement control factor - κ (line 40)
1.0
(42) Out-of-Straightness at midspan normalized by L - δp/L
(line 42)
0.0094
(44) Material stress-strain parameters - α, η, β, and γ
(line 44)
3.0,0.3,5.0,6
(46) Identify Three Layers by Their Numbers for Recording
Their Strain History (line 46)
1,11,80
(48) Value of Lower Bound for Neutral Axis Location (line
48)
50.
```

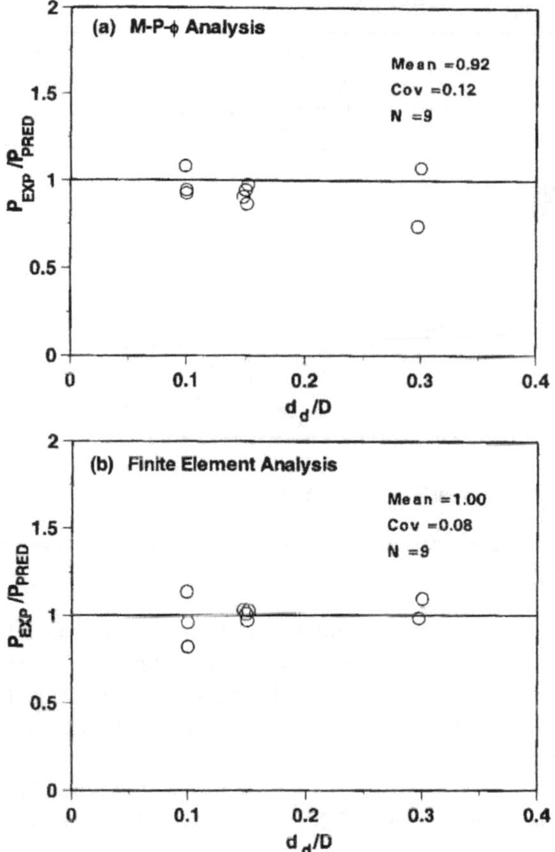

FIGURE 7.29 Comparison of experimental and predicted repaired member strength for selected specimens based on (a) M-P-ϕ integration and (b) finite element analysis.

```
(50) Load increment at which to print out the M-P-Phi
relationship - nprint (line 50)
30
```

During its execution, the user will be asked to provide a name for the output file that is created by DGROUT. Written to this file are values of the axial load, axial shortening, lateral deflection at midspan, and the strain history corresponding to converged solutions. In addition, the M-P-ϕ relationship at the selected axial load increment is also saved in the output file.

7.6 SOLUTIONS FOR MEMBER BEHAVIOR

The computer program DGROUT was used to conduct analysis of several dent-damaged test specimens studied in the laboratory. Some of the analysis results are

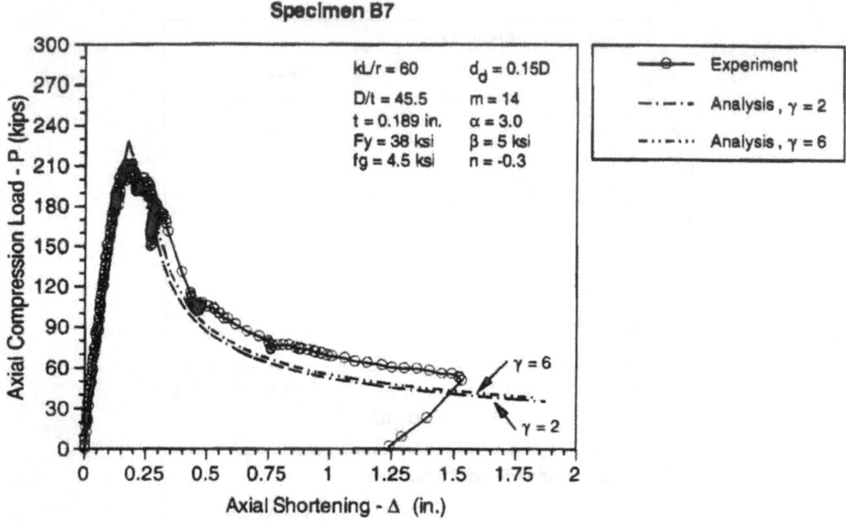

FIGURE 7.30 Effect of stress-strain parameter γ on predicted load-deformation response of an internal, grout-repaired cylindrical.

presented here in order to illustrate the use and accuracy of DGROUT in predicting the behavior of a grout-repaired dented cylindrical brace member. Also presented are the results of a parametric study of the repaired strength of dented bracing in order to illustrate the sensitivity to input data that describes the cylinder's material and geometrical characteristics.

Solutions (Fan, 1994) for the section analysis of a grout-filled undented and dented cylinder are shown in Figs. 7.20 to 7.23. The M-P-ϕ and ϵ-P-ϕ relationships

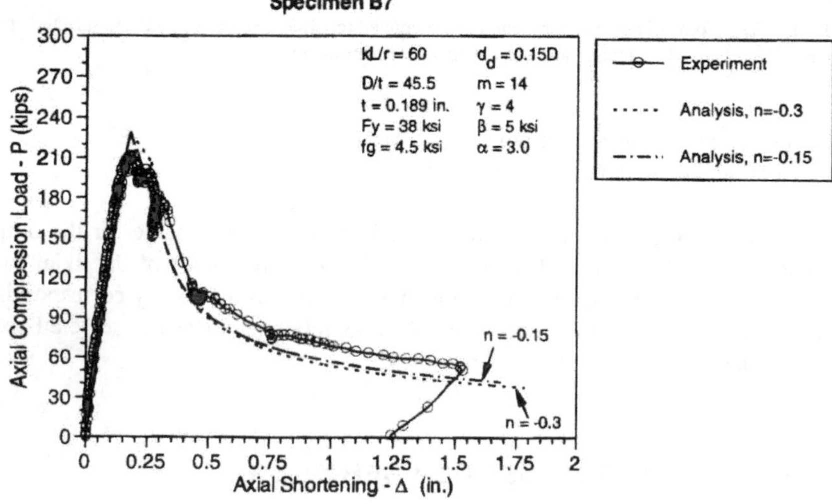

FIGURE 7.31 Effect of stress-strain parameter n on predicted load-deformation respsonse of an internal, grout-repaired cylindrical.

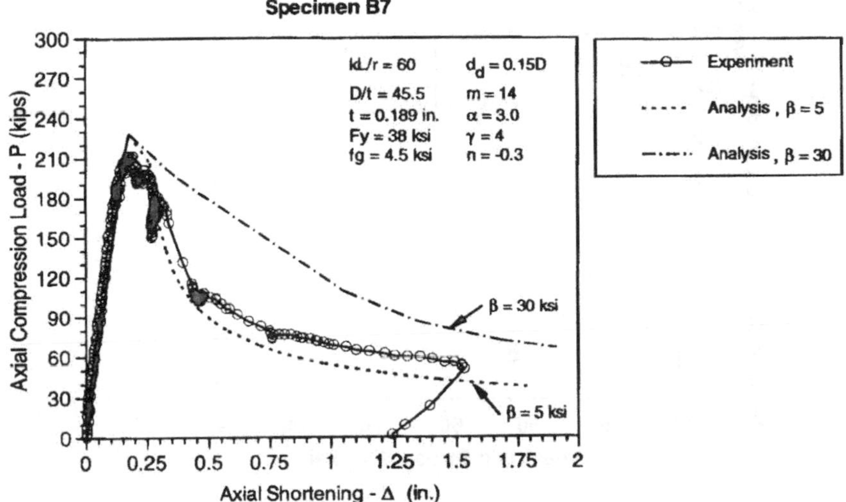

FIGURE 7.32 Effect of stress-strain parameter *n* on predicted load-deformation response of an internal, grout-repaired cylindrical.

shown were numerically generated for various levels of concentric axial force P between 50 and 250 kips using DGROUT. Pertinent input data is noted in these figures. In these examples CFAC1 and CFAC2 were both equal to 100, where a total of 80 layers (10 in both Groups 1 and 3 and 60 in Group 2) were used in the analysis. A Type 1 stress-strain curve was simulated by specifying for the stress-strain parameters the values of $n = -0.3$, $\beta = 0.01$, and $\gamma = 100$. A comparison of the undented and dented cross section M-P-ϕ and ϵ-P-ϕ relationships indicates that the capacity is notably less in the dented cylinder and that the axial strain ϵ associated with axial shortening Δ_s is somewhat larger in the dented cylinder compared to the undented cylinder.

Two test specimens (B7 and A9) were analyzed (Bruin et al., 1995) for member behavior. The loading condition, length, boundary conditions, and out-of-straightness are shown in Fig. 7.24 for these specimens. Specimens B7 and A9 had nominal dent depths of $d_d = 0.15D$ and $d_d = 0.3D$, respectively, with the dent located in both specimens at their mid-span. The input file for Specimen A9 was given previously as the example input file in Section 7.5. Solutions for member behavior in terms of the axial load-shortening and axial load–mid-span deflection predicted by DGROUT are compared with the experimental response in Figs. 7.25 and 7.26. Each analysis involved using 80 layers to describe the cross section (10 in Groups 1 and 3, and 60 in Group 2) and 21 segments (e.g., 22 stations) along the length of the specimen. Additional pertinent information related to the specimens is given in Figs. 7.25 and 7.26. The solutions from DGROUT are shown in Figs. 7.25 and 7.26 and compare reasonably well, particularly in the ascending branch of the axial load-displacement curves. The capacities predicted for Specimens B7 and A9 are within 8% and 6% of the experimental results. While the descending branch shows more discrepancy between the experimental response and DGROUT, the agreement is good compared

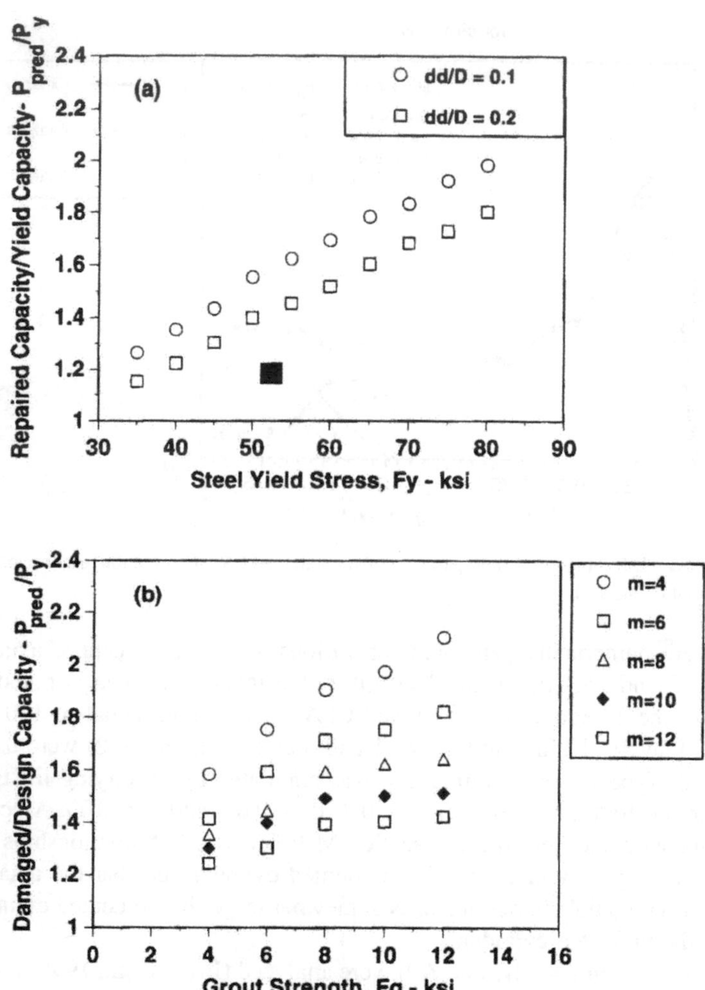

FIGURE 7.33 Effect of (a) steel yield strength and (b) grout strength on member-repaired strength.

to typical solutions using other analysis methods. The sensitivity of the predicted response to assumptions and variations of the input data will be discussed later.

Additional comparisons (Fan, 1994) with the axial-load capacity P_{exp} of experimental test specimens tested by various researchers (Boswell and D'Mello, 1990; Bruin et al, 1995; Parsanejad, 1987; Ricles et al, 1992, 1994b; Wimpy, 1986) with the predicted axial load capacity P_{pred} by DGROUT for grout-filled cylindrical bracing is shown in Fig. 7.27 as a function of dent depth d_d, and in Fig. 7.28 in the form of a histogram. A total of 73 specimens were analyzed, with dent depths ranging from d_d = 0D (i.e., no dent damage) to 0.3D. The mean and coefficient of variation associated with the ratio P_{exp}/P_{pred} are 1.075 and 0.16. These statistics indicate that DGROUT predicted reasonably well the capacity of the grout-filled

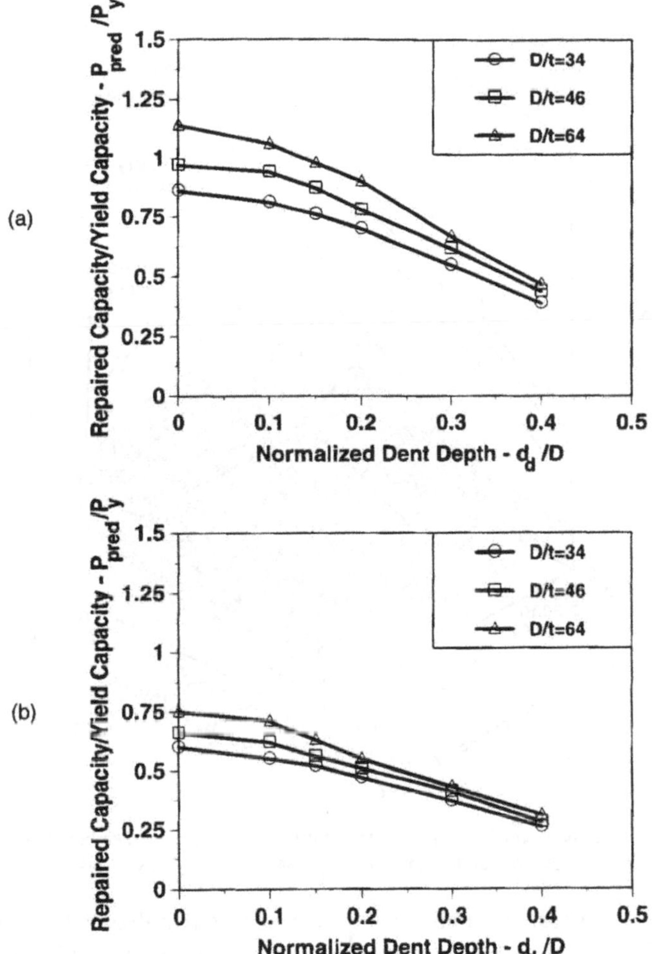

FIGURE 7.34 Effect of dent depth on member-repaired capacity for (a) $\delta_p = 0.01L$ and (b) $\delta_p = 0.02L$.

specimens. In these specimens the out-of-straightness ranged from $\delta_p = 0.0005L$ to 0.01L, the diameter-to-thickness ratio from D/t = 17 to 72, the grout strength from $f_g = 2.3$ ksi to 12.2 ksi, and the steel cylinder yield strength from 26 ksi to 85.1 ksi. In the analysis, a Type 1 stress-strain curve was simulated by specifying $n = -0.3$, $\beta = 0.01$, and $\gamma = 100$, since no local buckling in the cylinder was reported during the tests prior to the specimens achieving P_{exp}. Convergence in the analysis was based on specifying a value of 50 for both CFAC1 and CFAC2. Each analysis used 80 layers to define the cross section (10 in 1 and 3, and 60 in Groups 2) with 21 segments specified along the length. All specimens involved concentric load, except for three specimens which had an end eccentricity of e = 0.2D at both ends to cause single curvature (e.g., R = −1.0).

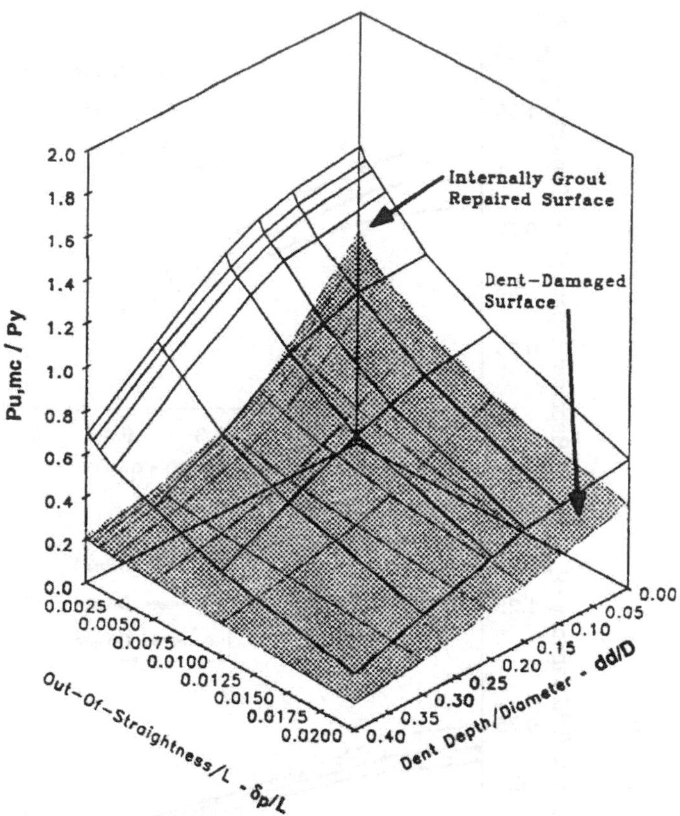

FIGURE 7.35 Comparison of nonrepaired and internal grout-repaired member strength as a function of dent depth and damaged out-of-straightness D/t = 34, fg = 5 ksi, σy = 34.8 ksi.

Some of the test specimens were also analyzed utilizing the finite element method (Ricles et al., 1994b), involving the use of shell and solid brick elements to model the steel tube and grout, respectively, in conjunction with plasticity and large deformation theory. The results are shown in Fig. 7.29, where analysis results from DGROUT are also shown for comparison. These results correspond to dent-damaged members having D/t ratios ranging from 34 to 69. In general, Fig. 7.29 indicates that, excluding a dent depth of d_d = 0.3D, the M-P-ϕ integration procedure provides solutions that are as reliable as the finite element method. At a dent depth of d_d = 0.3D the M-P-ϕ integration procedure has a tendency to diverge from the experimental results for members of larger D/t, where in Fig. 7.29, P_{EXP}/P_{PRED} = 0.73 for a test specimen having D/t = 69. For a specimen having D/t = 34 and d_d = 0.3D, better agreement is reached between the M-P-ϕ integration procedure and experimental results, where P_{EXP}/P_{PRED} = 1.07. The explanation for this phenomenon is that the Poisson effect is more significant for members of greater dent and higher D/t ratios, where the steel tube tends to separate from the grout at an earlier stage of loading during the tests. This effect is ignored in the M-P-ϕ integration procedure, and hence over-predicts the capacity of such members.

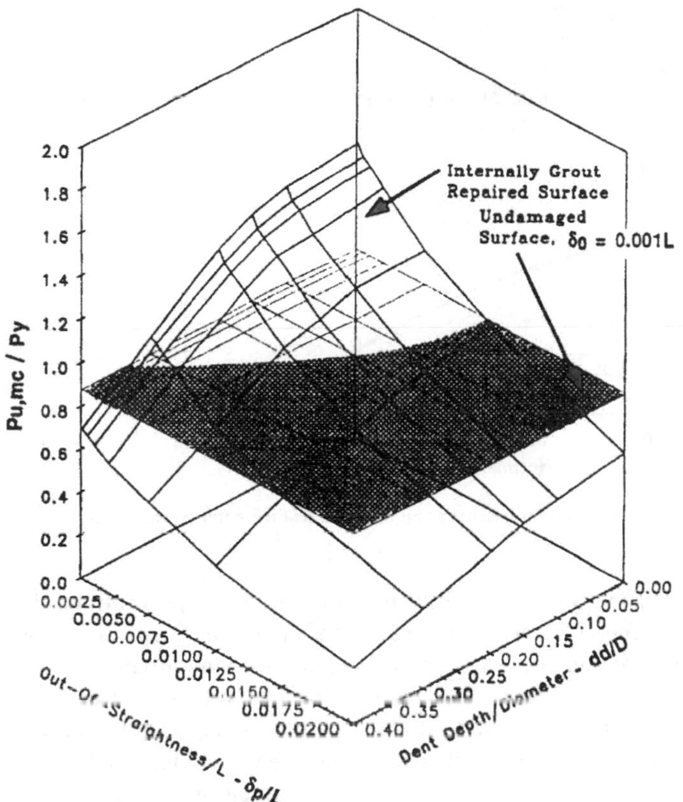

FIGURE 7.36 Comparison of undamaged and internal, grout-repaired member strength as a function of dent depth and damaged out-of-straightness D/t = 34, fg = 5 ksi, σy = 34.8 ksi.

The effects of varying the stress-strain parameters is illustrated in Figs. 7.30 to 7.32, where different values for n, β, and γ have been systematically varied and analyses performed to determine member load-deformation response. These results show that by varying γ from 2 to 6, the change in a member's capacity is not significant, while the post-ultimate load response has only a slightly slower rate of descent for analyses involving larger values of γ (Fig. 7.30). As noted previously, values for γ ranging from 2 to 6 were found to be consistent with experimental data associated with grout-filled dented cylinders (Ricles et al., 1992), where γ was approximately equal to 2 and 6 for specimens having a nominal ratio of D/t = 64 and 34.5. Specifying a larger value for the exponent n results in a similar trend, as shown in Fig. 7.31, where the capacity is not significantly different and the rate of descent is only slightly smaller in the post-ultimate load response region for the analysis using $n = -0.15$ compared to that using $n = -0.3$. The value of $n = -0.3$ was determined by Sato and Suzuki (1992) to fit experimental data for concrete-filled steel cylinders, and also works well in the M-P-ϕ analysis of grouted cylinders (Fan 1994), whose results were presented previously in Figs. 7.25 to 7.29. A large

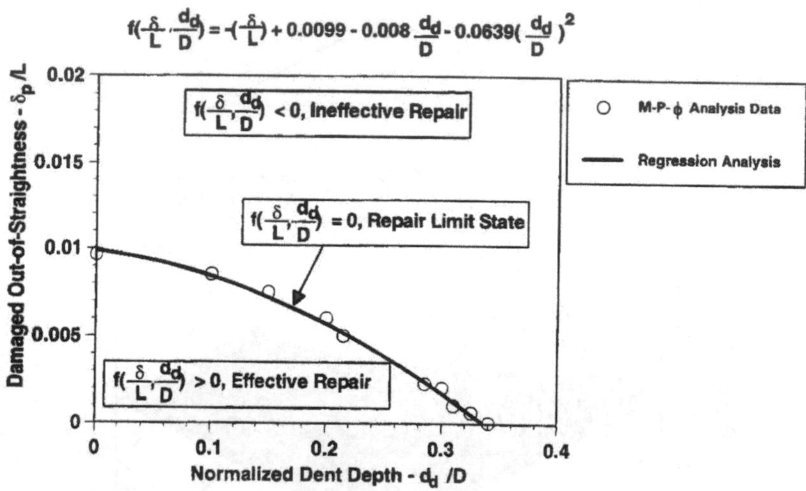

FIGURE 7.37 Feasibility of internal grout repair to restore member strength based on M-P-ϕ integration analysis.

variation in the value of β causes some difference in the post-ultimate load response, as shown in Fig. 7.32, where an increased amount of axial shortening occurs due to the failure and greater strength determination of the grout.

Other material properties that can effect the M-P-ϕ solution include the steel yield stress σ_y, grout strength f_g, and grout stiffness E_g. These effects are illustrated in Figs. 7.33a and b for member capacity, where in Fig. 7.33b, $m = E_s/E_g$, and represents the modular ratio, where E_s is constant and equal to 29,000 ksi. The quantity P_y in Fig. 7.33 is the yield strength based on the yield stress and cross-sectional area of the steel tube. The member capacity P_{pred} in Fig. 7.33 is shown to increase by an average amount of $0.066P_y$ for a unit decrease in the modular ratio, compared to a $0.038P_y$ and $0.016P_y$ increase, respectively, for a 1 ksi increase in f_g and σ_y. These results illustrate that a good estimate of the grout strength and stiffness, as well as the steel's yield strength σ_y for the member is recommended in the M-P-ϕ integration analysis in order to achieve reliable results.

The relationship of P_{pred} and dent depth d_d for two different values of out-of-straightness δ_p equal to 0.01L and 0.02L are shown in Figs. 7.34a and b. A comparison of these two figures indicates that the out-of-straightness can have a significant effect on P_{pred}; hence accurate values for δ_p should also be used in an M-P-ϕ integration analysis.

The relationship between member capacity, out-of-straightness δ_p, and dent depth d_d resulting from damage is illustrated in Fig. 7.35 for both an internal grout-repaired and non-repaired cylindrical brace having $D/t = 34$. Other member properties are identified in Fig. 7.35. The repaired member was analyzed using DGROUT, whereas the non-repaired member was analyzed using similar M-P-ϕ methods discussed in Chapter 6. Both δ_p and d_d are shown to have a significant effect on member capacity. The fact that in situ damage due to collision can cause both global bending and local dent damage, inducing δ_p and d_d, requires that both of these parameters be input as accurately as possible in order to establish a reliable prediction for member capacity.

The computational efficiency of the M-P-ϕ integration analysis procedure in providing load-deformation member response makes the method very practical for assessing internal grout-repair effectiveness. Such an assessment is shown in Fig. 7.36, where the repaired capacity as a function of δ_p and d_d is compared to the undamaged capacity of a similar, but nongrouted tubular having D/t = 34. On the basis that the undamaged capacity represents the required design strength, the intersection of the two surfaces associated with the capacities of the repaired and undamaged tubulars establishes when the repair is effective in reinstating a damaged member's capacity, as shown in Fig. 7.37.

REFERENCES

Boswell, L. F. and D'Mello, C. A. (1990) Residual and Fatigue Strength of Grout-Filled Damaged Tubular Members, *OTH 89 314, Offshore Technology Report,* U. K. Department of Energy, London, England.

Bruin, W. M., Ricles, J. M., and Sooi, T. K. (1995) Residual Strength and Repair of Dent-Damaged Tubulars—Experimental and Analytical Assessment, *ATLSS Report,* Advanced Technology for Large Structural Systems Research Center, Lehigh University, Bethlehem, PA to be published.

Chen, W. F. and Atsuta, F., (1976) *Theory of Beam-Columns,* Vol. 1: *In-Plane Behavior* and Design, McGraw-Hill, New York.

Ellinas, C. P. and Valsgard, S. (1985) Collision and Damage of Offshore Structures: A State-of-the-Art, *Journal of Energy Resource Technology,* 107(9), 297–314.

Fan, C. P. (1994) Assessment and Prediction of the Behavior of Dent Damaged and Internally Grout Repaired Tubular Steel Bracings Using Moment Curvature-Based Integration Methods, Thesis submitted to the Department of Civil Engineering, Fritz Engineering Laboratory, Lehigh University, Bethlehem, PA.

Kent, D. C. and Park, R. (1971) Flexural Members with Confined Concrete, *Journal of Structural Engineering,* ASCE, 97(7), 1969–90.

Mander, J. B., Priestley, M. J. N., and Park, R. (1988) Theoretical Stress-Strain Model for Confined Concrete, *Journal of Structural Engineering,* ASCE, 114(8), 1804–26.

Newmark, N. M. (1943) Numerical Procedure for Computing Deflection, Moments, and Buckling Loads, *Transactions,* ASCE, 108, 1161–88.

Parsanejad, S. (1987) Strength of Grout-Filled Damaged Tubular Members, *Journal of Structural Engineering,* ASCE, 113(3), 590–603.

Parsanejad, S. and Gusheh, P. (1988) Test on Partially Grout-Filled Damaged Tubular Members, *Civil Engineering Transactions,* Institution of Engineers, Sydney, Australia, CE30(5), 34–51.

Ricles, J. M., Gillum, T. E., and Lamport, W. (1992) Residual Strength and Grout Repair of Dented Offshore Tubular Bracing, *ATLSS Report No. 92-14,* Advanced Technology for Large Structural Systems Research Center, Lehigh University, Bethlehem, PA.

Ricles, J. M., Gillum, T. E., and Lamport, W. (1994a) Grout Repair of Dented Offshore Tubular Bracing–Experimental Behavior, *Journal of Structural Engineering,* ASCE, 120(7), 2086–2107.

Ricles, J. M., Sooi, T. K., Bruin, W. M., and Fan, C. P. (1994b) Behavior, Analysis and Repair of Dented Marine Platform Bracing Members, *Proceedings of the Structural Stability Research Council Annual Technical Session,* Bethlehem, PA.

Sato, T. and Suzuki, K. (1992) Proposal on Story Drift Limit Based on Axial Behavior of Column Under Seismic Loads, *Proceedings of the Second Engineering Foundation Composite Construction in Steel and Concrete,* ASCE.

U. K. Department of Energy (1988) Grout and Grouting for Construction and Repair of Offshore Structures—A Summary Report, *OTH 88 289,* Offshore Technology Report, London, England.

Wimpy Offshore, (1986) Static Testing of Grout-Filled Joints and Members, *Document 2/86a,* Report to Exxon Production Research Company, Southwest Research.

Index

A

ABAQUS computer program, 260–261
ADMCOL computer program
average flow moment method, 99–103
exact moment-curvature method, 89–99
input, entry of, 86–87
modified plastic hinge method, 89
sample calculations, 87–89
APCYCL computer program
cyclic analysis, application for, 133–137
Approximation
elastic-perfectly plastic, 122, 125, 177
stress-strain curves, 129
symmetric, 122, 125, 176
Assumed deflection method
beam columns, 57
Atsuta, F., 273
Average flow moment method
load deflection analysis, 85
load shortening analysis, 85
Axial forces
compressive, stages of, 144–150
strength interaction, 27
tensile, 150–153
values, 20
Axial load
hydrostatic pressure, 30
plane of bending, influence on, 30
Axial shortening
axial strain, 62, 67, 78–81
beam columns, 61–62
components, 97
lateral deflection, 67–68, 81
Axial stiffness, 85–86
Axial strain
axial shortening, 62, 67, 78–81
Axial strain increment
circumferential strain change, interaction with, 12

B

BCDENT computer program, 243
Beam columns
analysis, 169, 272
assumed deflection method, 57
axial shortening, 61–62
cyclic behavior, 137
cylindrical shape, 167
deflected shape, 179
dented, 261
distribution of curvature, 180
fixed-ended, 65–67, 69–70, 77–78, 85
load deflection, 172
pin-ended, 69, 76–77, 85
Beam theory
simple kinematic model, 167
Bending moment
right-hand screw rule, 30
values, 20
Biaxial stress interaction
plastic materials, 13
Biaxially loaded columns
tangent stiffness method, 7
Birkemoe, P. C., 260
BMCYCL computer program
cyclic analysis, application for, 129–133
Boswell, L. F., 271
BRACE computer program
cyclic analysis, application for, 167
Buckling
post-local, 169
pre-local, 169
Buckling load
out-of-straightness, effects of, 37

C

CFAC1 convergence factor, 284, 297
CFAC2 convergence factor, 297
Chen, W. F., 258, 259, 273
Circular cross section
interaction curve, 28
Circular cylindrical section
exact moment-thrust-axial relations, 221–224
Circular tubes
manufacturing processes, 1
Circumferential strain change
axial strain increment, interaction with, 12
Circumferential stresses
poisson effects, 10
Closed-form expressions
moment-thrust-curvature relationships, 61
Column failures
types, 6
Column strength curve
calculations, 40
Columns
axial shortening, 156

Compatibility relation
  matrix form, 22
Compression
  elastic loading, 146–147
  elastic unloading, 149–150
  hinges formation, 148–149
Compressive axial forces
  stages, 144–150
Convergence, 284–285, 299
Convergence of deflection
  specification requirements, 38
Cray X-MP/24 supercomputer, 261
Cross sectional distortion
  kinematic model, use of, 168
Cross sections
  calculations, 9
Curvature curves
  moment curves, 18
CYCL computer program
  cyclic analysis, application for, 133–137
Cyclic analysis
  BMCYCL computer program, application of, 129–133
  BRACE computer program, application of, 167
Cyclic axial compression-tension
  loading, 126
Cyclic behavior
  beam columns, 137
  fix-ended columns, 164
  load-displacement curves, 164
  pin-ended columns, 139–141
Cyclic loading
  moment-thrust-curvature relationships, 108–111
Cylindrical beam columns
  analysis of, 167
  distortion, causes of, 167

D

D'Mello, C. A., 271
Deflection function
  types, 60
Deflection shapes
  elastic beam columns, 60
Deflections
  fix-ended columns, 179
  lateral, 176
  load, 139
  pin-ended columns, 179
Deflections, computed
  assumed, 282
  corrected, 281
  uncorrected, 281
Dent damage
  calculation steps, 228–231
  description, 226–228
  equivalent concentrated load, 232–233
  moment-thrust-axial strain expressions, 221–225
  notations, 209–210
  undented cylindrical sections, 212–215
Dented cantilever member, 243
Dented columns
  beam columns, 261
  pin-ended, 260
Dented cylindrical beam columns
  loading cases, 226–228
Dented cylindrical sections
  moment-thrust-curvature, 212–225
DGROUT computer code
  transverse loading, effects of, 290
Diameter-to-thickness ratio
  fixed-ended columns, 201
Distribution of curvature
  beam columns, 180
Doubly symmetric curves
  approximation of, 174–176

E

Eccentricity, 292, 299
Elastic beam columns
  deflection shapes, 60
Elastic limit pressure
  equations, 23
Elastic loading
  compression, 146–147
Elastic tensioning, 153
Elastic-perfectly plastic approximation
  reversed loading, 177
Elastic-perfectly plastic curves
  approximation of, 122, 125, 177
End moments
  pin-ended beam columns, 203–204
Epsilon-P-phi, 274
Equilibrium
  equations of, 21
Equivalent concentrated load, 232–233
Exact moment-curvature method
  load deflection, 76–78
  load shortening, 78–81
Exact moment-thrust-axial relations
  circular cylindrical section, 221–224
Expressions
  closed-form, 123
  symmetric curve, 122
  three-regime, 123

F

Fabricated cylindrical columns
  design, 5
Fiber model, 274
Finite element analysis, 272
Finite segment method, 258

FIXCYCL computer program
cyclic analysis, application for, 158–163
Fixed-ended columns
cyclic behavior, 164
deflection, 179
diameter-to-thickness ratio, 201
hinge-by-hinge method, application of, 142
local buckling, 201
Fortran 77 computer programming language, 243, 283

G

Gosheh, P., 271
Gould NP1 computer, 263
Grouting
benefits, 269
internal, 271
steel clamps, 271

H

Hinge-by-hinge method
fixed-ended columns, application to, 142
Hydrostatic pressure
axial load, 30
out-of-roundness, effects on, 16

I

Imperfect cylinder
hydrostatic pressure, external, 16
Imperfections
maximum strength, effects on, 41
types, 2
Inelastic cyclic behavior, 126
Interaction curves
circular cross section, 28
computation, 28
end moment for, 41
wide flange section, 28
Intermediate curvature state
calculations, 10

K

Kent, D. C., 275
Kinematic model
cross sectional distortion, application for, 168

L

Landet and Johnsen's Test, 243, 260–261
Large deformation theory, 273, 300
Lateral deflection
axial shortening, 67–68, 81
Lateral load
pin-ended beam columns, effects on, 204–207
Lehigh tests
residual stress for, 39
Load deflection
beam columns, 172
behavior of, 139
tracing, 3
Load deflection analysis
average flow moment method, 85
exact moment-curvature method, 76
local buckling, 179
modified plastic hinge method, 70
moment-thrust-curvature relationships, 108
Load displacement curves
cyclic behavior, 164
Load history
plastic behavior, dependence of, 9
Load shortening analysis
average flow moment method, 85
exact-moment-curvature method, 78
local buckling, 187–188
modified plastic hinge method, 71
Loading
cyclic axial compression-tension, 126
monotonic, 125, 172
stages of, 128
Local buckling
computer implementation, 188–191
fixed-ended columns, 201
load deflection analysis, 179, 181–188
load shortening analysis, 187–188
pin-ended columns, effects on, 193–201
undented columns, 258
Longitudinal residual stress
patterns, 27
welding process, 14
Longitudinal stresses
poisson ratio effects, 10

M

M-P-phi curves, 259
M-P-phi integration method, 273
Maximum strength curve
out-of-roundness, effects of, 28
McIntyre, J., 260
Member load-displacement response, 279
Mesh sizes
tube circumference, 24
Modified assumed deflection method, 179–181
Modified plastic hinge method
load deflection analysis, 70
load shortening analysis, 71
Moment capacity
hydrostatic pressure, effects of, 32
tubular cross section, 29
Moment curves
curvature curves, 18

Moment-curvature behavior
welding location, effects of, 29, 29–30
Moment-curvature slope
stiffness, effects on, 5
Moment-thrust-axial strain expressions
dent damage, 221–225
Moment-thrust-curvature
dented cylindrical sections, 212–225
Moment-thrust-curvature integration analysis, 272–273
Moment-thrust-curvature relationships
closed-form expressions, 61, 73–76
cyclic loading, 108–111
load deflection analysis, 108
plastic hinge method, 68–69
simplification of, 61
Moments
thrusts, 7
Monotonic loading, 125, 172
MPCYCL computer program
input, entry of, 111–115

N

Newmark computer program
sample calculations, 49
Newmark's integration method
beam column strength, 34
computer implementation, 126–128
pin-ended columns, use for, 125
NSEG segments, 279
Numerical integration procedure, 278–282
Numerical iteration, 128

O

Offshore platforms, 269
One hinge formation
tensions, 154–155
Out-of-roundness imperfection
hydrostatic pressure, 16
measurements, 15
Out-of-straightness
buckling load, 37

P

Park, R., 275
Parsanejad, S., 271–272
Partially yielded section
matrices, 9
Perpendicular diameters
percentage of difference, 16
Pin-ended beam columns
cyclic analysis, 126
cyclic behavior, 139–141
deflection, 179
elastic regime, 181–183
end moments, effects of, 203–204
hinge-by-hinge method, application of, 144
lateral load, effects of, 204–207
Newmark's method, application of, 125
post-local-buckling regime, 183
primary yield regime, 183
reversed loading regime, 185
secondary yield regime, 183
Pipeline hydrostatic collapse
limit solution, 22
Plane cross section
wall thickness, 21
Plane of bending
axial load, influence of, 30
Plastic behavior
load history, dependence on, 9
Plastic hinge method
load deflection analysis, 70–71
load shortening analysis, 71
moment-thrust-curvature relationships, 68–69
Plastic materials
biaxial stress interaction, 13
Plastic moments, 173
Plastic range
sinusoidal function, 60
Plastic solution
strain vector, 11
Plasticity theory, 273, 300
Poisson effects, 300
circumferential stresses, 10
stress-strain relationships, 275
Post-local buckling, 169
pin-ended beam columns, 183
Pre-local buckling, 169

R

Residual stress
Lehigh tests, 39
types, 14
Reversed loading
analysis of, 170, 174
elastic-perfectly plastic approximation, 177
stages of, 142–144
symmetric approximation, 176
Right-hand screw rule
bending moment, 30

S

Sato, T., 274, 301
Seam welding location
direction of buckling, effects on, 38
Secondary yielded range, 125
Simple kinematic model
beam theory, 167
Sinusoidal function
plastic range, for, 60
Slenderness ratio
fixed-ended columns, effects on, 201–203

Sohal, I. S., 259
Steel clamps
  grouting, 271
Steel tubes
  linearized distributions, 15
Stiffness of cross section
  residual stress, effects on, 25
Strain vector
  plastic solution, 11
Strains
  axial compressive, 156
Strength interaction
  axial forces, 27
Stress distribution
  moment-curvature curve, effects on, 27
Stress paths
  wall thickness, 18
Stress-strain curves
  approximation of, 129
Stress-strain relationships
  effects of confinement, 275
  iterative computational procedure, 275
  Poisson effects, 275, 300
Sugimoto, H., 258
Suzuki, K., 274, 301
Symmetric curves
  approximation, 122, 125, 176

T

Taby, J., 261
Tangent stiffness method
  biaxially loaded columns, 7
Tensile axial forces, 150–153
Tensions
  one hinge formation, 154–155
  two hinge formation, 155–156
Three-regime expressions, 123
Thrusts
  moments, 7
Transverse loading
  DGROUT computer code, 290
Tresca yield curves
  stress axes, 11
Tube circumference
  mesh sizes, 24
Tube sizes
  structural properties, differences of, 2
Tubular cross section
  moment capacity, 29

U

Undented columns
  local buckling, 258–259
Undented cylindrical sections, 212–215
Undented member behavior, solutions of, 258

V

Values
  bending moment, 20

W

Wall thickness
  plane cross section, 21
  stress paths, 18
Welding location
  moment-curvature behavior, effects on, 29–30
Welding process
  longitudinal residual stress, 14
Wide flange section
  interaction curves, 28

Y

Yield moments, 173
Young, T., 276
Young's modulus
  linear variation, 11